T0134462

Multi-Objective Optimization

Jyotsna K. Mandal · Somnath Mukhopadhyay
Paramartha Dutta

Editors

Multi-Objective Optimization

Evolutionary to Hybrid Framework

Editors
Jyotsna K. Mandal
University of Kalyani
Kalyani, West Bengal, India

Paramartha Dutta
Visva Bharati University
Bolpur, Santiniketan, West Bengal, India

Somnath Mukhopadhyay
Assam University
Silchar, Assam, India

ISBN 978-981-13-4639-2   ISBN 978-981-13-1471-1   (eBook)
https://doi.org/10.1007/978-981-13-1471-1

© Springer Nature Singapore Pte Ltd. 2018
Softcover re-print of the Hardcover 1st edition 2018
This work is subject to copyright. All rights are reserved by the Publisher, whether the whole or part
of the material is concerned, specifically the rights of translation, reprinting, reuse of illustrations,
recitation, broadcasting, reproduction on microfilms or in any other physical way, and transmission
or information storage and retrieval, electronic adaptation, computer software, or by similar or dissimilar
methodology now known or hereafter developed.
The use of general descriptive names, registered names, trademarks, service marks, etc. in this
publication does not imply, even in the absence of a specific statement, that such names are exempt from
the relevant protective laws and regulations and therefore free for general use.
The publisher, the authors and the editors are safe to assume that the advice and information in this
book are believed to be true and accurate at the date of publication. Neither the publisher nor the
authors or the editors give a warranty, express or implied, with respect to the material contained herein or
for any errors or omissions that may have been made. The publisher remains neutral with regard to
jurisdictional claims in published maps and institutional affiliations.

This Springer imprint is published by the registered company Springer Nature Singapore Pte Ltd.
The registered company address is: 152 Beach Road, #21-01/04 Gateway East, Singapore 189721,
Singapore

Foreword

Multi-objective optimization problems have two or more (usually conflicting) objectives that we aim to solve simultaneously. The solution to these problems involves finding a set of solutions (rather than only one) representing the best possible trade-offs among the objectives, such that no objective can be improved without worsening another. In spite of the existence of a number of mathematical programming techniques that have been explicitly designed to solve multi-objective optimization problems, such techniques have several limitations, which has motivated the use of evolutionary algorithms, giving rise to an area known as *evolutionary multi-objective optimization*.

The first evolutionary multi-objective algorithm was published in 1985, but it was until the late 1990s that this research area started to gain popularity. Over the last 20 years, this discipline has given rise to a wide variety of algorithms, methodologies and applications that span practically all areas of knowledge.

This book brings together a very interesting collection of applications of multi-objective evolutionary algorithms and hybrid approaches in a variety of disciplines, including bioinformatics, networking, image processing, medicine and finance. This book should be of interest to researchers and students with or without experience in evolutionary multi-objective optimization, who will certainly benefit from the novel applications and concepts discussed in this volume.

Mexico City, México
April 2018

Carlos A. Coello Coello
CINVESTAV-IPN

Editorial Preface

In our day-to-day real life, we have to take decisions on the basis of various associated factors. Arriving at a decision becomes even more challenging if one has to deal with factors apparently contradictory to one another. In such a situation, addressing one factor falls short of being meaningful without addressing the other factors even though conflicting. The issue of multi-/many-objective framework typically deals with an environment where we have to consider simultaneous optimization of more than one factor/objective to arrive at a comprehensive conclusion/inference. In such cases, we try to achieve a collection of solutions, typically referred to as Pareto front, where a typical solution element in the front does neither dominate nor is dominated by another.

This edited book volume entitled *Multi-Objective Optimization: Evolutionary to Hybrid Framework* is a collection of fourteen chapters contributed by leading researchers in the field. The chapters were initially peer-reviewed by the Editorial Review Board members spanning over many countries across the globe. In that sense, the present endeavour is a resultant of contributions from serious researchers in the relevant field and subsequently duly peer-reviewed by pioneer scientists. A brief description of each of the chapters is as follows:

Chapter "Non-dominated Sorting Based Multi/Many-Objective Optimization: Two Decades of Research and Application" gives an exhaustive analysis and picture of non-dominated sorting-based multi-/many-objective optimization algorithms proposed in the last two decades of research and application. Authors have mentioned that for more than two decades, non-dominated sorting has been a cornerstone in most successful multi-/many-objective optimization algorithms. In this chapter, they have discussed the effect of non-dominated sorting in multi- and many-objective scenarios. Thereafter, they have presented some of the most widely used optimization algorithms involving non-dominated sorting, where they have discussed their extent and ubiquity across many scientific disciplines. Finally, they have gone over some of the state-of-the-art combinations of non-dominated sorting with other optimization techniques.

Chapter "Mean-Entropy Model of Uncertain Portfolio Selection Problem" deals with the portfolio selection problem, which is a single-period invest model where an investor has to select and distribute available capital among various securities to achieve the target investment. Authors have proposed in this study a bi-objective portfolio selection model, which maximizes the average return and minimizes the investment risk of the securities. In the proposed model, the average return and the risk are represented, respectively, by the mean and entropy of the uncertain securities. The expected value and the triangular entropy of the uncertain securities are determined to represent the mean and entropy, respectively. The proposed model is solved with two different classical multi-objective solution techniques: (i) weighted sum method and (ii) weighted metric method. Both the techniques generate a single compromise solution. To generate a set of non-dominated solutions, for the problem, two different multi-objective genetic algorithms (MOGAs)— (i) non-dominated sorting genetic algorithm II (NSGA-II) and (ii) multi-objective evolutionary algorithm based on decomposition (MOEA/D)—are used. Finally, the performances of the MOGAs are analysed and compared based on different performance metrics.

In Chapter "Incorporating Gene Ontology Information in Gene Expression Data Clustering Using Multiobjective Evolutionary Optimization: Application in Yeast Cell Cycle Data", authors have described how microarray technology has made it possible to simultaneously monitor the expression levels of a large number of genes over different experimental conditions or time points. In this chapter, authors have presented an approach for combining experimental gene expression information and biological information in the form of gene ontology (GO) knowledge through multi-objective clustering. The method combines the expression-based and GO-based gene dissimilarities. Moreover, the method simultaneously optimizes two objective functions—one from gene expression point of view and another from GO point of view. Authors have demonstrated the performance of the proposed technique on real-life gene expression dataset of yeast cell cycle. They have also studied here the biological relevance of the produced clusters to demonstrate the effectiveness of the proposed technique.

In Chapter "Interval-Valued Goal Programming Method to Solve Patrol Manpower Planning Problem for Road Traffic Management Using Genetic Algorithm", an interval-valued goal programming (IVGP) method is proposed for modelling and solving patrolmen deployment problem in traffic control system in an inexact environment. Here, the objective functions are to be optimized and represented as goals by assigning target intervals for the achievement of objective values and incorporating interval coefficients to objective parameter sets to reach satisfactory solution in decision horizon. Authors have also defined a performance measuring function to represent different kinds of objectives that are inherently fractional in form in decision premises by transforming it into linear equivalent to avoid computational difficulty with fractional objectives in course of searching solution to the problem. After that they have converted interval arithmetic rules, interval-valued goals into goals as in conventional GP to formulate standard model of the problem. Authors have designed the executable model by an extended GP methodology for solving the traffic control problem. They have demonstrated the proposed approach via a case example of the metropolitan city, Kolkata, West Bengal, in India.

In Chapter "Multi-objective Optimization to Improve Robustness in Networks", authors have proposed a new approach to address the budget-constrained multi-objective optimization problem of determining the set of new edges (of given size) that maximally improve multiple robustness measures. Authors have presented the experimental results to show that adding the edges suggested by their approach significantly improves network robustness, compared to the existing algorithms. The networks which they have presented can maintain high robustness during random or targeted node attacks.

Chapter "On Joint Maximization in Energy and Spectral Efficiency in Cooperative Cognitive Radio Networks" addresses a joint spectral efficiency (SE)–energy efficiency (EE) optimization problem in a single secondary user (SU) and primary user (PU) network under the constraints of sensing reliability, cooperative SE for primary network and transmission power constraints. Differential evolution (DE) is explored by the authors to handle this nonlinear optimization problem and to find the optimal set of values for the sensing duration, cooperation and transmission power of SU. The trade-off between EE–SE is shown through the simulation.

In Chapter "Multi-Objective Optimization Approaches in Biological Learning System on Microarray Data", authors have provided a comprehensive review of various multi-objective optimization techniques used in biological learning systems dealing with the microarray or RNA sequence data. In this regard, the task of designing a multi-class cancer classification system employing a multi-objective optimization technique is addressed first. Next, they have discussed how a gene regulatory network can be built from the perspective of multi-objective optimization problem. The next application deals with fuzzy clustering of categorical attributes using a multi-objective genetic algorithm. After this, how microarray data can be automatically clustered using a multi-objective differential evolution is addressed. Then, the applicability of multi-objective particle swarm optimization techniques in identifying the gene markers is explored. The next application concentrates on feature selection for microarray data using a multi-objective binary particle swarm optimization technique. Thereafter, a multi-objective optimization approach is addressed for producing differentially coexpressed module during the progression of the HIV disease. In addition, they have represented a comparative study based on the literature along with highlighting the advantages and limitations of the methods. Finally, they have depicted a new direction to bio-inspired learning system related to multi-objective optimization.

The main goal of the Chapter "Application of Multi-Objective Optimization Techniques in Biomedical Image Segmentation—A Study" is to give a comprehensive study of multi-objective optimization techniques in biomedical image analysis problem. This study mainly focusses on the multi-objective optimization techniques that can be used to analyse digital images, especially biomedical images. Here, some of the problems and challenges related to images are diagnosed and analysed with multiple objectives. It is a comprehensive study that consolidates some of the recent works along with future directions.

In Chapter "Feature Selection Using Multi-Objective Optimization Technique for Supervised Cancer Classification", a new multi-objective blended particle swarm optimization (MOBPSO) technique is proposed for the selection of significant and informative genes from the cancer datasets. To overcome local trapping, authors have integrated here a blended Laplacian operator. The concept is also implemented for differential evolution, artificial bee colony, genetic algorithm and subsequently multi-objective blended differential evolution (MOBDE), multi-objective blended artificial bee colony (MOBABC) and multi-objective blended genetic algorithm (MOBGA) to extract the relevant genes from the cancer datasets. The proposed methodology utilizes two objective functions to sort out the genes which are differentially expressed from class to class as well as provides good results for the classification of disease.

In Chapter "Extended Non-dominated Sorting Genetic Algorithm (ENSGA-II) for Multi-Objective Optimization Problem in Interval Environment", authors have proposed an efficient algorithm for solving a multi-objective optimization problem with interval objectives. For this purpose, they have developed an extended version of existing NSGA-II algorithm (ENSGA-II) for fixed objectives in an interval environment with the help of interval ranking and interval metric. In this connection, they have proposed non-dominated sorting based on interval ranking, interval-valued crowding distance and crowded tournament selection of solutions with respect to the values of interval objectives.

In Chapter "A Comparative Study on Different Versions of Multi-Objective Genetic Algorithm for Simultaneous Gene Selection and Sample Categorization", authors have studied different versions of multi-objective genetic algorithm for simultaneous gene selection and sample categorization. Here, authors have selected optimal gene subset, and sample clustering is performed simultaneously using multi-objective genetic algorithm (MOGA). They have employed different versions of MOGA to choose the optimal gene subset, where the natural number of optimal clusters of samples is automatically obtained at the end of the process. The proposed methods use nonlinear hybrid uniform cellular automata for generating initial population, tournament selection strategy, two-point crossover operation and a suitable jumping gene mutation mechanism to maintain diversity in the population. They have used mutual correlation coefficient, and internal and external cluster validation indices as objective functions to find out the non-dominated solutions.

In Chapter "A Survey on the Application of Multi-Objective Optimization Methods in Image Segmentation", authors have provided a comprehensive survey on multi-objective optimization (MOO), which encompasses image segmentation problems. Here, the segmentation models are categorized by the problem formulation with relevant optimization scheme. The survey they have done also provides the latest direction and challenges of MOO in image segmentation procedure.

In Chapter "Bi-objective Genetic Algorithm with Rough Set Theory for Important Gene Selection in Disease Diagnosis", a bi-objective genetic algorithm with rough set theory has been proposed for important gene selection in disease diagnosis. Here, two criteria are combined, and a novel bi-objective genetic algorithm is proposed for gene selection, which effectively reduces the dimensionality of the huge volume of gene dataset without sacrificing any meaningful information. The proposed method uses nonlinear uniform hybrid cellular automata for generating initial population and a unique jumping gene mechanism for mutation to maintain diversity in the population. It explores rough set theory and Kullback–Leibler divergence method to define two objective functions, which are conflicting in nature and are used to approximate a set of Pareto optimal solutions.

Chapter "Multi-Objective Optimization and Cluster-wise Regression Analysis to Establish Input-Output Relationships of a Process" deals with an approach which is used to establish input–output relationships of a process utilizing the concepts of multi-objective optimization and cluster-wise regression analysis. At first, an initial Pareto front is obtained for a given process using a multi-objective optimization technique. Then, these Pareto optimal solutions are applied to train a neuro-fuzzy system (NFS). The training of the NFS is implemented using a meta-heuristic optimization algorithm. Now, for generating a modified Pareto front, the trained NFS is used in MOEA for evaluating the objective function values. In this way, a new set of trade-off solutions is formed. These modified Pareto optimal solutions are then clustered using a clustering algorithm. Cluster-wise regression analysis is then carried out to determine input–output relationships of the process. These relationships are found to be superior in terms of precision to those of the equations obtained using conventional statistical regression analysis of the experimental data.

Contributions available in the fourteen chapters, after being meticulously reviewed, reflect some of the latest sharing of some serious researches of the concerned field. The editors want to avail this opportunity to express their sincere gratitude to all the contributors for their efforts in this regard without which this edited volume could have never come to a reality. The editors sincerely feel that the success of such an effort in the form of edited volume can be academically meaningful only when it is capable of drawing significant contributions from good researchers in the relevant field.

The editors also thank the reviewers who are leading researchers in the domain for spending their time from their busy schedules to give valuable suggestions to improve the quality of the contributed articles. Last but not least, the editors are inclined to express their sincere thanks to Springer Nature, Singapore, for being the publishing partner. But for their acceptance to publish this volume would never have been possible in the present standard.

Enjoy reading it.

Kalyani, India   Jyotsna K. Mandal
Silchar, India   Somnath Mukhopadhyay
Bolpur, Santiniketan, India   Paramartha Dutta

Contents

About the Editors

Dr. Jyotsna K. Mandal received his M.Tech. in computer science from the University of Calcutta and his Ph.D. from Jadavpur University in the field of data compression and error correction techniques. Currently, he is Professor of computer science and engineering and Director of IQAC at the University of Kalyani, West Bengal, India. He is Former Dean of Engineering, Technology and Management (2008–2012). He has 29 years of teaching and research experience. He has served as Professor of computer applications, Kalyani Government Engineering College, for two years and as Associate and Assistant Professor at the University of North Bengal for 16 years. He has been Life Member of the Computer Society of India since 1992. Further, he is Fellow of the IETE and Member of the AIRCC.

He has produced 146 publications in various international journals and has edited twenty volumes as Volume Editor for Science Direct, Springer, CSI, etc. and has successfully executed five research projects funded by the AICTE, Ministry of IT, Government of West Bengal. In addition, he is Guest Editor of *Microsystem Technology Journal* and Chief Editor of the *CSI Journal of Computing*.

Somnath Mukhopadhyay is currently Assistant Professor in the Department of Computer Science and Engineering, Assam University, Silchar, India. He completed his M.Tech. and Ph.D. in computer science and engineering at the University of Kalyani, India, in 2011 and 2015, respectively. He has co-authored one book and has five edited books to his credit. He has published over 20 papers in various international journals and conference proceedings, as well as three chapters in edited volumes. His research interests include digital image processing, computational intelligence and pattern recognition. He is Member of IEEE and IEEE Computational Intelligence Society, Kolkata Section; Life Member of the Computer Society of India; and currently Regional Student Coordinator (RSC) of Region II, Computer Society of India.

Dr. Paramartha Dutta completed his bachelor's and master's degrees in statistics at the Indian Statistical Institute, Calcutta, in 1988 and 1990, respectively. He received his Master of Technology in computer science from the same institute in

1993 and a Ph.D. in engineering from Bengal Engineering and Science University, Shibpur, in 2005. He is currently Professor in the Department of Computer and System Sciences, Visva Bharati University, West Bengal, India. Prior to this, he served Kalyani Government Engineering College and College of Engineering in West Bengal as a Full-Time Faculty Member.

He has co-authored eight books and has five edited books to his credit. He has published over 180 papers in various journals and conference proceedings, both international and national, as well as several book chapters in books from respected international publishing houses like Elsevier, Springer-Verlag, CRC Press and John Wiley. He has also served as Editor of special volumes of several prominent international journals.

Non-dominated Sorting Based Multi/Many-Objective Optimization: Two Decades of Research and Application

Haitham Seada and Kalyanmoy Deb

1 Introduction

The date Vector Evaluated Genetic Algorithms (VEGA) (Schaffer 1985) was proposed is the birthdate of Evolutionary Multi-objective Optimization (EMO), despite a few earlier suggestions on the importance of handling multiple objectives within an evolutionary algorithm. Since then, the classical trend of combining all objectives into one fitness function started to fade. The history of this relatively new field can be viewed from several different perspectives. Here, we are more concerned about the role non-dominated sorting played throughout this history. The concept of non-domination is related to the concept of "Pareto Optimality" first proposed by the Italian economist "Vilfredo Pareto", hence the naming. "Pareto Optimality" is simply a state of resource allocation among multiple criteria, where it is impossible to reallocate resources so as to make one criterion better without degrading one or more other criteria. In this context, a *non-dominated* solution is a solution that is not possible to be outperformed in all criteria simultaneously. Unlike "Pareto Optimality", "Non-domination" can be relative to a set of supplied solutions. A set of solutions are non-dominated with respect to each other if none of them outperforms any of the others in all criteria simultaneously, even if none of them is actually Pareto optimal. Figure 1 shows how the two terms are related yet slightly different.

H. Seada
Brown University, 70 Ship, Providence, RI 02903, USA
e-mail: haitham_seada@brown.edu; seadahai@msu.edu

K. Deb (✉)
Michigan State University, 428 South Shaw Lane, 2320 EB, East Lansing, MI 48824, USA
e-mail: kdeb@egr.msu.edu
URL: http://www.coin-laboratory.com

© Springer Nature Singapore Pte Ltd. 2018
J. K. Mandal et al. (eds.), *Multi-Objective Optimization*,
https://doi.org/10.1007/978-981-13-1471-1_1

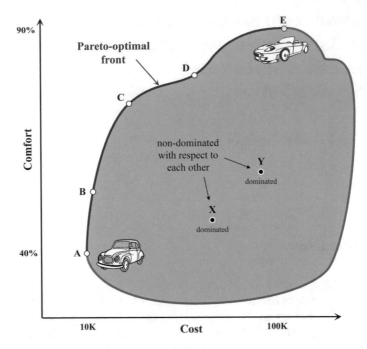

Fig. 1 Pareto optimality and non-domination

Everyone uses these principles in her/his daily life. If you are buying a new car, you will *ideally* be willing to get the cheapest yet most comfortable and luxurious car! This is called the *ideal* solution. In such a situation, the ideal solution does not exist, because your objectives are conflicting. This situation is a multi-objective optimization problem. Let us assume—for the sake of the argument—that the ideal solution exists! This means that your objectives are aligned and combining them into one or even optimizing just one of them is sufficient, which in turn means that your problem should have been formulated as a single-objective optimization problem in the first place, which is not the case in our car example. Figure 1 explains Pareto optimality and non-domination in the context of our car example. Notice that we are trying to minimize cost (x-axis) and maximize comfort (y-axis), thus for a car x to dominate a car y, the price of x must be lower and it needs to be more comfortable.[1] Consequently, any solution in the top left corner of another dominates it. This explains why the red line represents the "Pareto front", because any point (or, car) on this line can never be dominated, as it lies on the top left border of the entire search space. So, cars A, B, C, D, and E are Pareto optimal. On the other hand, notice that X is dominated by B, C, and D, while Y is dominated by C and D, yet they

---

[1] For simplicity, we only consider *strong* domination. However, according to the definition, domination can be *weak* as well. In our example, weak domination is when two cars have exactly the same price but one of them is more comfortable than the other. The more comfortable car is said to *weakly* dominate the other.

are both non-dominated with respect to each other.[2] Solutions A, X, Y, and E are all non-dominated to each other. With the knowledge of Pareto optimal solutions, we observe that solutions A, and E are Pareto optimal, but solutions are X and Y are not.

> In a minimization context, a vector $X = (x_1, x_2 \ldots x_m)$ is said to be non-dominated with respect to $Y = (y_1, y_2 \ldots y_m) \iff x_i \leq y_i \ \forall \ i$ and $x_i < y_i$, for at least one i. In this case, Y is said to be dominated-by or inferior-to X.

Although VEGA did not use non-dominated sorting, it used the non-domination principle to identify dominated solutions and penalize them. After VEGA, many successful evolutionary multi-objective optimization (EMO) algorithms were proposed. MOGA (Fonseca et al. 1993), NPGA (Horn et al. 1994), and NSGA (Srinivas and Deb 1994) are some of most notable initial studies in this filed. In these algorithms, researchers studied several potential approaches to handle more than one objective—mainly two objectives—through achieving a careful balance between convergence and diversity preservation. For achieving convergence to Pareto optimal solutions, these algorithms continued to further incorporate the concept of non-domination (Deb 2001). MOGA was the first algorithm to introduce the idea of grouping solutions into different layers based on their non-dominated ranking. A true Pareto optimal solution will never be dominated, and consequently it will always be ranked 1. In terms of diversity preservation, this family of algorithms continued to use several diversity preservation methods studied in the context of single-objective evolutionary computation methods (Goldberg and Richardson 1987; DeJong 1975; Deb and Goldberg 1989).

A second wave of algorithms followed. These algorithms incorporated elite-preservation concept which ensures the survival of non-dominated and well-diversified solutions from one generation to the next. Some of the most prominent studies of this wave are NSGA-II (Deb et al. 2002), SPEA (Zitzler and Thiele 1999), SPEA2 (Zitzler et al. 2001), and PAES (Knowles and Corne 1999), among others (Deb 2001; Coello et al. 2002; Goh and Tan 2009; Tan et al. 2006; Knowles et al. 2007).

Out of all these algorithms, NSGA-II has been the most widely used in the field (Jeyadevi et al. 2011; Dhanalakshmi et al. 2011; Fallah-Mehdipour et al. 2012; Bolaños et al. 2015; Arora et al. 2016). As showin in Fig. 2, in NSGA-II, elitism is ensured by merging both parents and their offspring into one double-sized combined population, sorting them, and then keeping the better half of the combined population. For diversity preservation, the authors of NSGA-II proposed an efficient approach where each member is assigned a crowding distance value. This value is simply an approximation of the perimeter of the cuboid whose edges are those pairs of solutions surrounding the designated solution in each dimension as shown in Fig. 3. The larger the crowding distance value, the more important the solution is.

---

[2]Unlike the car example, throughout the rest of this chapter, we assume a minimization context unless otherwise stated.

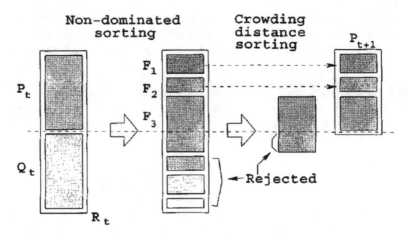

Fig. 2 NSGA-II algorithm (taken from (Deb et al. 2002))

Fig. 3 Crowding distance in
NSGA-II (taken from (Deb
et al. 2002))

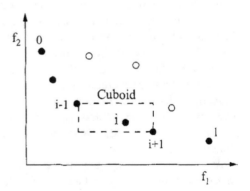

This is because such an individual[3] is considered the only representative of a less
crowded portion of the objective space. By continuously emphasizing lower ranks
than larger crowding distance values (in this specific order), NSGA-II achieves high
levels of both convergence and diversity eventually. The whole NSGA-II procedure
is shown in Fig. 2. On the other hand, SPEA and SPEA2 maintain an external archive
of outstanding solutions (in terms of both convergence and diversity) and use these
archived solutions in the creation of new populations. Another approach was adopted
by PAES which performs a competition between a parent and its child to enforce
elite preservation.

All the aforementioned algorithms suffer—in general—their inability to scale
up to more than two objectives.[4] And although most of them are applicable—
in theory—to any number of objectives, their performance obviously degrades
significantly (Khare et al. 2003). Only recently, researchers were able to devise new

---

[3] The terms "solution", "individual" and "points" are used interchangeably throughout this chapter.
[4] It is worth noting that some of the later ones were shown to solve up to three objectives as well.

scalable algorithms that can solve problems involving *many* objectives (Deb and Jain 2014; Jain and Deb 2014; Asafuddoula et al. 2015; Zhang et al. 2015; Li et al. 2015; Garza-Fabre et al. 2009; Adra and Fleming 2011; Khare et al. 2003; Ishibuchi et al. 2008; Hadka and Reed 2013; Singh et al. 2011; Aguirre and Tanaka 2009; Sato et al. 2010; Zou et al. 2008; Wickramasinghe and Li 2009; Purshouse and Fleming 2007; Köppen and Yoshida 2007; Zhang and Li 2007; Sindhya et al. 2013). Later, we will discuss some of the most prominent-based many-objective optimization algorithms.

Multi-objective   This term usually refers to scenarios involving two or three objectives at most.

Many-objective   This recently coined term refers to problems having more than three objectives, and possibly up to 20 objectives and in rare conditions even more.

Today, the field is more active than ever. Hundreds of researchers from a wide range of scientific disciplines have joined the field since its advent 30 years ago (Schaffer 1985). And it is easy to tell how far the field has gone.

The rest of this chapter is organized as follows. Section 2 shows the differences in effect and behavior of non-dominated sorting across different objective dimensions. Some of the most recent successful-based EMO algorithms are discussed in Sect. 3. Section 4 is dedicated to dive into the details of one of the state-of-the-art algorithms that combine non-dominated sorting with both point-to-point local search and a recently proposed theoretical convergence metric. Sections 3 and 4 capture briefly some of the results included in the original respective studies. Finally, Sect. 5 concludes our discussion.

2 Across Different Scenarios

Although the definition of non-dominated sorting is the same across all dimensions, its effect can be very different from one to the other. In this section, we will discuss these differences and explain the reasons behind them across the three main dimensional categories, namely, multi-objective, many-objective, and even single-objective.

2.1 Multi/Many-Objective Optimization

Visualization is an important tool in optimization. A visual inspection of your results can help identify gaps, missing extremes, and more importantly, any unexpected behavior of the algorithm under investigation. In multi-objective optimization, visualization is possible. In bi-objective problems, the entire objective space can be plotted in a two-dimensional plot where the Pareto front is represented by a line,

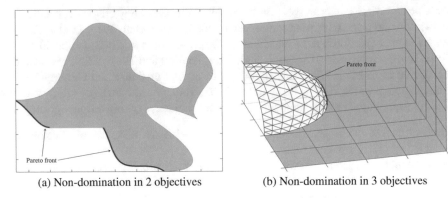

(a) Non-domination in 2 objectives   (b) Non-domination in 3 objectives

Fig. 4 Non-domination in multi-objective optimization

Fig. 5 Number of
non-dominated solutions is
directly proportional to the
number of objectives

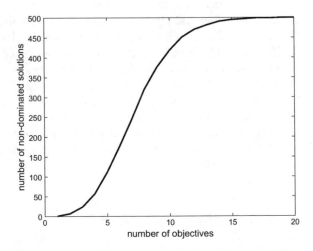

see Fig. 4. In cases involving three objectives, the objective space can be plotted in a three-dimensional plot, and the Pareto front will be a surface, see Fig. 4a. Obviously, beyond three objectives, visualization becomes a problem. Researchers have proposed numerous techniques to mitigate this problem.

Another problem arises from the fact that virtually all non-domination-based algorithms use non-dominated sorting (or at least the non-domination principle itself) as their main convergence pressure utility. This is not a problem in multi objective optimization per se; however, as the number of objectives increase, the selection pressure starts to fade! Figure 5 shows how the number of non-dominated solutions in a randomly generated population of 500 solutions increases at higher dimensions. For example, the figure shows that at 20 objectives, almost all solutions will be non-dominated with respect to each!

2.2 Single-objective Optimization

It is worth noting that researchers have utilized non-dominated sorting in single-objective optimization problems as well (Murugan et al. 2009; Deb and Tiwari 2008; Seada and Deb 2016; Sharma et al. 2008). These efforts can be roughly classified into two different approaches. The first approach is to add artificial objectives to an originally single-objective optimization problem. For example, some studies add constraint(s) violation minimization as an additional objective (Murugan et al. 2009). This optimization problem now becomes technically a multi-objective optimization problem, with a domination relationship among its solutions. Other researchers realized that even if non-dominated sorting is applied to a problem in its single-objective status, it degenerates to an ordinal comparison operator, which is an essential operation for progressing toward the optimum solution for a single-objective optimization problem. Consequently, the second approach appeared where researchers apply multi-objective optimization algorithms—involving non-dominated sorting—directly to single-objective optimization problems (Sharma et al. 2008; Deb and Tiwari 2008; Seada and Deb 2016).

3 Recent Non-dominated Sorting Based Algorithms

As mentioned in Sect. 2, researchers continue to employ non-dominated sorting in their algorithm even in the era of many objectives. One of the most notable efforts in this regard is NSGA-III, proposed by Deb and Jain in 2014 (Deb and Jain 2014; Jain and Deb 2014). They built on top of NSGA-II to create an algorithm capable of handling many objectives. The original study investigated a wide range of dimensionalities (3 to 15 objectives). Aside from those parameters required by standard evolutionary operations (e.g., population size, crossover probability, etc.), NSGA-III does not need any parameters, and thus dubbed "parameterless". It uses the same general framework of NSGA-II, but with a modified diversity-preserving operator that is based on a decomposition concept similar to that of MOEA/D (Zhang and Li 2007). Since many of the current lines of research in EMO uses NSGA-III or its extensions, we devote the next section to discuss its general philosophy and some of its details. For detailed explanations and extensive simulation results, the reader is advised to consult (Deb and Jain 2014; Jain and Deb 2014).

3.1 NSGA-III and Its Variants

NSGA-III starts with generating N random solutions (initial population) and a set of H prespecified M-dimensional *reference points*[5] that are evenly distributed on

---

[5]Although the original study used the notion of a "reference point", here we will mostly use the notion of a reference direction, which is the vector connecting the ideal point to the "reference

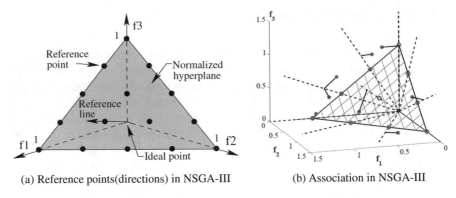

(a) Reference points(directions) in NSGA-III   (b) Association in NSGA-III

Fig. 6 NSGA-III operation

a unit hyperplane having a normal vector of ones covering the entire R_+^M region. The way the hyperplane is placed ensures that it intersects each objective axis at 1 (see Fig. 6a (Deb and Jain 2014)). $H = \binom{M+p-1}{p}$ reference points $((p + 1)$ along each boundary) are placed in the hyperplane using Das and Dennis's method (Das and Dennis 1998) N and H are chosen to be equal (or as close as possible). The justification behind this is that eventually one solution is expected to be found for each reference direction.

At generation t, recombination and mutation operators are applied to P_t (parent population) to generate Q_t (offspring population). Since the algorithm targets only one population member per reference direction, there should be no competition among reference directions. To avoid this inter-direction competition, NSGA-III does not have selection among feasible solutions. P_t and Q_t are then combined $R_t = P_t \cup Q_t$, and the whole combined population is sorted into several non-domination levels, and in the same way it is done in NSGA-II. Starting at the first front (non-domination level), solutions are included in P_{t+1} (the next parent population) one front at a time. Typically, the algorithm will reach a front that has more individuals than the remaining slots in the next parent population, i.e., a front that cannot be fully accommodated in P_{t+1}. Let us denote this last front as F_L. In such a case, a niche-preserving operator is used to select the subset of F_L that will be included in P_{t+1}. The niche-preserving operator works as follows. Each member of P_{t+1} (the new parent population that is partially full so far) and F_L (the last un-accommodated front) is *normalized* using extreme values of the current population (representing spread). After normalization, all objective vectors and reference directions should have commensurate values. Thereafter, each member of P_{t+1} and F_L is associated with its closest reference direction (in terms of perpendicular distance). Finally, in order to completely fill P_{t+1}, NSGA-III uses a careful *niching* strategy where it chooses those F_L members that represent the most sparse regions (reference directions) in P_{t+1}. The goal of

---

point". We found this notion more conceivable and avoids the confusion that may arise between the two terms "point" (an actual solution) and "reference point" (a hypothetical point in the objective space).

niching is to select a population member for as many supplied reference directions as possible. Niching prefers solutions attached to under-represented/un-represented reference directions. Because NSGA-III maintains a constant preference of non-dominated solutions along with the aforementioned niching strategy, the algorithm is expected to reach a Pareto optimal solution in the close vicinity of each supplied reference direction (provided that genetic variation operators can produce respective solutions), hence, an well-distributed sample representing the true Pareto optimal front (see Fig. 6b (Deb and Jain 2014)).

ALGORITHM 1 Generation t of NSGA-III procedure

Input: H structured reference directions Z^s or supplied aspiration points Z^a, parent population P_t
Output: P_{t+1}
1: $S_t = \emptyset, i = 1$
2: $Q_t =$ Recombination+Mutation(P_t)
3: $R_t = P_t \cup Q_t$
4: $(F_1, F_2, \ldots) =$ Non-dominated-sort(R_t)
5: **repeat**
6:     $S_t = S_t \cup F_i$ and $i = i + 1$
7: **until** $|S_t| \geq N$
8: Last front to be included: $F_l = F_i$
9: **if** $|S_t| = N$ **then**
10:     $P_{t+1} = S_t$, break
11: **else**
12:     $P_{t+1} = \cup_{j=1}^{l-1} F_j$
13:     Points to be chosen from F_l: $K = N - |P_{t+1}|$
14:     Normalize objectives and create reference set Z^r: `Normalize`$(\mathbf{f}^n, S_t, Z^r, Z^s, Z^a)$
15:     Associate each member $\mathbf{s}$ of S_t with a reference point: $[\pi(\mathbf{s}), d(\mathbf{s})] =$`Associate`$(S_t, Z^r)$
     % $\pi(\mathbf{s})$: closest reference point, d: distance between $\mathbf{s}$ and $\pi(\mathbf{s})$
16:     Compute niche count of reference point $j \in Z^r$: $\rho_j = \sum_{\mathbf{s} \in S_t/F_l} ((\pi(\mathbf{s}) = j) ? 1 : 0)$
17:     Choose K members one at a time from F_l to construct P_{t+1}: `Niching`$(K, \rho_j, \pi, d, Z^r, F_l, P_{t+1})$
18: **end if**

The original NSGA-III study (Deb and Jain 2014) showed very promising results in the many-objective realm. It is worth emphasizing that NSGA-III and its extension (Jain and Deb 2014) (for constraints handling) do not need any additional hard-coded parameters to achieve good performance.[6,7] That study has also introduced a computationally fast approach by which reference directions are adaptively updated on the fly based on the association status of each of them over a number of generations. The algorithm is outlined in Algorithm 1.

Instead of solving the reducing selection pressure problem at higher dimensions, NSGA-III views it as an advantage! According to the authors, in higher dimensions

---

[6]Other than the standard evolutionary parameters, e.g., population size, number of Solution evaluations(SEs)/generations, recombination/mutation probability, etc.

[7]One of the most important resources in optimization is the number of function evaluations (FEs) consumed to reach a solution. In a multi-objective optimization scenario, we use the term solution evaluation (SE) instead, as evaluating a single solution involves evaluating more than one function.

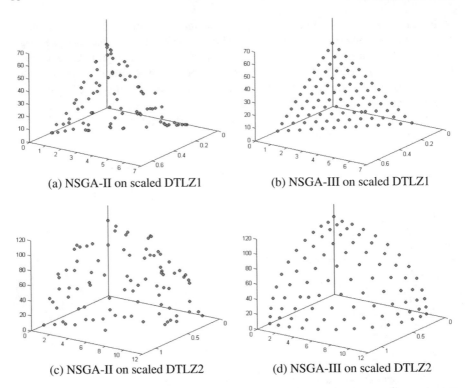

(a) NSGA-II on scaled DTLZ1   (b) NSGA-III on scaled DTLZ1

(c) NSGA-II on scaled DTLZ2   (d) NSGA-III on scaled DTLZ2

Fig. 7 Performance of NSGA-II and NSGA-III on scaled unconstrained three-objective DTLZ problems

the search space becomes too complex and the algorithm needs to slow down convergence to get enough opportunity to explore the whole search space. Otherwise, the algorithm might be trapped in a local Pareto front, or miss parts of the true Pareto front.

Figure 7 shows how NSGA-III achieves significantly better results than NSGA-II in three-objective scaled DTLZ1 and DTLZ2 problems (Deb et al. 2005). Obviously, NSGA-II fails to a well distribution completely.

NSGA-III is now being extensively used in a diverse set of fields including software engineering (Mkaouer et al. 2014, 2015), quality control (Tavana et al. 2016), renewable energy (Yuan et al. 2015), and feature selection (Zhu et al. 2017), among others. It was also extended to create new algorithms with specific desired features. One such extension is U-NSGA-III which is a parameterless extension of NSGA-III that improves on its ability to handle single-objective optimization problems, while maintaining the same outstanding performance in higher dimensions. θ-NSGA-III is another parameterized extension that is intended to solve the reducing selection pressure problem and maintain a better balance between convergence and diversity. Other extensions can also be found in literature (Ibrahim et al. 2016; Yuan et al. 2016a).

3.2 Other Successful Algorithms

So far, we have focused on non-domination-based EMO algorithms; however, there have been other successful attempts that are did not depend on non-domination. We devote this section to discuss some of these efforts. The most important effort in this line is Zhang and Li's MOEA/D (Zhang and Li 2007), which is considered one of the first multi/many-objective optimization algorithms. They proposed a decomposition-based method, guided by a set of predefined evenly distributed set of reference directions in the objective space. The original multi-objective optimization problem is decomposed into a set of single-objective optimization subproblems, one problem for each reference direction. Each subproblem utilizes information from its neighboring subproblems only. The authors proposed several approaches for formulating these subproblems. The most successful approach is the penalty-based boundary intersection (PBI) approach shown in Fig. 8 (notice that the original study was proposed in a maximization context). In this approach, for a given point x, objective vector $F(x)$, reference point z^*, and reference direction λ, the weighted sum of the two distances d_1 and d_2—in the objective space—is minimized. d_1 represents the Euclidean distance between z^* and the projection of $F(x)$ on λ, while d_2 represents the Euclidean distance between $F(x)$ and its projection on λ. Other formulation approaches including simple weighted sum, Tchebycheff, and non-penalized boundary intersection (BI) have also been discussed.

Although their original study showed better results compared to NSGA-II and other algorithms up to 4 objectives, their algorithms have been used successfully elsewhere up to 10 objectives (Deb and Jain 2014).

Researchers extended MOEA/D in many ways as well. Wang et al. proposed a modification of its replacement strategy (Global replacement) (Wang et al. 2014). The original version of MOEA/D simply assumes that a solution coming out of a specific subproblem can only replace another solution in the vicinity of this subprob-

Fig. 8 MOEA/D PBI (taken from (Zhang and Li 2007))

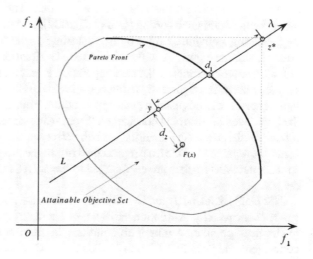

lem. Their approach on the other hand looks for the subproblem at which the new solution fits best and performs replacement in the vicinity of this subproblem instead of the original one, hence the name "Global replacement". They showed how this simple modification performs better on a set of test problems. Another recent study by Yuan et al. (2016b) proposed a steady-state modified version of MOEA/D that achieves better balance. In their study, they use a slightly relaxed mating restriction (controlled by a probability parameter δ). Their approach—MOEA/D-DU—utilizes the perpendicular distances between solutions and directions (d_2) for enhanced diversity. Another interesting extension is the one proposed by Li et al. (2014) where they used a stable matching selection model (Gale and Shapley 1962) to coordinate the selection process in MOEA/D.

Some researchers tried to get the best of both worlds by combining both non-dominated sorting used in NSGA-II and NSGA-III with the notion of sub-regions and mating selection from MOEA/D (Li et al. 2015).

4 State-of-the-Art Combinations

In this section, we will dive into the details of one of the most recent algorithmic developments of non-domination-based EMO algorithms, namely B-NSGA-III. B-NSGA-III is particularly interesting because it is a combination of EMO, point-to-point local search, and the recently proposed Karush Kuhn Tucker proximity measure. It shows the ability of carefully designed hybrid algorithms to handle complex problems with a significantly reduced number of solution evaluations.

B-NSGA-III retains the general outline of U-NSGA-III (Seada and Deb 2016) which in turn retains most of the features of NSGA-III. Starting with a randomly generated initial population, B-NSGA-III generates an equal number of offspring individuals (solutions/points) using niche-based tournament selection, simulated binary crossover, and polynomial mutation (Deb and Agrawal 1995). The two populations are then combined and the ideal point is updated. The combined population goes through non-dominated sorting (Deb et al. 2002), and the next population is formed by collecting individuals front by front starting at the first front. Since population size is fixed, the algorithm will typically reach a situation where the number of individuals needed to complete the next population is less than the number of individuals available in the front currently being considered. B-NSGA-III collects only as many individuals as it needs using a *niching* procedure. This niching procedure normalizes the objectives of all fronts considered. Then, using a fixed set of evenly distributed reference directions (in the normalized objective space), preference is given to those solutions representing the least represented reference directions in the objective space so far. Interested readers are encouraged to consult (Seada and Deb 2016) for more details.

NSGA-II, NSGA-III, and U-NSGA-III maintain a constant preference of convergence over diversity. A solution in front $n + 1$ will never be considered for inclusion in the next population unless all solutions in front n are already included. This convergence-always-first scheme has been recently criticized by several researchers

(Zhang and Li 2007; Li et al. 2010). B-NSGA-III breaks this constant emphasis on convergence, for every α generations. The algorithm changes its survival selection strategy, by favoring solutions solely representing some reference directions over other—possibly dominating—redundant solution. A redundant solution is a solution that is not the best representative of its niche, i.e., there exists another solution—in the same front—that is closer to the reference direction representing their niche.

Another difference between U-NSGA-III and B-NSGA-III is that U-NSGA-III treats all individuals/regions of the search space equally at all generations. And although it might seem better from a generic point of view, B-NSGA-III claims the opposite. Being fair the way U-NSGA-III is can lead to wasting SEs on easy sections of the Pareto front. Those wasted SEs could have been put into better use, if they were directed toward reaching more difficult sections of the front. Obviously, in order to achieve the maximum possible utilization of SEs, a truly dynamic algorithm that gives more attention to more difficult sections/points of the front is needed. However, designing such an algorithm needs an Oracle that knows deterministically the difficulty of attaining each point on the front relative to others. Unfortunately, for an arbitrary optimization problem, perfecting such an Oracle is a far-fetched dream so far, despite some studies (Liu et al. 2014). However, recent studies show some clues that can drive creating a nondeterministic version of the targeted Oracle. We summarize these clues in the following points:

1. Researchers have repeatedly shown that the important role normalization plays especially in achieving better coverage of the Pareto front. Usually, the extreme points of the current population dictate normalization parameters. During evolution, as new extreme points appear all previously normalized objective values become outdated, and normalization is repeated, and hence, the importance of extreme points in optimization. And as pointed in Talukder et al. (2017), the earlier we reach extreme points, the better normalization we have and the better coverage we attain.

2. Reaching some parts of the Pareto front may require more effort than others. Several tests as well as real-world problems exhibit such behavior (Zitzler et al. 2000; Deb et al. 2005). Usually, such difficult regions appear as gaps in the first front. In reference direction-based optimization algorithms (like MOEA/D, NSGA-III and U-NSGA-III), gaps can be identified by looking for reference directions having no associations so far.

3. In multi-objective optimization, all non-dominated solutions are considered equally good, thus deserving equal attention. This is not ideal though. Being non-dominated with respect to each other does not mean that two solutions are equally converged. The recently published approximate KKTPM now enables us to efficiently differentiate non-dominated solutions based on their proximity from local optima.

These clues are realized in B-NSGA-III through several phases. In α generations, the algorithm switches back and forth among three different phases. Each phase uses one some LS operator to fulfill its goal. The following subsection discusses these phases in greater detail.

4.1 Alternating Phases

One naive approach is to use sequential phases. Given their relative importance, the first phase may seek extreme points. Once found, the algorithm moves to subsequent phases and never looks back. But, as shown in Talukder et al. (2017), reaching true extreme points is not a trivial task. Even using LS, several optimizations might be needed to attain one extreme point. And since we can never safely assume that we have reached the true extreme points, this sequential design is not recommended. The same argument is valid for covering gaps. In an earlier generation, although your solutions may not provide the desired spread, it might be the case that there are no gaps within the small region they cover. As generations proceed and solutions expand, gaps may appear. This is a frequent pattern that is likely to repeat through an optimization run. Again, the sequential pattern is prone to failure, as we can never know if more gaps will appear in the future.

Another more involved yet simple approach is to move from one phase to the next after a fixed number of generations (or SEs). Once the algorithm reaches the final phase, it goes back to the first cyclically. This cyclic approach is more appealing, but how many generations (or SEs) to wait before switching from one phase to the next? Obviously, it is never easy to tell. In addition, using this rigid design obligates the algorithm to spend resources (SEs) in *possibly unnecessary* phases, just because it is their turn in the alternation cycle.

B-NSGA-III alternates among three phases *dynamically* and *adaptively*. Figure 9 shows the three phases in action. During *Phase-1*, the algorithm seeks extreme points. *Phase-2* is where an attempt is made to cover gaps found in the non-dominated front. Finally, during *Phase-3*, the focus is shifted toward helping poorly converged non-dominated solutions. In order to avoid the shortcomings of the two aforementioned

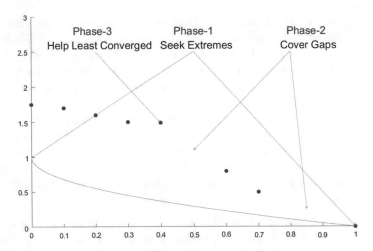

Fig. 9 *Phase-1*, *Phase-2*, and *Phase-3* in action

Fig. 10 Alternation of
phases

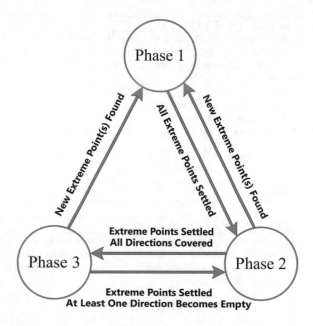

approaches, B-NSGA-III watches for specific incidents that trigger transitions from one phase to another. Those transitions are completely unrestricted, i.e., B-NSGA-III can move from any phase to the other if the appropriate trigger is observed. Figure 10 shows all possible transitions along with their triggers.

In the first α generation, the algorithm puts itself in *Phase-1*. Algorithm 2 shows the details of this phase. For an M objectives problem, *Phase-1* uses an extreme-LS operator (discussed later) to search for its M extreme points. If all extreme points remain unchanged from one α generation to the next, B-NSGA-III assumes *temporarily* that these settled points are the true extreme points, and moves to *Phase-2*. While being in phases 2 or 3, finding a better extreme point through evolution indicates that those extreme points previously settled are not the true ones. Consequently, B-NSGA-III returns to *Phase-1* in search for better extreme points again.

As mentioned earlier, B-NSGA-III gives a chance to possibly dominated solutions that solely represent their niche, every α generations. This is shown in Algorithm 3, line 3 and expanded in Algorithm 4. The points surrounding each *non-empty* reference direction are collected from the merged population (parents and offspring) (line 2), and the best ranked point is selected to represent this direction/niche (line 4). If more than one point share the same rank, the point closest to the direction is selected (line 5). Obviously, as opposed to U-NSGA-III points in B-NSGA-III compete only with their niche peers, which means that an inferiorly ranked point from one niche can be included because it is the best representative of its niche, while a superior point from another niche is left out because a better representative if its niche exists.

Once in *Phase-2*, B-NSGA-III looks for reference directions having no associations in the first front (*empty directions*). These directions represent gaps

ALGORITHM 2 Phase 1

Input: parent population (P), offspring (O) population size (N), reference directions (D), ideal point (I), intercepts (T), maximum number of function evaluations $(FeMax)$, maximum number of local search operations per iteration β, augmentation factor ε

Output: None

1: $All \leftarrow P \cup O$
2: $E \leftarrow getExtremePoints(All)$
3: **for** i = 1 to M **do**
4:   $E_i \leftarrow localSearch_{BWS}(E_i, I, T, FeMax, \varepsilon)$,
          $i = 1, \ldots, M$
5:   $O(randomIndex) \leftarrow E_i$,
          $1 \leq randomIndex \leq |O|$
6:   $i \leftarrow i + 1$
7: **end for**

ALGORITHM 3 Phases 2 and 3

Input: parents (P), offspring (O), number of objectives (M), population size (N), reference directions (D), ideal point (I), intercepts (T), maximum number of function evaluations $(FeMax)$, maximum number of local search operations per iteration β, last point used to cover direction d $(prev_d)$ for all d in D

Output: New Population $(\hat{P})$

1: $F \leftarrow nonDominatedSorting(All)$
2: $All \leftarrow P \vee O$
3: $\hat{P} \leftarrow getBestWithinNiche(d, O) \ \forall d \in D$
4: **if** $stagnant(E)$ **then**
5:   $D_{empty} \leftarrow \{d \in D \mid (\nexists x)[x \in F_1 \wedge x \in d_{surroundings}]\}$
6:   $s_{kktpm} \leftarrow calculateKKTPM(s) \ \ \forall s \in \hat{P}$
7:   **for** i = 1 to β **do**
8:     **if** $D_{empty} \neq \phi$ **then**   ► Phase-2
9:       $d \leftarrow randomPick(D_{empty})$
10:      $s \leftarrow \{x \in F_1 \mid$
              $(\nexists y)[y \in F_1 \wedge \perp_d (y) <\perp_d (x)]\}$
11:      **if** $prev_d = null$ or $\perp_d (s) <\perp_d (prev_d)$ **then**
12:        $prev_d \leftarrow s$
13:      **else**
14:        $s \leftarrow null$
15:        $D_{empty} \leftarrow D_{empty} \setminus \{d\}$
16:      **end if**
17:    **else**   ► Phase-3
18:      $s \leftarrow \{x \in \hat{P} \mid$
              $(\nexists y)[y \in \hat{P} \wedge y_{kktpm} > x_{kktpm}]\}$
19:    **end if**
20:    **if** $s \neq null$ **then**
21:      $\hat{s} \leftarrow localSearch_{ASF}(s, I, T, FeMax)$
22:      $\hat{P} \leftarrow \hat{P} \vee \{\hat{s}\}$
23:      $\beta \leftarrow \beta + 1$
24:    **end if**
25: **end for**
26: **end if**

in the non-dominated front (Algorithm 3, line 5). If several such directions exist, one is picked randomly (line 9) and the closest first front point to this direction is saved (line 10) to be used later as a starting point in local search. Notice that lines 11–16 ensure that B-NSGA-III will not retry to cover a gap until a closer starting point than the one previously used is found. If no empty directions exist, B-NSGA-III moves to *Phase-3*, looking for the least converged point among those selected so far. This point should have the highest KKTPM among all (line 18). An ASF-LS operator (discussed later) is employed in both cases (line 21), either to cover a gap (*Phase-2*) or to bring a poorly converged point closer to the front (*Phase-3*). In order to keep SEs as low as possible, a maximum of β LS operations are allowed, even if the number of gaps is more than β. Notice the ability of the algorithm to move directly from *Phase-1* to *Phase-3* if no gaps are found.

ALGORITHM 4 getBestWithinNiche(...)

Input: merged parents and offspring (All), reference directions (D)
Output: selected Individuals (one from each niche) ($\hat{P}$)
1: $\hat{P} \leftarrow \phi$
2: $S \leftarrow getSurroundings(d, All)$
3: **for all** $d \in D$ **do**
4:    $X_d \leftarrow \{x \in S \mid (\nexists y)[y \in S \land y_{rank} < x_{rank}]\}$
5:    $x_d \leftarrow \{x \in X_d \mid (\nexists y)[y \in S \land \perp_d (y) <\perp_d (x)]\}$
6:    $\hat{P} \leftarrow \hat{P} \cup \{x_d\}$
7: **end for**

ALGORITHM 5 fillUpPop(...)

Input: merged parents and offspring (All), population size (N), partially full new population ($\hat{P}$)
Output: completely full new population ($\hat{P}$)
1: $All \leftarrow All \setminus \hat{P}$
2: **while** $|\hat{P}| < N$ **do**
3:    $Z \leftarrow \{x \in All \mid (\nexists y)[y \in All \land y_{rank} < x_{rank}]\}$
4:    $z \leftarrow pickRandom(Z)$
5:    $All \leftarrow All \setminus \{z\}$
6:    $\hat{P} \leftarrow \hat{P} \vee \{z\}$
7: **end while**

It is worth noting that phases 2 and 3 may run simultaneously. If the number of empty directions (gaps) is less than β, B-NSGA-III moves to *Phase-3* and uses the remaining budget to help poorly converged solutions. Obviously, *Phase-1* has the highest priority followed by *Phase-2* then *Phase-3*. The following two subsections discuss both Extreme-LS and ASF-LS operators in detail.

Finally, since all the three phases are not guaranteed to completely fill the next population, a final pass is made to fill up the next population using points that B-NSGA-III has discarded so far. Algorithm 5 shows that the best ranked points—out of those not included yet in the next population—are given higher priority.

4.2 Two Local Search Operators

As mentioned earlier, B-NSGA-III uses two different local search operators. In each, all the objectives are combined into some aggregate function (scalarization). Any single-objective optimizer can be used to minimize these aggregate functions. Here, we chose to use Matlab's® `fmincon()` optimization routine, a point-to-point deterministic optimizer. Point-to-point optimizers use less function evaluations compared to set-based methods (e.g., evolutionary algorithms). But they are also less guaranteed to reach global optima. Yet, in an alternating multi-phased algorithm like B-NSGA-III the embedded single-objective optimizer is not expected to reach the global optimum in one shot. That is why `fmincon()` fits our criteria for an embedded single-objective optimizer. Earlier, we discussed the role of our two LS operators. Next, we discuss their formulations and how they fit into their designated roles.

4.2.1 Extreme-LS

Phase-1 uses Extreme-LS to find extreme points. This operator is formulated simply as a biased weighted sum (BWS) aggregate function of all objectives (Eq. 1). $\tilde{f}_k(x)$ represents the normalized value of objective k. When seeking the ith extreme point, the term *Biased* refers to the significantly smaller weight (we call it *augmentation factor*) multiplied by the ith objective, compared to the weights of all other objectives. Although weighted sum aggregate functions are straightforward and easy to implement, they (including ours) can only reach points lying on convex sections of the Pareto front. While this makes them less plausible as a generic formulation, they perfectly serve their purpose in B-NSGA-III, since extreme points by definition can never lie in a non-convex section of the Pareto front. Unlike (Seada et al. 2017), the operator we are proposing here is normalized based on the number of objectives. This allows using the same *augmentation factor* for different dimensions. It is important to note that adding the ith objective term to the formula helps avoiding weakly dominated points.

$$\underset{\mathbf{x}}{\text{Minimize}} \quad \text{BWS}_i(\mathbf{x}) = \varepsilon \tilde{f}_i(x) + \sum_{j=1,\ j\neq i}^{M} \frac{w_j \tilde{f}_j(x)}{M-1}, \tag{1}$$

where ε is set as one percent of $\min_{j=1,j\neq i}^{M} w_j$.

4.2.2 Achievement Scalarization Function LS (ASF-LS)

As mentioned earlier, a generic LS operator that is required to get an arbitrary Pareto point cannot rely on BWS. That is why we use ASF to formulate our second LS operator, ASF-LS. The formulation in Eq. 2 shows that ASF-LS targets the intersection between the provided direction and the Pareto front. Since B-NSGA-III is a reference direction-based algorithm, ASF-LS can follow these already existing directions. And because of its ability to reach points lying on both convex and non-convex sections of the Pareto front, this is the operator employed in both *Phase-2* and *Phase-3* of B-NSGA-III. It is worth noting that according to our earlier experiments, ASF-LS does not perform as well if used to find extreme points. This can be attributed to the steep gradient of the aggregate ASF function (at these points) on one side of the global optimum, which usually misleads the single-objective optimizer. Hence, we need both operators in B-NSGA-III, each playing its designated role.

$$\underset{\mathbf{x}}{\text{Minimize}} \quad \text{ASF}(\mathbf{x}, \mathbf{z}^r, \mathbf{w}) = \overset{M}{\underset{i=1}{\max}} \left(\frac{\tilde{f}_i(x) - u_i}{w_i} \right), \tag{2}$$

$$\text{subject to} \quad g_j(\mathbf{x}) \leq 0, \quad j = 1, 2, \dots, J.$$

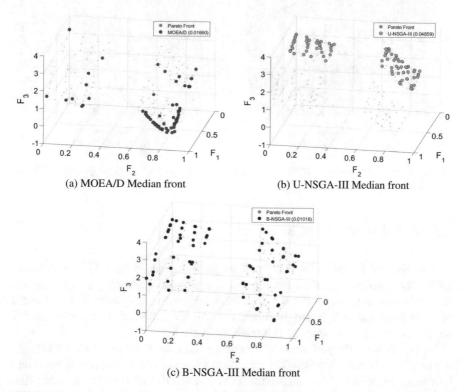

(a) MOEA/D Median front

(b) U-NSGA-III Median front

(c) B-NSGA-III Median front

Fig. 11 Median final fronts of MOEA/D, U-NSGA-III and B-NSGA-III on DTLZ7 (3 objectives) (Copied from (Seada et al. 2018))

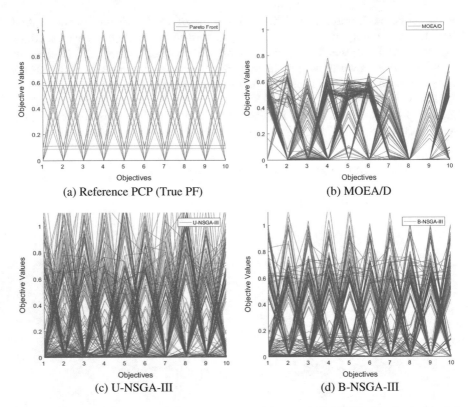

Fig. 12 PCPs of MOEA/D, U-NSGA-III, and B-NSGA-III on DTLZ4 (10 objectives) (Copied from (Seada et al. 2018))

4.3 B-NSGA-III Results

The authors of B-NSGA-III compared its performance to both U-NSGA-IIIand MOEA/D. Interested readers can consult the original study for an extensive set of simulations and results (Seada et al. 2018). Here, we will only show a sample of these results in both 3 and 10 objectives. All the following plots represent the median performance of their corresponding algorithms.

Figure 11 shows the performance of the three algorithms in three-objective DTLZ7 test problem (Deb et al. 2005). The true Pareto front of this problem is divided into four separated islands. Most algorithms fail to attain all islands and even if they did, they struggle to maintain a good distribution of solutions. Obviously, both MOEA/D and mine fail to cover the entire Pareto front. U-NSGA-III completely misses two of the four islands. MOEA/D was able to get some solutions at each island; however, the distribution is much worse than that of U-NSGA-III. On the other hand, B-NSGA-III is able to achieve a much better distribution of solutions all over the four islands.

Our second set of selected results shows the performance of the three algorithms on a 10-objective DTLZ4 problem (Deb et al. 2005). DTLZ4 is a variable density problem, where solutions are more dense toward one corner of the Pareto front. The results are shown in the form of parallel coordinate plots (PCPs). Obviously, B-NSGA-III median result (Fig. 12d) is the closest to the ideal performance (Fig. 12a).

5 Conclusions

In this chapter, we have discussed the role of non-dominated sorting in the EMO field over the last four decades. Non-dominated sorting has been one of the most useful utilities in this field since day one. The most widely used algorithms both in research and practice are non-domination dependent. Even in higher dimensions, non-dominated sorting still plays a significant role and researchers continue to devise new approaches that combine non-domination based sorting with decomposition, local search, and theoretical metrics.

Non-dominated sorting provides a more *global* picture of the best solutions in a population. Hence, an emphasis of non-dominated-based sorted points constitutes a more accurate search. The flip side is the computational requirement and the population issue, which may be prohibitive for it to be efficient for a distributed and parallel computing platform. Nevertheless, more than 25 years of research and application with non-dominated sorting based EMO methods have provided a solid foundation of its importance in evolutionary multi/many-objective optimization.

References

S.F. Adra, P.J. Fleming, Diversity management in evolutionary many-objective optimization. IEEE Trans. Evol. Comput. **15**(2), 183–195 (2011)

H. Aguirre, K. Tanaka, Many-objective optimization by space partitioning and adaptive ε-ranking on mnk-landscapes, in *Evolutionary Multi-criterion Optimization* (Springer, 2009), pp. 407–422

R. Arora, S. Kaushik, R. Kumar, R. Arora, Multi-objective thermo-economic optimization of solar parabolic dish stirling heat engine with regenerative losses using nsga-ii and decision making. Int. J. Electr. Power Energy Syst. **74**, 25–35 (2016)

M. Asafuddoula, T. Ray, R. Sarker, A decomposition-based evolutionary algorithm for many objective optimization. IEEE Trans. Evol. Comput. **19**(3), 445–460 (2015)

R. Bolaños, M. Echeverry, J. Escobar, A multiobjective non-dominated sorting genetic algorithm (nsga-ii) for the multiple traveling salesman problem. Decis. Sci. Lett. **4**(4), 559–568 (2015)

C.A.C. Coello, D.A. Van Veldhuizen, G.B. Lamont, *Evolutionary Algorithms for Solving Multi-Objective Problems*, vol. 242 (Springer, 2002)

I. Das, J.E. Dennis, Normal-boundary intersection: a new method for generating the pareto surface in nonlinear multicriteria optimization problems. SIAM J. Optim. **8**(3), 631–657 (1998)

K. Deb, *Multi-Objective Optimization Using Evolutionary Algorithms* vol.16 (Wiley, 2001)

K. Deb, R.B. Agrawal, Simulated binary crossover for continuous search space. Complex Syst. **9**(2), 115–148 (1995)

K. Deb, D.E. Goldberg (1989) An investigation of niche and species formation in genetic function optimization. in *Proceedings of the Third International Conference on Genetic Algorithms*, pp. 42–50

K. Deb, H. Jain, An evolutionary many-objective optimization algorithm using reference-point-based nondominated sorting approach, part i: solving problems with box constraints. IEEE Transa. Evol. Computat. **18**(4), 577–601 (2014)

K. Deb, S. Tiwari, Omni-optimizer: a generic evolutionary algorithm for single and multi-objective optimization. Eur. J. Oper. Res. **185**(3), 1062–1087 (2008)

K. Deb, A. Pratap, S. Agarwal, T. Meyarivan, A fast and elitist multiobjective genetic algorithm: Nsga-ii. IEEE Trans. Evol. Comput. **6**(2), 182–197 (2002)

K. Deb, L. Thiele, M. Laumanns, E. Zitzler (2005) Scalable test problems for evolutionary multiobjective optimization, in *Evolutionary multiobjective optimization theoretical advances and applications*, pp. 105–145

K.A. DeJong, An analysis of the behavior of a class of genetic adaptive systems. Ph.D. Thesis, Ann Arbor, MI, University of Michigan, dissertation Abstracts International 36(10), 5140B, 1975 (University Microfilms No. 76-9381)

S. Dhanalakshmi, S. Kannan, K. Mahadevan, S. Baskar, Application of modified nsga-ii algorithm to combined economic and emission dispatch problem. Int. J. Electr. Power Energy Syst. **33**(4), 992–1002 (2011)

E. Fallah-Mehdipour, O.B. Haddad, M.M.R. Tabari, M.A. Mariño, Extraction of decision alternatives in construction management projects: application and adaptation of nsga-ii and mopso. Expert Syst. Appl. **39**(3), 2794–2803 (2012)

C.M. Fonseca, P.J. Fleming et al., Genetic algorithms for multiobjective optimization: formulation, discussion and generalization. ICGA Citeseer **93**, 416–423 (1993)

D. Gale, L.S. Shapley, College admissions and the stability of marriage. Am. Math. Mon. **69**(1), 9–15 (1962)

M. Garza-Fabre, G.T. Pulido, C.A.C. Coello, Ranking methods for many-objective optimization, in *MICAI 2009: Advances in Artificial Intelligence* (Springer, 2009), pp. 633–645

C.K. Goh, K.C. Tan, *Evolutionary multi-objective optimization in uncertain environments* (Issues and Algorithms, Studies in Computational Intelligence, 2009), p. 186

D.E. Goldberg, J. Richardson, Genetic algorithms with sharing for multimodal function optimization, in *Genetic Algorithms and Their Applications: Proceedings of the Second International Conference on Genetic Algorithms* (1987), pp. 41–49

D. Hadka, P. Reed, Borg: An auto-adaptive many-objective evolutionary computing framework. Evol. Comput. **21**(2), 231–259 (2013)

J. Horn, N. Nafpliotis, D.E. Goldberg, A niched pareto genetic algorithm for multiobjective optimization, in *1994 Proceedings of the First IEEE Conference on Evolutionary Computation, IEEE World Congress on Computational Intelligence* (IEEE, 1994) pp. 82–87

A. Ibrahim, S. Rahnamayan, M.V. Martin, Deb K Elitensga-iii: an improved evolutionary many-objective optimization algorithm, in *2016 IEEE Congress on Evolutionary Computation (CEC)* (IEEE, 2016), pp. 973–982

H. Ishibuchi, N. Tsukamoto, Y. Nojima, Evolutionary many-objective optimization: a short review, in *IEEE Congress on Evolutionary Computation, Citeseer* (2008) pp. 2419–2426

H. Jain, K. Deb, An evolutionary many-objective optimization algorithm using reference-point based nondominated sorting approach, part ii: handling constraints and extending to an adaptive approach. IEEE Trans. Evol. Comput. **18**(4), 602–622 (2014)

S. Jeyadevi, S. Baskar, C. Babulal, M.W. Iruthayarajan, Solving multiobjective optimal reactive power dispatch using modified nsga-ii. Int. J. Electr. Power Energy Syst. **33**(2), 219–228 (2011)

V. Khare, X. Yao, K. Deb, Performance scaling of multi-objective evolutionary algorithms, in *Evolutionary Multi-Criterion Optimization* (Springer, 2003) pp. 376–390

J. Knowles, D. Corne, The pareto archived evolution strategy: a new baseline algorithm for pareto multiobjective optimisation, in *Proceedings of the 1999 Congress on Evolutionary Computation, CEC 99*, vol. 1 (IEEE, 1999)

J. Knowles, D. Corne, K. Deb, Multiobjective Problem Solving from Nature: from Concepts to Applications (Springer Science & Business Media, 2007)

M. Köppen, K. Yoshida, Substitute distance assignments in nsga-ii for handling many-objective optimization problems, in *Evolutionary Multi-Criterion Optimization* (Springer, 2007), pp. 727–741

K. Li, Q. Zhang, S. Kwong, M. Li, R. Wang, Stable matching-based selection in evolutionary multiobjective optimization. IEEE Trans. Evol. Comput. **18**(6), 909–923 (2014)

K. Li, K. Deb, Q. Zhang, S. Kwong, An evolutionary many-objective optimization algorithm based on dominance and decomposition. IEEE Trans. Evol. Comput. **19**(5), 694–716 (2015)

M. Li, J. Zheng, K. Li, Q. Yuan, R. Shen, Enhancing diversity for average ranking method in evolutionary many-objective optimization, in *Parallel Problem Solving from Nature, PPSN XI* (2010), pp. 647–656

H.L. Liu, F. Gu, Q. Zhang, Decomposition of a multiobjective optimization problem into a number of simple multiobjective subproblems. IEEE Trans. Evol. Comput. **18**(3), 450–455 (2014)

M.W. Mkaouer, M. Kessentini, S. Bechikh, K. Deb, M. Ó Cinnéide, High dimensional search-based software engineering: finding tradeoffs among 15 objectives for automating software refactoring using nsga-iii, in *Proceedings of the 2014 Annual Conference on Genetic and Evolutionary Computation* (ACM, 2014), pp. 1263–1270

W. Mkaouer, M. Kessentini, A. Shaout, P. Koligheu, S. Bechikh, K. Deb, A. Ouni, Many-objective software remodularization using nsga-iii. ACM Trans. Softw. Eng. Methodol (TOSEM) **24**(3), 17 (2015)

P. Murugan, S. Kannan, S. Baskar, Application of nsga-ii algorithm to single-objective transmission constrained generation expansion planning. IEEE Trans. Power Syst. **24**(4), 1790–1797 (2009)

R.C. Purshouse, P.J. Fleming, On the evolutionary optimization of many conflicting objectives. IEEE Trans. Evol. Comput. **11**(6), 770–784 (2007)

H. Sato, H.E. Aguirre, K. Tanaka, Pareto partial dominance moea and hybrid archiving strategy included cdas in many-objective optimization, in *2010 IEEE Congress on Evolutionary Computation (CEC)* (IEEE, 2010), pp. 1–8

J.D. Schaffer, Multiple objective optimization with vector evaluated genetic algorithms, in *Proceedings of the 1st International Conference on Genetic Algorithms* (L. Erlbaum Associates Inc., 1985), pp. 93–100

H. Seada, K. Deb, A unified evolutionary optimization procedure for single, multiple, and many objectives. IEEE Trans. Evol. Comput. **20**(3), 358–369 (2016)

H. Seada, M. Abouhawwash, K. Deb, Towards a better balance of diversity and convergence in nsga-iii: first results, *International Conference on Evolutionary Multi-Criterion Optimization* (Springer, Cham, 2017), pp. 545–559

H. Seada, K. Deb, M. Abouhawwash, Multi-phase balance of diversity and convergence in multi-objective optimization. Technical Report COIN Report Number 2018001, 3247 EB, (Michigan State University, East Lansing, Michigan, 48824, USA, 2018)

D. Sharma, K. Deb, N. Kishore, Towards generating diverse topologies of path tracing compliant mechanisms using a local search based multi-objective genetic algorithm procedure, in 2008 IEEE Congress on Evolutionary Computation, CEC 2008 (IEEE World Congress on Computational Intelligence) (IEEE, 2008) pp. 2004–2011

K. Sindhya, K. Miettinen, K. Deb, A hybrid framework for evolutionary multi-objective optimization. IEEE Trans. Evol. Comput. **17**(4), 495–511 (2013)

H.K. Singh, A. Isaacs, T. Ray, A pareto corner search evolutionary algorithm and dimensionality reduction in many-objective optimization problems. IEEE Trans. Evol. Comput. **15**(4), 539–556 (2011)

N. Srinivas, K. Deb, Muiltiobjective optimization using nondominated sorting in genetic algorithms. Evol. Comput. **2**(3), 221–248 (1994)

A.K.A. Talukder, K. Deb, S. Rahnamayan, Injection of extreme points in evolutionary multiobjective optimization algorithms, in *International Conference on Evolutionary Multi-Criterion Optimization* (Springer, 2017), pp. 590–605

K.C. Tan, E.F. Khor, T.H. Lee, *Multiobjective Evolutionary Algorithms And Applications* (Springer Science & Business Media, 2006)

M. Tavana, Z. Li, M. Mobin, M. Komaki, E. Teymourian, Multi-objective control chart design optimization using nsga-iii and mopso enhanced with dea and topsis. Expert Syst. Appl. **50**, 17–39 (2016)

Z. Wang, Q. Zhang, M. Gong, A. Zhou, A replacement strategy for balancing convergence and diversity in moea/d, in 2014 IEEE Congress on Evolutionary Computation (CEC) (IEEE, 2014), pp. 2132–2139

U.K. Wickramasinghe, X. Li, A distance metric for evolutionary many-objective optimization algorithms using user-preferences, in *AI 2009: Advances in Artificial Intelligence* (Springer, 2009), pp. 443–453

X. Yuan, H. Tian, Y. Yuan, Y. Huang, R.M. Ikram, An extended nsga-iii for solution multi-objective hydro-thermal-wind scheduling considering wind power cost. Energy Convers. Manage. **96**, 568–578 (2015)

Y. Yuan, H. Xu, B. Wang, X. Yao, A new dominance relation-based evolutionary algorithm for many-objective optimization. IEEE Trans. Evol. Comput. **20**(1), 16–37 (2016a)

Y. Yuan, H. Xu, B. Wang, B. Zhang, X. Yao, Balancing convergence and diversity in decomposition-based many-objective optimizers. IEEE Trans. Evol. Comput. **20**(2), 180–198 (2016b)

Q. Zhang, H. Li, Moea/d: a multiobjective evolutionary algorithm based on decomposition. IEEE Trans. Evol. Comput. **11**(6), 712–731 (2007)

X. Zhang, Y. Tian, Y. Jin, A knee point-driven evolutionary algorithm for many-objective optimization. IEEE Trans. Evol. Comput. **19**(6), 761–776 (2015)

Y. Zhu, J. Liang, J. Chen, Z. Ming, An improved nsga-iii algorithm for feature selection used in intrusion detection. Knowl.-Based Syst. **116**, 74–85 (2017)

E. Zitzler, L. Thiele, Multiobjective evolutionary algorithms: a comparative case study and the strength pareto approach. IEEE Trans. Evol. Comput. **3**(4), 257–271 (1999)

E. Zitzler, K. Deb, L. Thiele, Comparison of multiobjective evolutionary algorithms: empirical results. Evol. Comput. **8**(2), 173–195 (2000)

E. Zitzler, M. Laumanns, L. Thiele, *Spea2: Improving the Strength Pareto Evolutionary Algorithm* (2001)

X. Zou, Y. Chen, M. Liu, L. Kang, A new evolutionary algorithm for solving many-objective optimization problems. IEEE Trans. Syst. Man Cybern. Part B Cybern. **38**(5), 1402–1412 (2008)

Mean-Entropy Model of Uncertain Portfolio Selection Problem

Saibal Majumder, Samarjit Kar and Tandra Pal

1 Introduction

Investment is an important factor in many institutional sectors of the economy and is essentially a profit-making avenue. Making decision about where to invest in order to curtail the associated risk, involves intricate mathematics. In this chapter, we investigate a portfolio selection problem, which maximizes the return and minimizes the associated risk of the investor. The portfolio selection problem is a branch of economic and financial modelling which deals with selecting optimal portfolio. The pioneering work of optimal portfolio selection problem was first done by Markowitz (1952) which is regarded as the basis of financial theory. The author considered the trade-off relationship between investment return and risk, and eventually introduced the mean-variance (MV) model which has a vital role in the modern portfolio theory. Markowitz developed a bi-criteria optimization problem to maximize the expected return at a predetermined portfolio risk or to minimize the risk for a predetermined expected return. Markowitz's MV portfolio model is conflicting in nature. There does not exist any decision point (optimal solution) at which both the objectives attain their extremum values. Therefore, the better compromise solutions for both the objectives are taken into consideration.

S. Majumder · T. Pal (✉)
Department of Computer Science and Engineering, National Institute of Technology
Durgapur, Durgapur 713209, West Bengal, India
e-mail: tandra.pal@gmail.com

S. Majumder
e-mail: saibaltufts@gmail.com

S. Kar
Department of Mathematics, National Institute of Technology
Durgapur, Durgapur 713209, West Bengal, India
e-mail: kar_s_k@yahoo.com

© Springer Nature Singapore Pte Ltd. 2018
J. K. Mandal et al. (eds.), *Multi-Objective Optimization*,
https://doi.org/10.1007/978-981-13-1471-1_2

An inherent assumption of the seminal work of Markowitz is that the future security returns are influenced by their past performances. In other words, the security returns are considered as probabilistic in nature and represented as random variables. The characteristics (expected value, variance, entropy, etc.) of these random variables are determined by the samples of historical data. These situations become valid if there is an availability of abundant historical data in a financial market. However, in real-world scenarios, there are many parameters in the security markets, such as company performance, political factors, and market forces of supply and demand, which are associated with non-statistical uncertainty and cannot be determined using probability theory due to the lack of historical data. Moreover, quite often we come across some situations in the financial market, where new securities may occur. In such cases, the investors may find it difficult to gather enough samples due to the lack of historical data about the security returns. Under such circumstances, many researchers suggest estimation for security returns by inviting domain experts and represent those estimations using fuzzy set theory. Essentially, portfolio optimization under fuzziness depends on the possibility distribution, which can be determined by the predicted value associated with the returns of the securities, estimated by the experts.

Considering the fuzzy portfolio optimization, a thorough investigation reveals that paradoxes may appear if the subjective estimation of an investment security return is expressed as fuzzy variable. As an example, let us consider a security return which is represented as a triangular fuzzy number $\zeta = (0.7, 1.4, 1.9)$, and is characterized by a membership function. Therefore, based on the possibility measure (Liu 2002) of a fuzzy number, the occurrence of the event that *the return is exactly* 1.4 will certainly possess a belief degree 1. However, it is not acceptable as the belief degree of *exactly* 1.4 is almost 0. Moreover, according to possibility measure, the events, *exactly* 1.4 and *not exactly* 1.4, have the same belief. These facts essentially infer a paradoxical conclusion that the occurrence of the two events is equally likely. In order to deal with such improbable situations, Liu (2007) proposed uncertainty theory to effectively deal with subjective imprecise quantity by estimating beliefs of experts, and further developed it in Liu (2010, 2012). Nowadays, applications of uncertainty theory can be well observed in different research domains such as finance (Chen and Gao 2013; Guo and Gao 2017), economics (Yang and Gao 2016, 2017) and management (Gao and Yao 2015; Gao et al. 2017).

In the context of the studies related to portfolio selection problem, under uncertainty theory (Liu 2007) framework (cf. Sect. 2), we have observed that apart from the study of Kar et al. (2017), an uncertain portfolio problem is optimized either by maximizing the return at a particular preset value of risk or minimizing the risk at a predetermined value of expected return. Considering the framework of return-risk trade-off of the investors' portfolio selection, in this study, we have proposed a bi-objective mean-entropy portfolio selection problem under the framework of uncertainty theory. Here, the mean and the entropy respectively represent the expected return and the associated risk of the uncertain securities of the investors, which are determined by the expected value and the triangular entropy, respectively. The uncertain securities are also represented as uncertain variables. The proposed model is solved by two classical multi-objective solution techniques as well as by two multi-objective genetic algorithms (MOGAs). We have also proposed two theorems to verify the Pareto optimality of the compromise solution, generated by the proposed model, using the two classical multi-objective solution techniques. A dataset of 20 uncertain security returns, determined from expert's opinion, are used as input data of the proposed model. Finally, the performances of the MOGAs for solving the proposed model are also studied.

The rest of the chapter is organized as follows. The survey of the related literature of our study is presented in Sect. 2. The preliminaries, required for the study, are provided in Sect. 3. In Sect. 4, the concept of the uncertain multi-objective programming (Liu 2007) is discussed. Two multi-objective genetic algorithms: NSGA-II (Deb et al. 2002) and MOEA/D (Zhang and Li 2007), used to solve the portfolio selection model, are presented in Sect. 5. In Sect. 6, different performance metrics of the MOGAs are defined. The proposed bi-objective portfolio selection model, under the framework of the uncertainty theory, is formulation in Sect. 7. The results of the proposed model and their discussions are provided in Sect. 8. Finally, the chapter is concluded in Sect. 9.

2 Literature Study

In this section, we present a brief overview of different variants of portfolio optimization problem which are proposed in previous studies. The survey, by no means, encompasses all the related researches in the literature. However, some studies having significant contributions to portfolio selection problem are reviewed.

In 1952, Markowitz presented a precise mathematical model while representing the famous mean-variance (MV) portfolio selection problem. Further, the author observed that the returns of different securities of a portfolio follow stochastic distributions and therefore suggested that the securities of a portfolio can be considered as random variables. In addition, the author also represented the associated risk of the investors by determining the variance of the random variables. Later, motivated from the noteworthy research of Markowitz, many researchers put forward several modifications of the mean-variance portfolio model. The single index model or market model is one such modification which does not consider the covariance between the security returns (Sharpe 1964; Lintner 1965). Thereafter, several studies have been conducted in portfolio theory. Among those, Fishburn (1977) formulated the mean-risk portfolio selection model in which risk is measured by a probability-weighted function of deviation below a specific target return. Later, Konno and Yamazaki (1991) and Feinstein and Thapa (1993) demonstrated the mean-absolute deviation (MAD) portfolio optimization models. In their study, the authors modelled the associated risk as absolute deviation and minimized it for a given expected return. Lai (1991) defined the mean-variance-skewness model and solved it with goal programming method. The same model was also addressed by Konno and Suzuki (1995) and eventually solved it as a single objective optimization problem. Following the contribution of Konno and Yamazaki (1991), Simaan (1997) compared the risks of MV and MAD portfolio models for a given expected return. Afterwards, Jorion (1996) formulated the mean-value at risk (m-VaR) model. Besides, Konno and Wijayanayake (1999) formulated a convex maximization problem for MAD model with transaction cost. Rockafellar and Uryaser (2000) defined the mean-conditional value at risk (m-CVaR) model of portfolio selection problem. Oh et al. (2006) introduced a new risk measuring index, portfolio beta β_p which effectively measures the portfolio volatility relative to capital market. Furthermore, Soleimani et al. (2009) extended the Markowitz's MV portfolio model considering market capitalization as a constraint.

Since the development of fuzzy set theory (Zadeh 1965), many scholars and researchers have developed different fuzzy optimization models for fuzzy portfolio selection problem. Among them, Watada (1997) first extended the mean-variance portfolio selection problem in fuzzy decision principle, where the goals for expected return and the associated risk are represented by membership functions. Since then, many scholars have studied fuzzy portfolio selection problem. For examples, Tanaka and Gao (1999) used fuzzy probability and possibility distributions to optimized mean and variance of a portfolio. Arenas-Parra et al. (2001) proposed an optimal portfolio selection problem for a private investor by considering return, risk and liquidity, and eventually solved the problem by fuzzy goal programming. Carlsson et al. (2002) used the new definitions of the mean and the variance of fuzzy numbers (Carlsson and Fullér 2001), and determined the optimal portfolio for the investors by maximizing the utility score. León et al. (2002) proposed an algorithm which used fuzzy linear programming technique under the framework of return-risk trade-off to manage invertors' portfolio. Zhang and Nie (2004) proposed a fuzzy portfolio selection problem by assuming admissible errors of the expected return and associated risk. Subsequently, Huang (2008a) proposed two different fuzzy mean-semivariance models of portfolio selection and eventually solved those models with fuzzy simulation based genetic algorithm. Thereafter, Huang (2008b) proposed a mean-entropy fuzzy portfolio selection problem, where mean and entropy respectively represent the return and risk for the securities. Afterwards, Huang (2008c) defined a risk curve as a new alternative to represent the risk, and formulated a mean-risk curve fuzzy portfolio model. Qin et al. (2009) proposed the cross entropy minimization model for fuzzy portfolio selection problem. Li et al. (2010) introduced a mean-variance-skewness fuzzy portfolio selection model, which respectively models the return, the risk and the asymmetric behaviour of the fuzzy portfolio returns. Later, Qin et al. (2011), and Vercher and Bermúdez (2015) represented the associated risk of the fuzzy returns of the portfolio selection problem respectively by absolute deviation and semiabsolute deviation. However, some recent studies (Liu 2012; Huang 2012b; Chen et al. 2016) reveal paradoxical results using fuzzy set theory, making the use of fuzzy set theory incongruous in the context of portfolio selection problem. Hence, many researchers have considered uncertainty theory (Liu 2007) as a substitute of fuzzy set theory.

Huang (2010) first proposed a portfolio selection problem under the Liu's uncertainty theory framework. Since then, many researchers and scholars have significantly contributed to improve different variants of the problem. As for example,

Huang (2011) proposed a risk curve and presented a mean-risk portfolio selection model by considering the experts subjective estimation for security returns as uncertain variables. Huang (2012a) defined the mean-variance and the mean-semivariance models, for which the experts' estimation of security returns is considered as uncertain variables. Subsequently, Huang (2012b) defined a risk index and introduced a safe criterion for judging the investor's portfolio for uncertain portfolio selection. Motivated from the work of Huang (2012b), Bhattacharyya et al. (2012) presented the mean-variance-skewness uncertain portfolio model and solved it by fuzzy goal programming technique. Further, the study of Li and Qin (2014) put forward a mean-semiabsolute portfolio selection model by considering the expected returns of the securities as intervals and represented them as uncertain variables. Later, Zhang et al. (2015) presented expected-variance-chance and chance-expected-variance models of uncertain portfolio selection problem which are solved by a genetic algorithm. Of late, by considering the risk-return trade-off relationship of the portfolio selection problem, Qin et al. (2016) proposed the mean-semiabsolute deviation model for uncertain investment returns. Besides, Huang and Di (2016) considered the return and risk associated with background assets to develop an uncertain portfolio selection problem. Afterwards, Kar et al. (2017) proposed a mean-variance-cross entropy uncertain portfolio selection problem and applied the model on a Shenzhen Stock Exchange data set and solved the models with MOGAs. Zhai and Bai (2017) considered the background risk and the asset liquidity and thereby developed a mean-risk model along with background risk, transaction costs and liquidity for uncertain portfolio selection problem. Recently, Li et al. (2018) introduced an uncertain multi-period portfolio selection problem, which maximizes the final wealth and minimizes the investment risk, and generates a compromise solution of the problem using a genetic algorithm.

3 Preliminaries

In this section, some preliminary concepts of uncertainty theory (Liu 2007) are introduced, which are required to formulate the proposed model.

Definition 2.1 (Liu 2007) Let Γ be a nonempty set and $\mathcal{L}$ be a σ-algebra over. Each element $\Lambda \in \mathcal{L}$ is called an event. For each event Λ, it is necessary to assign a number $\mathcal{M}\{\Lambda\}$, which determines the chance of occurrence of each event Λ. In order to define $\mathcal{M}\{\cdot\}$ axiomatically, Liu (2007) proposed the following four axioms.

Axiom 1 (*Normality*) $\mathcal{M}\{\Lambda\} = 1$.

Axiom 2 (*Self-Duality*) For any event Λ, $\mathcal{M}\{\Lambda\} + \mathcal{M}\{\Lambda^c\} = 1$.

Axiom 3 (*Subadditivity*) $\mathcal{M}\left\{\sum_{i=1}^{\infty} \Lambda_i\right\} \leq \sum_{i=1}^{\infty} \mathcal{M}\{\Lambda_i\}$, for all countable sequence of events $\Lambda_1, \Lambda_2, \Lambda_3, \ldots$.

Subsequently, Liu (2010) presented the product measure axiom which is stated as below.

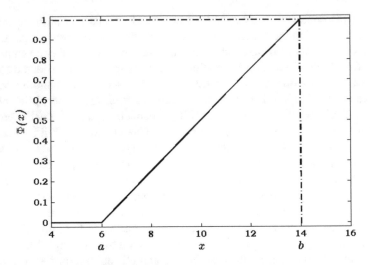

Fig. 1 Linear uncertainty distribution of $\mathcal{L}(6, 14)$

Axiom 4 (*Product*) Let $(\Gamma_i, \mathcal{L}_i, \mathcal{M}_i)$ be the uncertain space for $i = 1, 2, \ldots,$ then the product uncertain measure $\mathcal{M}$ becomes an uncertain measure such that $\mathcal{M}\{\prod_{i=1}^{\infty} \Lambda_i\} = \bigwedge_{\infty}^{i=1} \mathcal{M}_i\{\Lambda_i\}$, where Λ_i is the respective arbitrarily chosen event from $\mathcal{L}_i$ for all $i = 1, 2, \ldots$.

Definition 2.2 (Liu 2007) An uncertain variable ζ defined in (1) is a measurable function from $\mathfrak{R}$ to $[0, 1]$, and is characterized by an uncertainty distribution Φ.

$$\Phi(x) = \mathcal{M}\{\zeta \leq x\}, \forall x \in \mathfrak{R} \tag{1}$$

Example 2.1 Let ζ be an uncertain variable. Then, ζ is said to be a linear uncertain variable (LUV) if

$$\Phi(x) = \begin{cases} 0, & \text{if } x \leq a \\ \frac{x-a}{b-a}, & \text{if } a < x \leq b \\ 1, & \text{if } x > b. \end{cases}$$

An LUV is denoted as $\mathcal{L}(a, b)$, where $a, b \in \mathfrak{R}$ and $a < b$. The uncertainty distribution of an LUV is shown in Fig. 1.

Example 2.2 Let ζ be an uncertain variable. Then, ζ is said to be a normal uncertain variable (NUV) if

$$\Phi(x) = \left\{1 + exp\left(\frac{\pi(\mu - x)}{\sqrt{3}\sigma}\right)\right\}^{-1}.$$

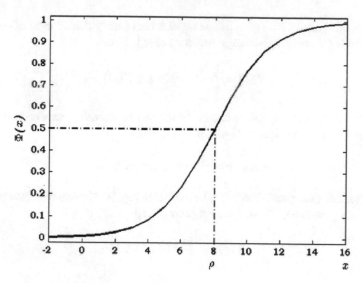

Fig. 2 Normal uncertainty distribution of $\mathcal{N}(8, 3)$

An NUV is denoted as $\mathcal{N}(\mu, \sigma)$, where $\mu, \sigma \in \mathfrak{R}$ and $\sigma > 0$. The uncertainty distribution of an NUV is displayed in Fig. 2.

Definition 2.3 (Liu 2010) Let $\Phi(x)$ be a regular uncertainty distribution of an uncertain variable ζ. Then, the inverse uncertainty distribution of ζ is an inverse function denoted by $\Phi^{-1}(\alpha)$, where $\alpha \in [0, 1]$.

Theorem 2.1 *(Liu 2010) Let* $\Phi_1, \Phi_2, \ldots, \Phi_n$ *be respectively the regular uncertainty distributions of the independent uncertain variables* $\zeta_1, \zeta_2, \ldots, \zeta_n$. *If the function* $f(\zeta_1, \zeta_2, \ldots, \zeta_n)$ *is strictly increasing with respect to* $\zeta_1, \zeta_2, \ldots, \zeta_m$ *and strictly decreasing with respect to* $\zeta_{m+1}, \zeta_{m+2}, \ldots, \zeta_n$, *then the inverse uncertainty distribution of an uncertain variable* ξ *is*

$$\Phi^{-1}(\alpha) = f\left(\Phi_1^{-1}(\alpha), \Phi_2^{-1}(\alpha), \ldots, \Phi_m^{-1}(\alpha), \Phi_{m+1}^{-1}(1-\alpha), \ldots, \Phi_n^{-1}(1-\alpha)\right), \quad (2)$$

where $\xi = f(\zeta_1, \zeta_2, \ldots, \zeta_n)$.

Definition 2.4 (Liu 2007) Let ζ be an uncertain variable. Then, the expected value of ζ is

$$\mathrm{E}(\zeta) = \int\limits_0^{+\infty} \mathcal{M}\{\zeta \geq r\}dr - \int\limits_{-\infty}^0 \mathcal{M}\{\zeta \leq r\}dr, \quad (3)$$

with at least one finite integral.

Remark 2.1 If ζ is an LUV and $\zeta = \mathcal{L}(a, b)$, then the corresponding expected value, $\mathrm{E}[\zeta] = \frac{(a+b)}{4}$. Whereas, if ζ is an NUV such that $\zeta = \mathcal{N}(\mu, \sigma)$, then $\mathrm{E}[\zeta] = \mu$.

Theorem 2.2 *(Liu 2010) Let* Φ *be the regular uncertainty distribution of an uncertain variable* ζ *having a finite expected value* $E[\zeta]$. *Then*

$$E[\zeta] = \int_0^1 \Phi^{-1}(r)dr, \; r \in [0, 1] \tag{4}$$

Theorem 2.3 *(Liu 2010) Let* ζ_1 *and* ζ_2 *be the two independent uncertain variables having finite expected values. Then, for any* $a, b \in \Re$

$$E[a\zeta_1 + b\zeta_2] = aE[\zeta_1] + bE[\zeta_2]. \tag{5}$$

Definition 2.5 (Tang and Gao 2012) Let Φ be the regular uncertainty distribution of an uncertain variable ζ. Then, the triangular entropy $T(\zeta)$ of ζ is

$$T(\zeta) = \int_{-\infty}^{+\infty} F(\Phi(x))dx, \tag{5}$$

where $F(u) = \begin{cases} u & ;\text{if } 0 \le u \le \frac{1}{2} \\ 1 - u & ;\text{if } \frac{1}{2} < u \le 1. \end{cases}$

Example 2.3 (Tang and Gao 2012) The triangular entropy of a LUV, $\zeta = \mathcal{L}(a, b)$, is represented as

$$T(\zeta) = \int_a^{\frac{(b-a)}{2}} \frac{x-a}{b-a}dx + \int_{\frac{(b-a)}{2}}^b \left(1 - \frac{x-a}{b-a}\right)dx = \frac{b-a}{4}. \tag{7}$$

Example 2.4 (Tang and Gao 2012) The triangular entropy of a NUV, $\zeta = \mathcal{N}(a, b)$, is expressed as

$$T(\zeta) = \int_{-\infty}^e \left(1 + \exp\left(\frac{\pi(e-x)}{\sqrt{3}\sigma}\right)\right)^{-1} dx + \int_e^{+\infty} \left(1 - \left(1 + \exp\left(\frac{\pi(e-x)}{\sqrt{3}\sigma}\right)\right)^{-1}\right) dx = \frac{2\sqrt{3}\ln 2}{\pi}\sigma. \tag{8}$$

Theorem 2.4 *(Tang and Gao 2012) Let* Φ^{-1} *be the inverse uncertainty distribution of an uncertain variable* ζ. *Then*

$$T(\zeta) = \int_{\frac{1}{2}}^1 \Phi^{-1}(r)dr - \int_0^{\frac{1}{2}} \Phi^{-1}(r)dr. \tag{9}$$

Theorem 2.5 *(Ning et al. 2014) Let* Φ_1 *and* Φ_2 *be respectively the regular uncertainty distributions of the two independent uncertain variables* ζ_1 *and* ζ_2. *Then, for any* $a, b \in \Re$

$$T[a\zeta_1 + b\zeta_2] = aT[\zeta_1] + bT[\zeta_2]. \tag{10}$$

4   Uncertain Multi-Objective Programming

Many real-world optimization problems presume a vector of conflicting objectives. These objectives are to be optimized simultaneously. To optimize such a vector of objectives, multi-objective programming techniques are widely applied in various application domains. However, the real-life multi-objective problems, where most of the associated input parameters are not precisely defined, necessarily require the support of different uncertainty theories, e.g. fuzzy sets (Zadeh 1965), rough sets (Pawlak 1982), uncertainty theory (Liu 2007), etc. Under the framework of uncertainty theory, Liu and Chen (2015) first modelled the uncertain multi-objective programming, presented in (11).

$$
\begin{cases}
\underset{x}{\textit{minimize}} \; \mathrm{E}\big[f_1(x, \zeta)\big], \mathrm{E}\big[f_2(x, \zeta)\big], \ldots, \mathrm{E}\big[f_r(x, \zeta)\big] \\
\textit{subject to} \\
\quad \mathcal{M}\{g_l(x, \zeta) \le 0\} \ge \alpha_l \\
\quad l = 1, 2, \ldots, p,
\end{cases}
\tag{11}
$$

where ζ is an uncertain vector, x is a decision vector, $f_i(x, \zeta)$s are the uncertain objective functions for $i = 1, 2, \ldots, r$, $\mathrm{E}[\cdot]$ determines the expected value of each uncertain objective function $f_i(x, \zeta)$, $g_l(x, \zeta)$s are the uncertain constraints and α_ls are the confidence levels for $l = 1, 2, \ldots, p$. Due to the existence of trade-off among the conflicting objectives, multiple solutions are generated by simultaneously minimizing all the objectives. Therefore, improving an objective becomes impossible without deteriorating at least one of the remaining objectives.

Similarly, we can design an uncertain multi-objective model by minimizing the triangular entropy $\mathcal{T}[\cdot]$ of each uncertain objective function $f_i(x, \zeta)$ as given below in (12).

$$
\begin{cases}
\underset{x}{\textit{minimize}} \; \mathcal{T}\big[f_1(x, \zeta)\big], \mathcal{T}\big[f_2(x, \zeta)\big], \ldots, \mathcal{T}\big[f_r(x, \zeta)\big] \\
\textit{subject to} \\
\quad \mathcal{M}\{g_l(x, \zeta) \le 0\} \ge \alpha_l \\
\quad l = 1, 2, \ldots, p.
\end{cases}
\tag{12}
$$

Combining the models, presented in (11) and (12), we formulate below another multi-objective model in (13).

$$\begin{cases} \underset{x}{minimize} \begin{cases} \mathrm{E}[f_1(x,\zeta)], \mathrm{E}[f_2(x,\zeta)], \dots, \mathrm{E}[f_u(x,\zeta)] \\ \mathcal{T}[f_{u+1}(x,\zeta)], \ \dots, \mathcal{T}[f_r(x,\zeta)] \end{cases} \\ subject\ to \\ \quad \mathcal{M}\{g_l(x,\zeta) \le 0\} \ge \alpha_l \\ \quad l = 1, 2, \dots, p. \end{cases} \tag{13}$$

Remark In our study, we have proposed an uncertain bi-objective portfolio selection problem, where the first objective determines the expected return and the second objective determines the risk of the uncertain security returns. Here, the expected return and the risk are respectively determined by expected values and the triangular entropy of the uncertain securities. Accordingly, we formulate model (13) by combining models (11) and (12). It is to be noted that model (13) is an uncertain multi-objective optimization problem for r uncertain objective functions, where u objectives optimize the expected values and $r - u$ uncertain objectives optimize the triangular entropy. However, in the context of the uncertain portfolio selection, the proposed model (cf. model (20)) is formulated by considering $r = 2$, among which the first objective maximizes the expected value and the second objective minimizes the triangular entropy. Moreover, the constraint functions of the proposed problem are considered deterministic.

Definition 4.1 (Liu and Chen 2015) A feasible solution x^* is considered as Pareto optimal to the uncertain multi-objective problem, defined in (11), if there does not exist any feasible solution x such that

(i) $\mathrm{E}[f_i(x,\zeta)] \le \mathrm{E}[f_i(x^*,\zeta)], \forall i \in \{1, 2, \dots, r\}$
(ii) $\mathrm{E}[f_q(x,\zeta)] < \mathrm{E}[f_q(x^*,\zeta)]$ for at least one index $q = 1, 2, \dots, r$.

Similarly, the Pareto optimal solution x^* for uncertain multi-objective programming, presented in (12), is defined below.

Definition 4.2 A feasible solution x^* is considered as Pareto optimal to the uncertain multi-objective problem, presented in (12), if there does not exist any feasible solution x such that

(i) $\mathcal{T}[f_i(x,\zeta)] \le \mathcal{T}[f_i(x^*,\zeta)], \forall i \in \{1, 2, \dots, r\}$
(ii) $\mathcal{T}[f_q(x,\zeta)] < T[f_q(x^*,\zeta)]$ for at least one index $q = 1, 2, \dots, r$.

The uncertain multi-objective problem in (13) can be converted to its equivalent single objective problem subject to the same set of chance constraints by aggregating all the $f_i(x,\zeta)$ with a real-valued preference function. The equivalent model is then referred as a compromise model whose solution is usually known as compromise solution. In order to formulate the compromise model, in subsections 4.1 and 4.2, we discuss two classical multi-objective solution techniques: (i) weighted sum method, and (ii) weighted metric method, respectively. Two related theorems are also proposed to prove that the solution of the compromised models (cf. model (14) and model (15)) is also Pareto optimal to the corresponding multi-objective model (13).

4.1 Weighted Sum Method

We apply weighted sum method (Zadeh 1963) on the model (13) to formulate its compromise model as given below.

$$
\begin{cases}
\underset{x}{minimize} \left(\begin{array}{l} \omega_1 E[f_1(x, \zeta)] + \ldots + \omega_u E[f_u(x, \zeta)] + \\ \omega_{u+1} T[f_{u+1}(x, \zeta)] + \ldots + \omega_r T[f_r(x, \zeta)] \end{array} \right) \\
subject\ to \\
\qquad \mathcal{M}\{g_l(x, \zeta) \leq 0\} \geq \alpha_l \\
\qquad l = 1, 2, \ldots, p,
\end{cases}
\tag{14}
$$

where $\sum_{m=1}^{r} \omega_m = 1$, $\forall \omega_m \in [0, 1]$.

The following theorem proves the Pareto optimal conditions of the solution generated by the model (14).

Theorem 4.1.1 *Let x^* be an optimal solution of the uncertain model (14), then x^* is Pareto optimal to the uncertain multi-objective model, presented in (13)*

Proof Let us assume that x^* is an optimal solution to (14) but is not Pareto optimal to (13). So there exists a feasible solution x such that

$$
x \preceq_E x^* \Rightarrow E[f_t(x, \zeta)] \leq E[f_t(x^*, \zeta)] \forall t = \{1, 2, \ldots, u, u+1, \ldots, r\} \text{ and}
$$
$$
x \preceq_T x^* \Rightarrow T[f_q(x, \zeta)] \leq T[f_q(x^*, \zeta)] \forall q = \{u+1, u+2, \ldots, r\}.
$$

Considering the objective functions, for at least one index value n and m, we have

$$
x \prec_E x^* \Rightarrow E[f_n(x, \zeta)] < E[f_n(x^*, \zeta)] \text{ and}
$$
$$
x \prec_T x^* \Rightarrow T[f_m(x, \zeta)] < T[f_m(x^*, \zeta)],
$$

where $n \in \{1, 2, \ldots, u\}$ and $m \in \{u+1, u+2, \ldots, r\}$.

Moreover,

$$
\omega_t E[f_t(x, \zeta)] \leq \omega_t E[f_t(x^*, \zeta)], \forall t = \{1, 2, \ldots, u\} \text{ and}
$$
$$
\omega_q T[f_q(x, \zeta)] \leq \omega_q T[f_q(x^*, \zeta)], \forall q = \{u+1, u+2, \ldots, r\},
$$

and for at least one index value of each n and m

$$
\omega_n E[f_n(x, \zeta)] < \omega_n E[f_n(x^*, \zeta)] \text{ and}
$$
$$
\omega_m T[f_m(x, \zeta)] < \omega_m T[f_m(x^*, \zeta)].
$$

Therefore

$$\left(\begin{array}{l} \sum_{t=1}^{u} \omega_t \mathrm{E}[f_p(x, \zeta)]+ \\ \sum_{q=u+1}^{r} \omega_q \mathcal{T}[f_q(x, \zeta)] \end{array} \right) \leq \left(\begin{array}{l} \sum_{t=1}^{u} \omega_p \mathrm{E}[f_p(x^*, \zeta)]+ \\ \sum_{q=u+1}^{r} \omega_q \mathcal{T}[f_q(x^*, \zeta)] \end{array} \right)$$

$\forall t = \{1, 2, \ldots, u\}$ and $\forall q = \{u+1, u+2, \ldots, r\}$.
Further, for at least one index value of n and m

$$\left(\begin{array}{l} \omega_n \mathrm{E}[f_n(x, \zeta)]+ \\ \omega_m \mathcal{T}[f_m(x, \zeta)] \end{array} \right) < \left(\begin{array}{l} \omega_n \mathrm{E}[f_n(x^*, \zeta)]+ \\ \omega_m \mathcal{T}[f_m(x^*, \zeta)] \end{array} \right).$$

It implies, for model (14), x^* is not the optimal solution, which contradicts with our initial assumption that x^* is an optimal solution of (14). Therefore, we conclude that x^* is the Pareto optimal solution of model (13).

4.2   Weighted Metric Method

An alternative approach to formulate the compromise model of (13) using weighted metric method (Asgharpour 1998; Deb 2001) is discussed here. In weighted metric method, the distance metric L_s, $s \in \{1, 2, \ldots, \infty\}$ is commonly associated with non-negative weights. We implement the weighted metric method with Euclidean distance metric L_2 to formulate the compromise model of (13), presented below in (15).

$$\begin{cases} \underset{x}{minimize} \quad \sqrt{\begin{array}{l} \omega_1 \left(\dfrac{(\mathrm{E}_1^* - \mathrm{E}[f_1(x,\zeta)])}{\mathrm{E}_1^*} \right)^2 + \ldots + \\[2ex] \omega_u \left(\dfrac{(\mathrm{E}_u^* - \mathrm{E}[f_u(x,\zeta)])}{\mathrm{E}_u^*} \right)^2 + \\[2ex] \omega_{u+1} \left(\dfrac{(\mathcal{T}[f_{u+1}(x,\zeta)] - \mathcal{T}_r^*)}{\mathcal{T}_{u+1}^*} \right)^2 + \ldots + \\[2ex] \omega_r \left(\dfrac{(\mathcal{T}[f_r(x,\zeta)] - \mathcal{T}_r^*)}{\mathcal{T}_r^*} \right)^2 \end{array}} \\[6ex] subject\ to \\ \mathcal{M}\{g_l(x, \zeta) \leq 0\} \geq \alpha_l, l = 1, 2, \ldots, p \\ \sum\limits_{m=1}^{r} \omega_m = 1, \forall \omega_m \in [0, 1], \end{cases}$$

$$(15)$$

where, $\mathrm{E}_t^* (t \in \{1, 2, \ldots, u\})$ and $\mathcal{T}_q^* (q \in \{u+1, u+2, \ldots, r\})$ are the optimal solutions of tth and qth objective functions when solved individually as a single objective problem.

The Pareto optimality of the compromise solution generated by the model (15) is proved in the following theorem.

Theorem 4.2.1 *Let x^* be an optimal solution of an uncertain programming model (15,) then x^* is Pareto optimal to the uncertain multi-objective programming model, defined in (13).*

Proof Let us assume that x^* is an optimal solution of (15) but is not Pareto optimal to (13). So there exists a feasible solution x such that

$$x \preccurlyeq_E x^* \Rightarrow \omega_t E[f_t(x, \zeta)] \leq \omega_t E[f_t(x^*, \zeta)] \forall t = \{1, 2, \ldots, u\} \text{ and}$$
$$x \preccurlyeq_T x^* \Rightarrow \omega_q T[f_q(x, \zeta)] \leq T[f_q(x^*, \zeta)] \forall q = \{u+1, u+2, \ldots, r\}.$$

For at least one index value of $n(\in \{1, 2, \ldots, u\})$ and $m(\in \{u+1, u+2, \ldots, r\})$, we have

$$x \prec_E x^* \Rightarrow \omega_n E[f_n(x, \zeta)] < \omega_n E[f_n(x^*, \zeta)] \text{ and } x \prec_T x^* \Rightarrow \omega_m T[f_m(x, \zeta)] < \omega_m T[f_m(x^*, \zeta)].$$

E_t^* and T_q^* are the optimized solutions of tth and qth objective functions respectively when solved individually as a single objective problem. Therefore, we have

$$\sqrt{\omega_t \left(\frac{(E_t^* - E[f_t(x, \zeta)])}{E_t^*}\right)^2} \leq \sqrt{\omega_t \left(\frac{E_t^* - (E[f_t(x^*, \zeta)])}{E_t^*}\right)^2} \text{ and } \sqrt{\omega_q \left(\frac{(T[f_q(x, \zeta)] - T_q^*)}{T_q^*}\right)^2} \leq \sqrt{\omega_q \left(\frac{(T[f_q(x^*, \zeta)] - T_q^*)}{T_q^*}\right)^2}$$

$\forall t = \{1, 2, \ldots, u\}$ and $\forall q = \{u+1, u+2, \ldots, r\}$, respectively.
Also for at least one index value of n and m

$$\sqrt{\omega_n \left(\frac{E_n^* - (E[f_n(x, \zeta)])}{E_n^*}\right)^2} < \sqrt{\omega_n \left(\frac{E_n^* - (E[f_n(x^*, \zeta)])}{E_n^*}\right)^2} \text{ and } \sqrt{\omega_m \left(\frac{(T[f_m(x, \zeta)] - T_m^*)}{T_m^*}\right)^2} < \sqrt{\omega_m \left(\frac{(T[f_m(x^*, \zeta)] - T_m^*)}{T_m^*}\right)^2}.$$

Thus, $x \preccurlyeq_E x^*$ and $x \preccurlyeq_T x^*$ together imply that, for model (15), x^* is not the optimal solution, which essentially contradicts with our initial assumption that x^* is an optimal solution of model (15). Hence, x^* is the Pareto optimal solution of model (13).

5 Multi-Objective Genetic Algorithm

Multi-objective genetic algorithms (MOGAs) provide a set of nondominated solutions of multi-objective optimization problem (MOP) in a single execution. MOGA simultaneously searches different unexplored regions of convex, non-convex and discontinuous solution spaces of different complex MOPs making it a potential candidate to explore a diverse set of solutions of MOPs. Furthermore, MOGAs do not require any decision-maker to prioritize the objectives. All these characteristics effectively make MOGAs suitable for solving MOPs. With generations, a MOGA aims to converge towards Pareto optimality by generating better nondominated solutions.

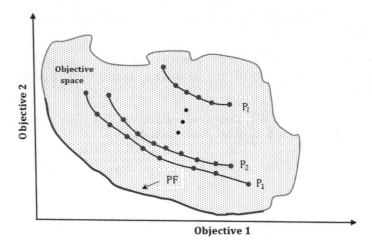

Fig. 3 Different nondominated fronts generated by MOGAs

Figure 3 depicts different nondominated fronts generated by a MOGA during executing an optimization problem with two objectives. In this figure, the shaded region represents the feasible objective space and the PF represents the Pareto front. The front closest to the PF is the first front P_1. Higher fronts from P_2 to P_l depicted in the figure are in the increasing order of their distance from the PF. Moreover, no solutions in the first front P_1 are dominated by any solution of other fronts. Similarly, no solution in second front P_2 is dominated by any solution of the fronts $P_3, P_4, \ldots,$ P_l. However, a solution in P_2 is dominated by at least one solution of P_1. The front P_l in Fig. 3 covers the spectrum of the PF much less compared to that of P_1 , i.e., the solutions of P_l are less diverse compared to those in P_1. During the optimization of MOP, the fronts approach towards PF with subsequent generation.

MOGA was first presented by Fonseca and Flaming (1993). Since then MOGA has drawn colossal interest among researchers and has become a well-known methodology for solving complex multi-objective optimization problems. Different variants of MOGAs (Srinivas and Deb 1995; Knowles and Corne 1999; Zitzler et al. 2001; Deb et al. 2002; Coello 2006; Zhang and Li 2007; Nebro et al. 2008; Nag et al. 2015, etc.) exist in the literature.

In this article, we consider NSGA-II and MOEA/D, developed respectively by Deb et al. (2002) and Zhang and Li (2007).

5.1 Nondominated Sorting Genetic Algorithm II (NSGA-II)

NSGA-II is an elitist model, which ensures retaining the fittest candidates in the next population to enhance the convergence. The algorithm starts by initializing a population P_0 of N randomly generated solutions. In a particular generation t, the genetic operators, i.e., selection, crossover and mutation are applied on the individuals of parent population P_t to generate an offspring population C_t with an equal number of candidate solutions. To ensure elitism, the parent and offspring population are mingled to produce a population S_t of size $2N$. In order to select N solutions from S_t for next generation, NSGA-II performs following two steps.

(1) Frontify the individuals of S_t, by assigning a rank to each individual with non-domination sorting. Based on their ranks, the solutions in S_t are frontified to different nondominated fronts $N_1, N_2, \ldots, N_l$. Each $N_k, k \in \{1, 2, \ldots, l\}$ represents a set of nondominated solutions of rank k. Solutions with the same rank value are placed in a nondominated front. Solutions with lower nondomination ranks are always preferred. In other words, if p and q are the two solutions in S_t, and if $p_{rank} < q_{rank}$, then p is superior to q. In order to build the next population P_{t+1}, the solutions belonging to N_1 are considered first. Here, if the size of N_1 is smaller than N, all the solutions of N_1 are inserted in P_{t+1}. The remaining solutions of P_{t+1} are considered from subsequent nondominated fronts in order of their ranking. In this way, the solutions of 2nd front, i.e., N_2 are moved next to P_{t+1} followed by solutions of N_3 and so on until all the solutions of a nondominated front N_k cannot be fully inserted to P_{t+1}.

(2) If all the solutions of N_k cannot be accommodated in P_{t+1}, the solutions are then sorted in descending order based on their corresponding values of crowding distance ($i_{distance}$) as proposed by Deb et al. (2002). Particularly, if p and q are the two nondominated solutions belong to N_k and p has better crowding distance than q, i.e. if $p_{rank} = q_{rank}$ and $p_{distance} > q_{distance}$, then p is preferred over q in N_k. Accordingly, the solutions with higher crowding distance are eventually selected from N_k to fill the remaining slots of P_{t+1}.

In this way, as the formation of P_{t+1} completes, the individuals of P_{t+1} replaces P_t for the next generation. This process continues until the termination condition (i.e. maximum function evaluations or generations) is reached. The working principle of NSGA-II is depicted in Fig. 4.

5.2 Multi-Objective Evolutionary Algorithm Based on Decomposition (MOEA/D)

MOEA/D decomposes a multi-objective problem (MOP) into N finite scalar optimization sub-problems and optimizes them simultaneously by generating a population of solutions for each of them. Each sub-problem is optimized by using information from its neighbourhood relations with several sub-problems. The neighbourhood relationship is defined by the distance between the aggregations of the coefficient vectors of the corresponding sub-problems. The sub-problems are scalarized to a single objective problem each having a weight vector $\lambda^k, k \in \{1, 2, 3, \ldots, N\}$ which are uniformly distributed. Since each solution is associated with λ^k, the population

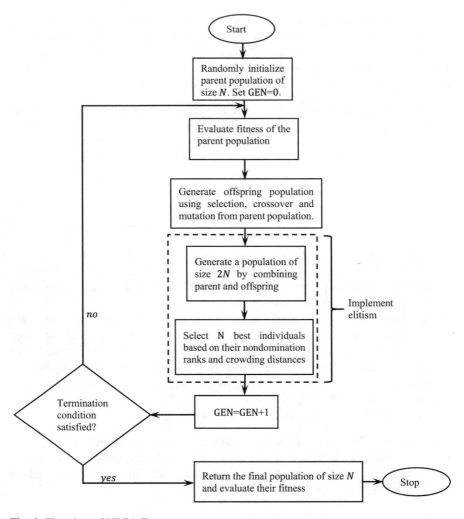

Fig. 4 Flowchart of NSGA-II

size becomes N. For each kth solution having weight vector λ^k, there exists a set of T closest neighbours, $\lambda^{k_1}, \lambda^{k_2}, \ldots, \lambda^{k_T}$ of λ^k. T closest neighbours of λ^k are determined by the Euclidian distances between λ^k and $\lambda^{k_q}, q \in [1, T]$. To keep track of the T closest neighbours of $\lambda^k (k = 1, 2, \ldots N)$, MOEA/D maintains a vector $B(k)$ of dimension T which contains the indices of the weight vectors, $\lambda^{k_1}, \lambda^{k_2}, \ldots, \lambda^{k_T}$. In MOEA/D, it is worth mentioning that a weight vector λ^k and its T closest neighbours correspond to a kth sub-problem along with its T sub-problems, respectively. In order to preserve the nondominated solutions found during the search process, an external population (EP) is maintained in MOEA/D.

At each generation, MOEA/D randomly selects two parent solutions from two different sub-problems among T closest neighbour of kth sub-problem and generate offspring using genetic operators. Considering all the T neighbouring sub-problems, if the newly generated solutions improve any of the kth sub-problem with respect to any other existing solution p, then the offspring replaces p in the EP. This process continues at every subsequent generation of MOEA/D until the termination criteria are reached. The value of T is very crucial to maintain the balance between exploration and exploitation during the execution of MOEA/D. If T is very large, the randomly selected parents may generate poor offspring which can reduce the exploitation of the algorithm. On the other hand, if T is very small then the generated offspring may be very similar to its parents which will ultimately degrade the exploitation capability of the algorithm. Therefore, the value of T should be neither too large nor too small.

The diversity among these sub-problems of a MOP is introduced by properly selecting the corresponding weight vector for the sub-problems and an appropriate decomposition method. There are several approaches for decomposing a MOP, e.g., weighted sum (Miettinen 1999), Tchebycheff (Miettinen 1999), normal boundary intersection (Das and Dennis 1998), normalized normal constraint (Messac et al. 2003), etc. Here, the individuals of a population are associated with different sub-problems. Therefore, the diversity between the sub-problems will eventually promote the diversity in the entire population. The possibility of generating a uniformly distributed approximation of the Pareto front increases if the sub-problems are optimized effectively. The flowchart of MOEA/D is displayed in Fig. 5.

The input parameters of MOEA/D are as follows.

(a) N—The number of sub-problems and uniformly distributed weight vectors to be considered for MOEA/D.
(b) T—The size of the closest neighbours for each weight vector.
(c) CR—Crossover rate.
(d) F—Mutation rate.
(e) δ—Probability that neighbouring solutions of a sub-problem are considered for mating.
(f) n_r—Number of rejected solutions from the population when a better quality offspring simultaneously improves quite a number of sub-problems.

The characteristics of MOEA/D are listed below.

(a) MOEA/D uses the decomposition techniques of mathematical programming in multi-objective evolutionary computing.
(b) MOEA/D promotes diversity preservation easily by optimizing scalar sub-problems of finite number with a uniformly distributed weight vectors for a MOP.
(c) For a small population, MOEA/D can generate a small number of evenly distributed solutions.
(d) MOEA/D maintains a mating restriction by allowing two solutions to mate only when the solutions belong to two different sub-problems.

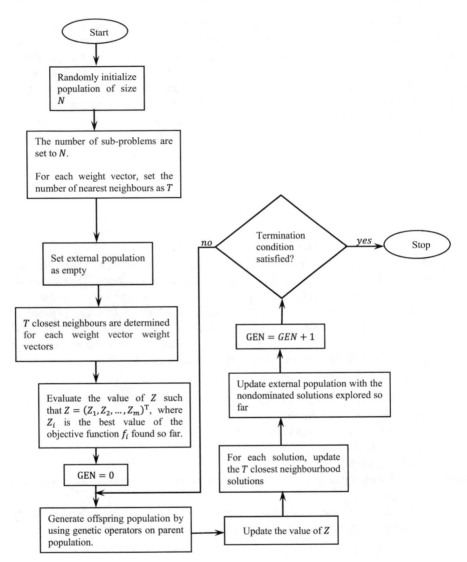

Fig. 5 Flowchart of MOEA/D

6 Performance Metrics

For the purpose of comparison between MOGAs, we have used following performance metrics.

Hypervolume (HV): Hypervolume (HV), presented by Zitzler and Thiele (1999), is the volume covered by the set of all nondominated solutions N_D in the approximate front. The volume is generally the region enclosed by all the elements of N_D in the

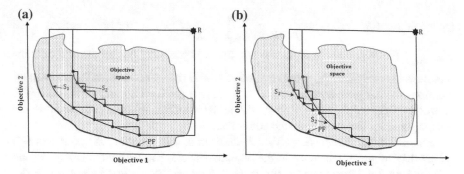

Fig. 6 HV for different optimized fronts S_1 and S_2 with respect to a reference solution R. **a** The front S_1 is closer to PF than S_2 hence shall have higher HV value, **b** solutions of S_2 is more diversely spread than S_1, hence shall have higher HV value

objective space with respect to a reference solution point R, where R is a vector of worst objective function values in the entire objective space of MOP. HV is calculated as

$$HV = volume\left(\cup_{e \in N_D} h_e\right), \tag{16}$$

where h_e is the hypercube for each $e \in N_D$. Larger values of HV are always desirable. The HV metric ensures both the convergence and diversity. More the value of HV, closer will be the approximate front to the Pareto front, generated by a MOGA.

As an example, let S_1 and S_2 be the two fronts generated by two different MOGAs for the same problem as depicted in Fig. 4. In the figure, the Pareto front of the MOP is represented by PF (red colour) in the corresponding objective space. In Fig. 6a, S_1 is more close to PF as well as more diverse compared to S_2. As a result, the HV for S_1 is more than S_2 with respect to reference solution R. If solutions of S_1 are not well diverse compared to S_2 as depicted in Fig. 6b, a better HV for S_2 is obtained with respect to R, since in this case, S_2 covers more spectrum of PF compared to S_1.

Spread (Δ): Spread determines the diversity in a nondominated front which measures the magnitude of diversity between the solutions, generated from a simulation. This metric was first proposed by Deb et al. (2002), which is capable to find the distance between a pair of closest solutions in a nondominated front for bi-objective problems. This metric is then extended to multi-objective problems by Zhou et al. (2006), where the distance between a nondominated solution and its nearest neighbour is measured. In the ideal situation, the value of spread is zero. It is expressed as

$$\Delta = \frac{\sum_{i=1}^{m} d(e_i, N_D) + \sum_{X \in \text{PF}} |d(X, N_D) - \bar{d}|}{\sum_{i=1}^{m} d(e_i, N_D) + |\text{PF}| \times \bar{d}} \forall e_i \in \text{PF} \tag{17}$$

such that $d(X, N_D) = \underset{Y \in N_D, Y \neq X}{min} \|F(X) - F(Y)\|^2$ and $\bar{d} =$
$\frac{1}{|PF|} \sum_{X \in PF} d(X, N_D)$, where F is the set of m objectives, N_D is the set of non-
dominated solutions in the approximate front, PF is the set of known Pareto optimal
solutions for the problem and $e_i(i = 1, 2, \ldots, m)$ are the extreme solutions in PF.
Lower the value of Δ, more diverse will be the solutions. Hence, Δ determines the
diversity among the nondominated solutions.

GenerationalDistance(GD): Van Veldhuizen and Lamont (1998) proposed *GD*
which determines the distance in the objective space, between the elements of the
nondominated solution vectors and their corresponding nearest neighbour in the
Pareto front. The formulation of *GD* metric is presented in (18).

$$GD = \sqrt{\frac{\sqrt{\sum_{i=1}^n d_i^2}}{n}}, \tag{18}$$

where in the objective space, the total number of solutions in the approximate front
is denoted by n and $d_i = \underset{j}{min} \|F(x_i) - PF(x_j)\|$ is the distance between the solution
x_i and the nearest member x_j in the Pareto front. This metric determines the distance
of the approximate front from the Pareto front (PF). In other words, the *GD* metric
ensures the convergence of a MOGA by measuring how far the approximate front of
a MOGA is from the Pareto front. In ideal condition $GD = 0$.

Inverted Generational Distance(IGD): Presented by Van Veldhuizen and Lamont (1998), *IGD* measures the distance of the Pareto front from an approximate front
and is represented as follows:

$$IGD = \frac{\sum_{i=1}^{|P|} d(\rho_i, N_D)}{|P|} \forall \rho_i \in PF, \tag{19}$$

where PF is the Pareto front, $d(\rho_i, N_D)$ is the Euclidian distance between ρ_i to its
closest neighbour in N_D and N_D is a set of nondominated solutions in the approximate
front. ρ_i represents a solution in PF. *IGD* assures both convergence and diversity. A
lower value of this metric can only be achieved if the solution set N_D is close to PF
and nearly covers the spectrum of PF. Therefore, smaller value of *IGD* is always
preferred.

7 Proposed Uncertain Bi-Objective Portfolio Selection Model

In this section, an uncertain bi-objective portfolio selection problem has been formulated. Here, a financial market with n risky assets is considered. Let x_i and
$\zeta_i(i = 1, 2, \ldots, n)$ be respectively the investment proportion and the uncertain return
of the ith security. The uncertainty involved with each ζ_i is represented under the

uncertainty theory framework (Liu 2007). The purpose of our model is to maximize the expected return and to minimize the investment risk. In this context, the expected value and the triangular entropy are used respectively to measure the return and the risk of the uncertain securities. We formulate a bi-objective portfolio selection problem by considering mean and triangular entropy under return-risk framework, and present the model in (20). An efficient portfolio[1] for uncertain returns is indeed a solution of the model (20). Essentially all efficient portfolios construct the efficient frontiers of model (20). Different investors can select their efficient portfolio from the efficient frontier generated by model (20).

$$
\begin{cases}
maximize\ E[\zeta_1 x_1 + \zeta_2 x_2 + \ldots + \zeta_n x_n] \\
minimize\ T[\zeta_1 x_1 + \zeta_2 x_2 + \ldots + \zeta_n x_n] \\
subject\ to \\
\quad x_1 + x_2 + \ldots + x_n = 1 \\
\quad x_i \geq 0, i = 1, 2, \ldots, n.
\end{cases}
\tag{20}
$$

Following the generalized models (14) and (15), the corresponding compromise models (21) and (22) are formulated below with deterministic constraint set for the proposed uncertain bi-objective portfolio selection problem.

$$
\begin{cases}
minimize \begin{pmatrix} -\omega_1 E[\zeta_1 x_1 + \zeta_2 x_2 + \ldots + \zeta_n x_n] + \\ \omega_2 T[\zeta_1 x_1 + \zeta_2 x_2 + \ldots + \zeta_n x_n] \end{pmatrix} \\
subject\ to \\
\quad x_1 + x_2 + \ldots + x_n = 1 \\
\quad x_i \geq 0, i = 1, 2, \ldots, n,
\end{cases}
\tag{21}
$$

where $\omega_1 + \omega_2 = 1$ and $\omega_1, \omega_2 \in [0, 1]$.

$$
\begin{cases}
minimize\ \sqrt{\omega_1 \left(\dfrac{\left(E^* - E_{[\zeta_1 x_1 + \zeta_2 x_2 + \ldots + \zeta_n x_n]}\right)}{E^*} \right)^2 + \omega_2 \left(\dfrac{(T[\zeta_1 x_1 + \zeta_2 x_2 + \ldots + \zeta_n x_n] - T^*)}{T^*} \right)^2} \\
subject\ to \\
\quad x_1 + x_2 + \ldots + x_n = 1 \\
\quad x_i \geq 0, i = 1, 2, \ldots, n. \\
\quad \omega_1 + \omega_2 = 1, \omega_1, \omega_2 \in [0, 1].
\end{cases}
\tag{22}
$$

---

[1]A portfolio is said to be efficient if it is not possible to achieve a higher expected return without increasing the value of triangular entropy or it is not possible to get a lower value of triangular entropy without decreasing the expected return.

Table 1 Twenty uncertain security returns (units per stock)

Security number	Security returns	Security number	Security returns
1	$\mathcal{L}(0.8490, 1.1072)$	11	$\mathcal{N}(0.8535, 0.2568)$
2	$\mathcal{L}(0.8521, 1.0851)$	12	$\mathcal{N}(1.0096, 0.3572)$
3	$\mathcal{L}(0.8532, 1.3129)$	13	$\mathcal{N}(1.0659, 0.3269)$
4	$\mathcal{L}(0.7143, 1.6147)$	14	$\mathcal{N}(0.9987, 0.4926)$
5	$\mathcal{L}(0.7952, 1.3284)$	15	$\mathcal{N}(1.1141, 0.6792)$
6	$\mathcal{L}(0.6582, 1.2271)$	16	$\mathcal{N}(1.1051, 0.6428)$
7	$\mathcal{L}(0.6866, 1.4207)$	17	$\mathcal{N}(0.9249, 0.5847)$
8	$\mathcal{L}(0.8724, 1.4712)$	18	$\mathcal{N}(1.0379, 0.4219)$
9	$\mathcal{L}(0.7496, 1.1958)$	19	$\mathcal{N}(1.0677, 0.2436)$
10	$\mathcal{L}(0.7998, 1.3947)$	20	$\mathcal{N}(1.1552, 0.6120)$

In (22), E^* and $\mathcal{T}^*$ are the optimized solutions of the first and the second objective functions of the model (20) when solved individually.

8   Results and Discussion

In this section, we demonstrate the proposed portfolio selection problem, presented in (20), with a numerical example on uncertain portfolio optimization problem. Purposefully, we have considered a problem, where an investor plans to invest the funds among twenty securities. Here, the future security returns are considered as uncertain variables among which the return of ten securities is expressed as linear uncertain variables and the remaining ten security returns are represented as normal uncertain variables. All these investment returns are presented in Table 1. In the proposed portfolio model (20), the mean return is determined by the expected value, and the risk is defined by triangular entropy for the security returns. The expected value (E) and the triangular entropy $(\mathcal{T})$ for all the investment returns (cf. Table 1) are displayed in Table 2. Accordingly, the expected security returns are calculated by Eq. (3) whereas, the triangular entropy of linear and normal security returns are determined using Eqs. (7) and (8) respectively.

The compromise models, (21) and (22) of model (20), which are formulated by using respectively the weighted sum method and the weighted metric method, are eventually solved by setting the weights of the objectives, i.e., ω_1 and ω_2, to 0.5 each for both the compromise models. These models are then solved by a standard optimization package, Lingo 11.0. The generated compromise solutions of these models are nondominated to each other and are listed in Table 3.

In order to determine the multiple nondominated solutions, we solve the model (20) using two MOGAs, i.e., NSGA-II (Deb et al. 2002) and MOEA/D (Zhang and Li 2007). For the purpose of comparison of the performances of the MOGAs with that of

Table 2 Expected value (E) and triangular entropy ($\mathcal{T}$) of 20 uncertain investment returns

Security number	Expected value (E)	Triangular entropy ($\mathcal{T}$)	Security number	Expected value (E)	Triangular entropy ($\mathcal{T}$)
1	0.9781	0.0645	11	0.8535	0.1963
2	0.9686	0.0582	12	1.0096	0.2730
3	1.0831	0.1149	13	1.0659	0.2499
4	1.1645	0.2251	14	0.9987	0.3765
5	1.0438	0.1423	15	1.1141	0.5191
6	0.9426	0.1422	16	1.1051	0.4913
7	1.0537	0.1835	17	0.9249	0.4469
8	1.1718	0.1497	18	1.0379	0.3225
9	0.9727	0.1115	19	1.0677	0.1862
10	1.0973	0.1487	20	1.1552	0.4678

Table 3 Compromised solutions obtained from weighted sum and weighted metric methods

Objective Functions	Weighted sum method	Weighted metric method
maximize $E[f(x, \zeta)]$	1.1718	1.1269
minimize $\mathcal{T}[f(x, \zeta)]$	0.1497	0.1294

the optimization of model (20), the quality of the nondominated solutions generated by NSGA-II and MOEA/D are analysed in terms of the performance metrics, hypervolume (HV), spread (Δ), generational distance (GD), and inverted generational distance (IGD). For most of the real-life problems, the set of optimal solutions in the Pareto front (P) is usually unavailable. Likewise, for the proposed model in (20), Pareto optimal solutions do not exist in the literature. So, we approximate the Pareto front by generating a reference front by collecting all the best quality solutions from every independent execution of NSGA-II and MOEA/D for 250 generations.

The parameter settings of NSGA-II and MOEA/D, used to optimize the proposed portfolio model in (20), are listed below.

(a) NSGA-II

Size of the population $= 100$, Crossover probability $(p_c) = 0.9$, Mutation probability $(p_m) = 0.03$, Maximum generation $= 250$.

(b) MOEA/D

Size of the population $= 100$, $T = 35$, $\delta = 0.97$, $n_r = 2$, Crossover Rate $(CR) = 0.9$, Mutation rate $(F) = 0.3$, Mutation probability $= 0.03$, Maximum generation $= 250$.

The nondominated solutions, generated by executing NSGA-II and MOEA/D on the proposed model (20), are depicted respectively in Fig. 7a, b. Each nondominated solution of the proposed uncertain portfolio model (20), shown in Fig. 7, determines the return and risk of the uncertain portfolios. It is observed that the investor can receive more expected return only if the investor is willing to withstand higher risk

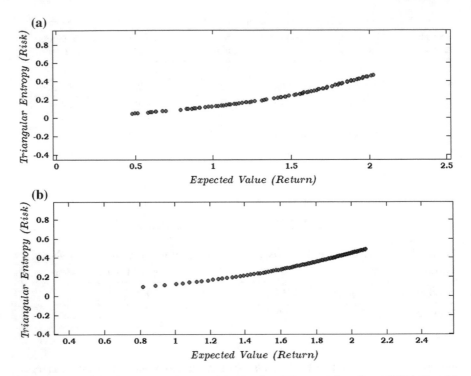

Fig. 7 Nondominated solutions of the model (20) after 250 generations for **a** NSGA-II and **b** MOEA/D

for the uncertain securities. In other words, minimization of the risk results in a progressive decrement of the expected returns. This fact implies a significant return-risk trade-off of the proposed portfolio selection problem, where the return and the risk of the uncertain securities are respectively represented by the expected value and the triangular entropy.

Figure 7 shows the return and risk of the portfolio which essentially constructs the nondominated front for the proposed uncertain portfolio model (20). It is observed that the investor can receive more expected return only if the investor is willing to withstand higher risk for the uncertain securities. In other words, minimization of the risk results in a progressive decrement of the expected returns. This fact implies a significant return-risk trade-off of the proposed portfolio selection problem, where the return and the risk of the uncertain securities are respectively represented by the expected value and the triangular entropy.

Due to stochastic characteristics of the MOGAs, every simulation of the results with the above-mentioned parameter settings is executed for 100 times. For each execution, different performance metrics, i.e., HV, Δ, GD and IGD are evaluated with respect to the optimized solutions obtained after 250 generations and the corresponding reference front. Here, jMetal 4.5 (Durillo and Nebro 2011) framework

Table 4 Mean and *sd* of HV, Δ, GD and IGD after 100 runs of NSGA-II and MOEA/D

MOGAs	HV		Δ		GD		IGD	
	mean	sd	mean	sd	mean	sd	mean	sd
NSGA-II	6.36e−01	4.03e−03	**4.74e−01**	3.50e−02	4.93e−04	**5.30e−05**	6.31e−04	3.05e−04
MOEA/D	**6.39e−01**	**2.90e−03**	6.34e−01	**1.20e−02**	**3.69e−04**	1.40e−04	**1.79e−04**	**3.60e−05**

Table 5 Median and *IQR* of HV, Δ, GD and IGD for 100 runs of NSGA-II and MOEA/D

MOGAs	HV		Δ		GD		IGD	
	median	IQR	median	IQR	median	IQR	median	IQR
NSGA-II	6.36e−01	4.40e−03	**4.73e−01**	4.02e−02	4.87e−04	**7.20e−05**	5.87e−04	2.2e−04
MOEA/D	**6.40e−01**	**3.20e−03**	6.31e−01	**3.40e−03**	**3.49e−04**	2.90e−04	**1.79e−04**	**2.90e−05**

has been used for simulation of the MOGAs. For all 100 observations of each of the performance metrics, we calculate two measures each, for central tendency (mean and median) and variability (standard deviation (*sd*) and interquartile range (*IQR*)). Table 4 reports the mean and *sd* and Table 5 summarizes the median and *IQR* of HV, Δ, GD and IGD. In each of these tables, the better results are shown in bold.

In Table 4, MOEA/D becomes superior to NSGA-II in terms of the performance measures, HV, GD and IGD for the *mean*. However, with respect to Δ, NSGA-II generates a better mean than MOEA/D. In Table 5, better median is obtained for MOEA/D compared to NSGA-II for HV, GD and IGD. Whereas for Δ, NSGA-II becomes better than MOEA/D by generating a relatively smaller *median*. Therefore, by studying the performance metrics in Tables 4 and 5 for both NSGA-II and MOEA/D, it can be observed that during execution, MOEA/D simultaneously maintains better exploration and exploitation compared to NSGA-II for the proposed model presented in (20). This fact can be well understood from the results of HV and IGD reported in Tables 4 and 5. As it has been mentioned above that the performance metrics, HV and IGD ensure convergence as well as diversity, the better result of both these performance metrics for MOEA/D essentially infers that a better balance between exploration and exploitation is maintained by MOEA/D compare to NSGA-II while optimizing the proposed model.

In order to have graphical interpretation of median and *IQR*, boxplots of HV, Δ, GD and IGD, for both the MOGAs, are depicted in Fig. 8a–d. The boxplots show that corresponding to HV, Δ and IGD, the deviation around the median is less for MOEA/D than NSGA-II. While with GD, the deviation around the median is less for NSGA-II compared to MOEA/D. This suggests that after executing model (20), the probabilistic fluctuations of MOEA/D are less compared to NSGA-II for all the performance measures except GD, for which the probabilistic fluctuation of NSGA-II is comparatively less.

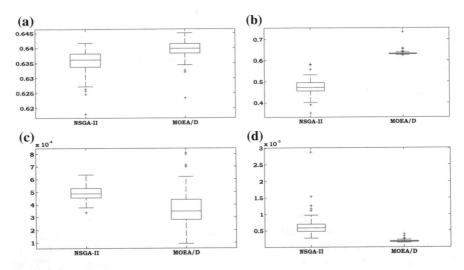

Fig. 8 Boxplots for **a** *HV* **b** Δ **c** *GD* and **d** *IGD*

9 Conclusion

In this article, a bi-objective uncertain portfolio selection model has been proposed under the paradigm of uncertainty theory. The proposed model maximizes the investment return and the risk of the uncertain security returns is, respectively, maximized and minimized by the proposed model. Here, the expected value and triangular entropy of the uncertain returns are, respectively, considered to determine the return and the risk. Twenty uncertain securities are considered for the proposed portfolio selection model. These investment returns are considered as linear uncertain variables and normal uncertain variables. It is worth mentioning that the proposed portfolio selection model can also be considered for more or less securities. The proposed model is solved by two compromise multi-objective solution techniques: (i) weighted sum approach, and (ii) weighted metric method. Under the uncertainty theory framework, except the contribution of Kar et al. (2017), there is no further study of multi-objective portfolio selection problem, where MOGAs are used as solution methodologies. Therefore, we have considered two different MOGAs, NSGA-II and MOEA/D, for solving the proposed model. The quality of solutions obtained from these MOGAs is analysed in terms of the performance metrics HV, Δ, GD and IGD. Moreover, to analyse the performance metrics, we have considered two summary statistics: (i) central tendency and (ii) variability. All these analyses show that MOEA/D outperforms NSGA-II while solving our proposed portfolio selection model.

In future, a large number of securities can be used to study the performance of the model. The proposed model can also be extended in fuzzy-random, uncertain-random and other hybrid uncertain environments.

Acknowledgements This work has been supported by INSPIRE fellowship (No. DST/INSPIRE Fellowship/2015/IF150410), Department of Science & Technology (DST), Ministry of Science and Technology, Government of India.

References

M. Arenas-Parra, A. Bilbao-Terol, M.V. Rodríguez-Uría, A fuzzy goal programming approach to portfolio selection. Eur. J. Oper. Res. **133**, 287–297 (2001)

M.J. Asgharpour, *Multiple Criteria Decision Making* (Tehran University Press, Tehran, 1998)

R. Bhattacharyya, A. Chatterjee, S. Kar, Mean-variance-skewness portfolio selection model in general uncertain environment. Indian J. Ind. Appl. Math. **3**, 44–56 (2012)

C. Carlsson, R. Fullér, On possibilistic mean value and variance of fuzzy numbers. Fuzzy Sets Syst. **122**, 315–326 (2001)

C. Carlsson, R. Fullér, P. Majlender, A possibilistic approach to selecting portfolios with highest utility score. Fuzzy Sets Syst. **131**, 13–21 (2002)

X. Chen, J. Gao, Uncertain term structure model of interest rate. Soft Comput. **17**(4), 597–604 (2013)

L. Chen, J. Peng, B. Zhang, I. Rosyida, Diversified models for portfolio selection based on uncertain semivariance. Int. J. Syst. Sci. **48**, 637–648 (2016)

C.A.C. Coello, Evolutionary multi-objective optimization: A historical view of the field. IEEE Comput. Intell. Mag. **1**, 28–36 (2006)

I. Das, J.E. Dennis, Normal-boundary intersection: A new method for generating Pareto optimal points in multi-criteria optimization problems. SIAM J. Optim. **8**, 631–657 (1998)

K. Deb, *Multi-objective Optimization Using Evolutionary Algorithms* (Wiley, New York, NY, 2001)

K. Deb, S. Agrawal, A. Pratap, T. Meyarivan, A fast and elitist multiobjective genetic algorithm: NSGA-II. IEEE Trans. Evol. Comput. **6**, 182–197 (2002)

J.J. Durillo, A.J. Nebro, jMetal: A Java framework for multi-objective optimization. Adv. Eng. Softw. **42**, 760–771 (2011)

C.D. Feinstein, M.N. Thapa, A reformulation of a mean-absolute deviation portfolio optimization model. Manage. Sci. **39**, 1552–1553 (1993)

D.C. Fishburn, Mean-risk analysis with risk associated with below-target returns. American Economical Review **67**, 117–126 (1977)

C.M. Fonseca, P.J. Fleming, Genetic algorithms for multi-objective optimization: Formulation, discussion and generalization, in *Proceedings of the fifth International Conference on Genetic Algorithms* (1993), pp. 416–423

J. Gao, K. Yao, Some concepts and theorems of uncertain random process. Int. J. Intell. Syst. **30**(1), 52–65 (2015)

J. Gao, X. Yang, D. Liu, Uncertain Shapley value of coalitional game with application to supply chain alliance. Appl. Soft Comput. **56**, 551–556 (2017)

C. Guo, J. Gao, Optimal dealer pricing under transaction uncertainty. J. Intell. Manuf. **28**(3), 657–665 (2017)

X. Huang, Mean-semivariance models for fuzzy portfolio selection. J. Comput. Appl. Math. **217**, 1–8 (2008a)

X. Huang, Mean-entropy models for fuzzy portfolio selection. IEEE Trans. Fuzzy Syst. **16**, 1096–1101 (2008b)

X. Huang, Risk curve and fuzzy portfolio selection. Comput. Math Appl. **55**, 1102–1112 (2008c)

X. Huang, *Portfolio Analysis: From Probabilistic to Credibilistic and Uncertain Approaches* (Springer, Berlin, 2010)

X. Huang, Mean-risk model for uncertain portfolio selection. Fuzzy Optim. Decis. Mak. **10**, 71–89 (2011)

X. Huang, Mean-variance models for portfolio selection subject to expert's estimations. Expert Syst. Appl. **39**, 5887–5893 (2012a)

X. Huang, A risk index model for portfolio selection with return subject to experts' evaluations. Fuzzy Optim. Decis. Making **11**, 451–463 (2012b)

X. Huang, H. Di, Uncertain portfolio selection with background risk. Appl. Math. Comput. **276**, 284–296 (2016)

P.H. Jorion, Value at Risk: A New Benchmark for Measuring Derivatives Risk (Irwin Professional Publishers, 1996)

M.B. Kar, S. Majumder, S. Kar, T. Pal, Cross-entropy based multi-objective uncertain portfolio selection problem. J. Intell. Fuzzy Syst. **32**, 4467–4483 (2017)

Knowles,J. D. Corne, The pareto archived evolution strategy: a new baseline algorithm for multi-objective optimization, in *Proceedings of the 1999 Congress on Evolutionary Computation CEC 99* (1999), pp. 9–105

H. Konno, K. Suzuki, A mean-variance-skewness optimization model. J. Oper. Res. Soc. Jpn. **38**, 137–187 (1995)

H. Konno, A. Wijayanayake, Mean-absolute deviation portfolio optimization model under transaction costs. J. Oper. Res. Soc. Jpn. **42**, 422–435 (1999)

H. Konno, H. Yamazaki, Mean-absolute deviation portfolio optimization model and its applications to Tokyo stock market. Manage. Sci. **37**, 519–531 (1991)

T.Y. Lai, Portfolio selection with skewness: A multi-objective approach. Rev. Quant. Financ. Acc. **1**, 293–305 (1991)

T. León, V. Liern, F. Vercher, Viability of infeasible portfolio selection problems: A fuzzy approach. Eur. J. Oper. Res. **139**, 178–189 (2002)

X. Li, Z. Qin, Interval portfolio selection models within the framework of uncertainty theory. Econ. Model. **41**, 338–344 (2014)

X. Li, Z. Qin, S. Kar, Mean-variance-skewness model for portfolio selection with fuzzy returns. Eur. J. Oper. Res. **202**, 239–247 (2010)

B. Li, Y. Zhu, Y. Sun, G. Aw, K.L. Teo, Multi-period portfolio selection problem under uncertain environment with bankruptcy constraint. Appl. Math. Model. **56**, 539–550 (2018)

J. Lintner, The valuation of risk assets and the selection of risky investments in stock portfolios and capital budgets. Rev. Econ. Stat. **41**, 13–37 (1965)

B. Liu, *Theory and practice of uncertain programming* (Springer, Berlin, Heidelberg, 2002)

B. Liu, *Uncertainty Theory*, 2nd edn. (Springer, Berlin, Heidelberg, 2007)

B. Liu, *Uncertainty theory: A Branch of Mathematics for Modeling Human Uncertainty* (Springer, Berlin, 2010)

B. Liu, Why is there a need for uncertainty theory? J. Uncertain Syst. **6**, 3–10 (2012)

B. Liu, X.W. Chen, Uncertain multi-objective programming and uncertain goal programming. J. Uncertain. Anal. Appl. **3**, 1–8 (2015)

S. Majumder, S. Kar, Multi-criteria shortest path for rough graph. J. Ambient Intell. Humaniz. Comput. (2017). https://doi.org/10.1007/s12652-017-0601-6

S. Majumder, P. Kundu, S. Kar, T. Pal, Uncertain Multi-Objective Multi-Item Fixed Charge Solid Transportation Problem with Budget Constraint. Soft. Comput. (2018). https://doi.org/10.1007/s00500-017-2987-7

H. Markowitz, Portfolio selection. J. Financ. **7**(1), 77–91 (1952)

A. Messac, A. Ismail-Yahaya, C. Mattson, The normalized normal constraint method for generating the Pareto frontier. Structural and Multidisciplinary Optimization **25**, 86–98 (2003)

K. Miettinen, *Nonlinear Multiobjective Optimization* (Kluwer, Norwell, MA, 1999)

K. Nag, T. Pal, N.R. Pal, ASMiGA: an archive-based steady-state micro genetic algorithm. IEEE Trans. Cybern. **45**, 40–52 (2015)

A.J. Nebro, F. Luna, E. Alba, B. Dorronsoro, J.J. Durillo, A. Beham, AbYSS: Adapting scatter search to multiobjective optimization. IEEE Trans. Evol. Comput. **12**, 439–457 (2008)

Y. Ning, H. Ke, Z. Fu, Triangular entropy of uncertain variables with application to portfolio selection. Soft. Comput. **19**, 2203–2209 (2014)

K.J. Oh, T.Y. Kim, S.-H. Min, H.Y. Lee, Portfolio algorithm based on portfolio beta using genetic algorithm. Expert Syst. Appl. **30**(3), 527–534 (2006)

Z. Pawlak, Rough sets. Int. J. Inf. Comput. Sci. **11**, 341–356 (1982)

Z. Qin, X. Li, X. Ji, Portfolio selection based on fuzzy crossentropy. J. Comput. Appl. Math. **228**, 139–149 (2009)

Z. Qin, M. Wen, C. Gu, Mean-absolute deviation portfolio selection model with fuzzy returns. Iran. J. Fuzzy Syst. **8**, 61–75 (2011)

Z. Qin, S. Kar, H. Zheng, Uncertain portfolio adjusting model using semiabsolute deviation. Soft. Comput. **20**, 717–725 (2016)

T.R. Rockafellar, S.P. Uryaser, Optimization of conditional value-at-risk. J. Risk **2**, 21–41 (2000)

W.F. Sharpe, Capital asset prices: A theory of market equilibrium under conditions of risk. J. Financ. **19**, 425–442 (1964)

Y. Simaan, Estimation risk in portfolio selection: the mean variance model versus the mean absolute deviation model. Manage. Sci. **43**, 1437–1446 (1997)

H. Soleimani, H.R. Golmakani, M.H. Salimi, Markowitz-based portfolio selection with minimum transaction lots, cardinality constraints and regarding sector capitalization using genetic algorithm. Expert Syst. Appl. **36**(1), 5058–5063 (2009)

N. Srinivas, K. Deb, Multiobjective function optimization using nondominated sorting genetic algorithms. Evol. Comput. **2**, 221–248 (1995)

H. Tanaka, P. Guo, Portfolio selection based on upper and lower exponential possibility distributions. Eur. J. Oper. Res. **114**(1), 115–126 (1999)

W. Tang, W. Gao, Triangular entropy of uncertain variables. Information **15**, 411–414 (2012)

D.A. Van Veldhuizen, G.B. Lamont, Multiobjective evolutionary algorithm research: a history and analysis, Technical Report TR-98-03, Department of Electrical and Computer Engineering, Graduate School of Engineering, Air Force Institute of Technology, Wright-Patterson, AFB, OH 1998

E. Vercher, J.D. Bermúdez, Portfolio optimization using a credibility mean-absolute semi-deviation. Expert Syst. Appl. **42**, 7121–7131 (2015)

J. Watada, Fuzzy portfolio selection and its application to decision making. Tatra Mountains Math. Publ. **13**, 219–248 (1997)

X. Yang, J. Gao, Linear quadratic uncertain differential game with application to resource extraction problem. IEEE Trans. Fuzzy Syst. **24**(4), 819–826 (2016)

X. Yang, J. Gao, Bayesian equilibria for uncertain bimatrix game with asymmetric information. J. Intell. Manuf. **28**(3), 515–525 (2017)

L.A. Zadeh, Optimality and nonscalar-valued performance criteria. IEEE Trans. Autom. Control **8**, 59–60 (1963)

L.A. Zadeh, Fuzzy Sets. Information Control **8**, 338–353 (1965)

J. Zhai, M. Bai, Uncertain portfolio selection with background risk and liquidity constraint. Math. Probl. Eng. **2017**, 1–10 (2017)

Q. Zhang, H. Li, MOEA/D: a multi-objective evolutionary algorithm based on decomposition. IEEE Trans. Evol. Comput. **11**, 712–731 (2007)

W.G. Zhang, Z.K. Nie, On admissible efficient portfolio selection problem. Appl. Math. Comput. **159**, 357–371 (2004)

B. Zhang, J. Peng, S. Li, Uncertain programming models for portfolio selections with uncertain returns. Int. J. Syst. Sci. **46**, 2510–2519 (2015)

A. Zhou, Y. Jin, Q. Zhang, B. Sendho, E. Tsang, Combining model-based and genetics-based offspring generation for multiobjective optimization using a convergence criterion, *in 2006 IEEE Congress on Evolutionary Computation* (Sheraton Vancouver Wall Center Vancouver, BC, Canada, 2006), pp. 3234–3241

E. Zitzler, L. Thiele, Multi-objective evolutionary algorithms: a comparative case study and the strength Pareto approach. IEEE Trans. Evol. Comput. **3**, 257–271 (1999)

E. Zitzler, M. Laumanns, L. Thiele, SPEA2: Improving the Strength Pareto Evolutionary Algorithm, Computer Engineering and Networks Laboratory Technical Report, Department of Electrical Engineering (2001), 103

Incorporating Gene Ontology Information in Gene Expression Data Clustering Using Multiobjective Evolutionary Optimization: Application in Yeast Cell Cycle Data

Anirban Mukhopadhyay

1 Introduction

The advancement in the area of microarray technology allows simultaneously study of the expression levels of several genes across a number of experimental conditions or time points (Alizadeh et al. 2000; Eisen et al. 1998; Bandyopadhyay et al. 2007; Lockhart and Winzeler 2000). The gene expression levels are measured at different time points during a biological experiment. A microarray gene expression data, which consists of g number of genes and t number of time points, is usually organized in the form of a 2D matrix $E = [e_{ij}]$ of size $g \times t$. Each element e_{ij} of the matrix represents the expression level of the ith gene at the jth time point. Clustering (Jain and Dubes 1988; Maulik and Bandyopadhyay 2002), a popular microarray analysis technique, is utilized to find the sets of genes having similar expression profiles. Clustering algorithms aim to partition a set of n genes into K clusters depending on some gene-to-gene distance metric.

Traditional clustering algorithms usually optimize a single cluster quality measure while performing clustering on a dataset. However, a single cluster quality measure is seldom equally applicable to different data properties. In this regard, several attempts have been made in optimizing more than one cluster quality measure simultaneously in order to obtain better robustness to different data properties (Mukhopadhyay and Maulik, in press; Bandyopadhyay et al. 2007; Mukhopadhyay et al., in press; Mukhopadhyay et al. 2015). These methods have used popular Multiobjective Optimization (MOO) (Deb 2001; Coello 2006; Mukhopadhyay et al. 2014a, b) tools as the underlying optimization strategy. Many of these multiobjec-

On leave from Department of Computer Science & Engineering, University of Kalyani, West Bengal, India. Email: anirban@klyuniv.ac.in.

A. Mukhopadhyay (✉)
Department of Computer Science, Colorado State University, Fort Collins, CO 80523-1873, USA
e-mail: anirban.mukhopadhyay@colostate.edu

© Springer Nature Singapore Pte Ltd. 2018
J. K. Mandal et al. (eds.), *Multi-Objective Optimization*,
https://doi.org/10.1007/978-981-13-1471-1_3

tive clustering algorithms have been applied for clustering genes in microarray gene expression data and have been shown to have improved performance compared to the traditional single-objective clustering algorithms (Bandyopadhyay et al. 2007; Maulik et al. 2009).

Validating clustering solutions for gene expression data is a challenging task. The validation measures mainly try to find how good is the clustering from either statistical point of view or biological point of view. In the later case, biological knowledge like Gene Ontology (GO) (Ashburner et al. 2000) information are utilized for analyzing gene clusters. GO data have been used to find the functional enrichment of a group of genes. In recent time, several researchers have attempted to include the GO information directly for clustering to yield more biologically relevant gene clusters.

Previously, a GO-driven clustering approach for gene expression data using genetic algorithm-based optimization was proposed in (Mukhopadhyay et al. 2010). However, as far as our knowledge goes, none of the previous attempts has used this GO information under the framework of multiobjective clustering. This motivates to make an attempt in incorporating the knowledge from GO into the distance (dissimilarity) between genes. The aim is to balance the gene expression information and GO information. Therefore, expression-based distance and GO-based distance are merged to form a combined distance metric. A well-known genetic algorithm for multiobjective optimization called Non-dominated Sorting Genetic Algorithm-II (NSGA-II) (Deb et al. 2002) has been employed as the underlying optimization framework. Two objective functions, one from gene expression point of view and another from GO point of view, have been optimized simultaneously. The performance of the proposed technique has been demonstrated for two real-life microarray gene expression datasets, viz., Heidelberg cell lines and yeast cell cycle. The validation of the clusters obtained by various distance matrices has been conducted using both GO (Ashburner et al. 2000)-based and KEGG pathway (Kanehisa and Goto 2000) based enrichment studies.

The rest of the chapter is organized as follows. In Sect. 2, the concept of gene ontology and different semantic similarity measures are discussed. Section 3 contains a discussion on the basic ideas of multiobjective optimization. In Sect. 4, the proposed multiobjective clustering algorithm is discussed in detail. Section 5 describes the dataset for experiments and the experimental results with illustrations. Finally, Sect. 6 concludes the chapter.

2 Gene Ontology and Similarity Measures

Gene Ontology (GO) (Ashburner et al. 2000) is a tool for associating a gene product with some ontological terms from three vocabularies, viz., Biological Process (BP), Molecular Function (MF), and Cellular Component (CC). Each of these BP, MF, and CC terms is arranged in the form of a Directed Acyclic Graph (DAG) where the nodes represent the GO terms and their relationships are represented through edges. A GO term thus can have multiple offsprings as well as more than one parent.

From the DAG structure of GO, the similarity between two gene products can be computed based on the GO terms associated with them. In this work, three popular metrics to compute GO-based similarity have been used. These are Resnik's measure (Resnik 1995), Lin's measure (Lin 1998), and Weighted Jaccard measure (Pesquita et al. 2007). These three measures are defined below (Mukhopadhyay et al. 2010).

2.1 Resnik's Measure

The basis of computing the Resnik's similarity metric is the Information Content (IC) in the concerned GO terms. The IC of a GO term is calculated based on the how many times the term is appearing in annotations. It is assumed that a term which is used rarely has greater IC. The probability $p(t)$ of observing a term t is computed as

$$p(t) = \frac{Frequency(t)}{MaxFreq}, \tag{1}$$

where $Frequency(t)$ represents how many times the term t or any of its descendents occur in the gene products annotated with the terms. $MaxFreq$ represents the frequency which is maximum among all GO terms. Based on this, the IC of a term t is then defined as (Mukhopadhyay et al. 2010; Resnik 1995)

$$IC(t) = -log_2 p(t). \tag{2}$$

The IC of a GO term can be calculated for all the three vocabularies, viz., BP, MF, and CC.

Resnik's similarity between two terms t and t', denoted as $Resnik(t, t')$, is then defined as the IC of the Minimum Subsumer (MS) or Lowest Common Ancestor (LCA) of the t and t' in the GO DAG (Mukhopadhyay et al. 2010; Resnik 1995).

$$Resnik(t, t') = IC_{LCA}(t, t') = \max_{\tau \in CA(t,t')} IC(\tau). \tag{3}$$

Here, $CA(t, t')$ represents the set of all common ancestors (direct and indirect) of t and t'.

Now, the similarity score of two genes (or their products) g and g' is computed as the largest of the pairwise similarities of the associated GO terms of the gene products. Let us consider two genes g and g' that are associated with GO terms $t_1, \ldots, t_n$ and $t'_1, \ldots, t'_m$, respectively. The Resnik's functional similarity $RSim(g, g')$ between the genes g and g' is then defined as (Mukhopadhyay et al. 2010; Resnik 1995)

$$RSim(g, g') = \max_{i=1,\ldots,n,\ j=1,\ldots,m} \{Resnik(t_i, t'_j)\}. \tag{4}$$

2.2 Lin's Measure

Lin similarity measure is also based on IC of the terms. Unlike Resnik's measure, Lin's measure considers how distant the terms are from their common ancestor. For this purpose, Lin's measure relates the IC of the LCA to the IC of the GO terms which are being compared. The Lin's similarity measure $Lin(t, t')$ between two GO terms t and t' is defined as (Mukhopadhyay et al. 2010; Lin 1998)

$$Lin(t, t') = \frac{2IC_{LCA}(t, t')}{IC(t) + IC(t')}.$$ (5)

The similarity between two genes is computed as the largest pairwise similarity of the associated GO terms of the genes. Therefore, if two genes g and g' are annotated with GO terms $t_1, \ldots, t_n$ and $t'_1, \ldots, t'_m$, then as per Lin's measure, the functional similarity $LSim(g, g')$ between g and g' is defined as (Mukhopadhyay et al. 2010; Lin 1998)

$$LSim(g, g') = \max_{i=1,\ldots,n,\ j=1,\ldots,m} \{Lin(t_i, t'_j)\}.$$ (6)

2.3 Weighted Jaccard Measure

This measure employs a set theoretic concept for computing the similarity of two genes or their products. Let us assume that gene g is annotated with n GO terms of the set $G = \{t_1, \ldots, t_n\}$ and gene g' is annotated with m GO terms of the set $G' = \{t'_1, \ldots, t'_m\}$. The Jaccard similarity measure $Jaccard(g, g')$ between g and g' is calculated as (Mukhopadhyay et al. 2010; Pesquita et al. 2007)

$$Jaccard(g, g') = \frac{|G \cap G'|}{|G \cup G'|}.$$ (7)

If each term is weighted with its IC, then the Weighted Jaccard (WJ) measure of similarity $WJSim(g, g')$ between g and g' is computed as (Mukhopadhyay et al. 2010; Pesquita et al. 2007)

$$WJSim(g, g') = \frac{\sum_{\tau_i \in G \cap G'} IC(\tau_i)}{\sum_{\tau_j \in G \cup G'} IC(\tau_j)}.$$ (8)

To compute the distance $D(g, g')$ between genes g and g', first the similarity $Sim(g, g')$ between the genes is computed using any of the above methods and normalized between 0 and 1. Thereafter, $D(g, g')$ is computed as $1 - Sim(g, g')$.

2.4 Combining Expression-Based and GO-Based Distances

In this work, for clustering genes, a combination of distance metrics obtained from gene expression data and GO is exploited. This leads to extraction of more biologically relevant gene clusters.

In this approach, a convex combination of distance measures is performed. The combined distance $\mathcal{D}_{exp+GO}(g, g')$ between the genes g and g' is computed as (Mukhopadhyay et al. 2010)

$$\mathcal{D}_{exp+GO}(g, g') = w \cdot D_{exp}(g, g') + (1 - w) \cdot D_{GO}(g, g'), \tag{9}$$

where $D_{exp}(g, g')$ denotes the distance computed from the expression data and $D_{GO}(g, g')$ is the distance calculated based on GO knowledge. The component distances, $D_{exp}(g, g')$ and $D_{GO}(g, g')$, are normalized to have values from 0 to 1. The weight w, a value between 0 and 1, determines the importance of each component on the combined measure. Here, w is set to 0.5 signifying equal importance to both the component distance metrics.

In this work, the well-known Pearson correlation-based distance measure is utilized to compute the distance between two genes based on their expression values. For computing distance from GO information, Resnik's, Lin's, and Weighted Jaccard distance measures are considered at a time. This means correlation-based distance is merged with one of Resnik's, Lin's, or Weighted Jaccard similarity based distances using Eq. (9). The distance metric values are normalized between 0 and 1 before combination.

3 Multiobjective Optimization and Clustering

In different real-life applications, there may be a requirement to optimize multiple objectives simultaneously to solve a certain problem. In multiobjective optimization, there is no notion of existence of a single solution that optimizes all the objectives simultaneously. Therefore, it is not easy to compare one solution with another. Generally, these problems produce a set of competent solutions instead of a single solution each of which is considered equally good when the relative importance of the objectives is unknown. In this context, the best solution is often subjective and it depends on the trade-off between the multiple conflicting objectives.

3.1 Formal Definitions

The multiobjective optimization can be formally stated as follows (Coello 1999): Find the vector $\bar{x}^* = [x_1^*, x_2^*, \ldots, x_n^*]^T$ of decision variables which will satisfy the m

inequality constraints:

$$g_i(\bar{x}) \geq 0, \quad i = 1, 2, \ldots, m, \tag{10}$$

the p equality constraints

$$h_i(\bar{x}) = 0, \quad i = 1, 2, \ldots, p, \tag{11}$$

and optimizes the vector function

$$\bar{f}(\bar{x}) = [f_1(\bar{x}), f_2(\bar{x}), \ldots, f_k(\bar{x})]^T. \tag{12}$$

The constraints shown in Eqs. (10) and (11) define the feasible region $\mathcal{F}$ that contains all the admissible solutions. Any solution falling outside this region is treated as inadmissible since it does not satisfy one or more constraints. The vector $\bar{x}^*$ provides an optimal solution in $\mathcal{F}$ satisfying these equality and inequality constraints. In the context of multiobjective optimization, multiple solutions may be evolved as the competent solutions. Hence, the difficulty lies in the definition of optimality, since it is very unlikely to generate a single vector $\bar{x}^*$ representing the optimum solution with respect to all the objective functions.

The concept of *Pareto-optimality* is useful in the field of multiobjective optimization. A formal definition of Pareto optimality from the viewpoint of a minimization problem may be stated as follows: A decision vector $\bar{x}^*$ is called Pareto-optimal if and only if there is no $\bar{x}$ that dominates $\bar{x}^*$, i.e., there is no $\bar{x}$ such that

$$\forall i \in \{1, 2, \ldots, k\}, f_i(\bar{x}) \leq f_i(\bar{x}^*)$$

and

$$\exists i \in \{1, 2, \ldots, k\}, f_i(\bar{x}) < f_i(\bar{x}^*).$$

In other words, $\bar{x}^*$ is called Pareto-optimal if there does not exist any feasible vector $\bar{x}$ which causes a decrease in some objective function without a simultaneous increase in at least another.

There exists a spectrum of approaches for multiobjective optimization problems (Deb 2001; Coello 1999), e.g., aggregating, population-based non-Pareto and Pareto-based techniques, etc. In aggregating techniques, the diverse objective functions are generally integrated into one using weighting or goal-based techniques. Among the population-based non-Pareto approaches, Vector Evaluated Genetic Algorithm (VEGA) is a technique in which different subpopulations are used for various objectives. Multiple-Objective GA (MOGA), Non-dominated Sorting GA (NSGA), Niched Pareto GA (NPGA), etc. are the most popular state-of-the-art methods under the Pareto-based non-elitist approaches (Deb 2001). On the other hand, NSGA-II (Deb et al. 2000), SPEA (Zitzler and Thiele 1998), and SPEA2 (Zitzler et al. 2001) are the examples of some recently developed multiobjective elitist techniques (Mukhopadhyay et al. 2014a, b).

3.2 Multiobjective Clustering

A number of approaches have been reported in recent literature that employs multi-objective optimization tool for data clustering (Mukhopadhyay and Maulik, in press; Bandyopadhyay et al. 2007; Mukhopadhyay et al., in press; Maulik et al. 2009). Some of them have also been used for clustering genes in microarray gene expression data (Bandyopadhyay et al. 2007; Maulik et al. 2009). These approaches mainly involve searching for suitable cluster centers using multiobjective optimization and optimizing some cluster validity measures simultaneously. Many of these approaches have used NSGA-II as the underlying optimization technique. The validity measures used for optimization have been computed based on some expression-based distance measure such as correlation or Euclidean distance. However, as far as our knowledge goes, no multiobjective clustering technique has incorporated the gene ontological knowledge in clustering gene expression data. With this motivation, a multiobjective clustering algorithm which utilizes both expression-based and GO-based knowledge for clustering microarray gene expression data has been developed here. NSGA-II has been used as the underlying optimization tool. The next section describes the proposed technique in detail.

4 Incorporating GO Knowledge in Multiobjective Clustering

In this section, the proposed multiobjective clustering technique which incorporates GO information in gene clustering is described. The proposed method uses NSGA-II as the underlying multiobjective optimization tool. The processes of chromosome representation, initialization of population, computation of fitness, selection, crossover, mutation, and elitism have been discussed. Finally, how a single solution is obtained from the Pareto-optimal front is stated.

4.1 Chromosome Representation and Initialization of Population

The main goal is to yield a suitable set of cluster centers through multiobjective optimization. Therefore, each chromosome of GA encodes K cluster centers, K being the number of clusters. In Maulik et al. (2009), real-valued encoding of cluster centers, each chromosome having a length of $K \times d$, was used. Here, d denotes the number of dimensions (time points) of the dataset. However, the cluster centers (arithmetic mean of feature vectors of expression values representing genes) may not represent actual genes. Therefore, if the complete pairwise gene-to-gene distance matrix is there in hand, then it is better to encode cluster medoids (most centrally located point in a cluster) instead of cluster centers. Unlike cluster centers (means), cluster medoids are actual points present in the dataset. Therefore, they represent actual genes. Here, the length of each chromosome is equal to K. Each position of the chromosome has a value selected randomly from the set $\{1, 2, \ldots, n\}$, where n is the number of points (genes). Hence, a chromosome is composed of the indices of the points (genes) in the dataset. Each point index in a chromosome represents a cluster medoid. A chromosome is valid if no point index is repeated in the chromosome. The initial population is generated by randomly creating P such chromosomes, where P is the user-defined population size. The value of P is kept unchanged over all the generations.

4.2 Computation of Fitness Functions

As stated in Sect. 2.4, three distance matrices are considered: the distance matrix computed based on Pearson correlation measure from gene expression values (D_{exp}), the distance matrix calculated based on one of Resnik's, Lin's, or Weighted Jaccard measures from GO information (D_{GO}), and the combined distance metric matrix ($\mathcal{D}_{exp+GO}$) as defined in Eq. (9). The combined distance is used for updating a chromosome, whereas the individual distance metrics are utilized for computing the fitness values.

A chromosome is decoded by forming clusters considering the genes encoded in it as the medoid genes and assigning other genes in the dataset to their closest medoids. Thereafter, new medoids for each cluster are computed by choosing the most centrally located point of the cluster and the chromosome is updated with the indices of those medoids. The most centrally located point of a cluster is the point from which the sum of the distances (dissimilarities) to the other points of the cluster is minimum. Note that all these computations are performed using the combined distance measure $\mathcal{D}_{exp+GO}$.

Two similar fitness functions, calculated using two different distance measures (D_{exp} and D_{GO}), are optimized simultaneously. The two objective functions f_1 and f_2 are given as

$$f_1 = \frac{\sum_{i=1}^{K} \sum_{x \in C_i} D_{exp}(x, \alpha_i)}{n \times \min_{p,q=1,\ldots,K,\ p \neq q} \{D_{exp}(\alpha_p, \alpha_q)\}}, \tag{13}$$

and

$$f_2 = \frac{\sum_{j=1}^{K} \sum_{y \in C_j} D_{GO}(y, \beta_j)}{n \times \min_{u,v=1,\ldots,K,\ u \neq v} \{D_{GO}(\beta_u, \beta_v)\}}. \tag{14}$$

Here, α_i denotes the medoid gene of the cluster C_i. To find the genes $x \in C_i$ (i.e., all genes whose nearest cluster medoid is α_i), the distance D_{exp} is used. Similarly, β_j denotes the medoid gene of the cluster C_j, and to obtain the genes $y \in C_j$ (i.e., all genes whose nearest cluster medoid is β_j), the distance D_{GO} is used. K and n denote the number of clusters and the number of genes in the dataset.

Note that both the objective functions are basically crisp version of well-known Xie–Beni (XB) validity index. The numerators represent the summed up dissimilarities of the genes from their respective cluster medoids. This is the global cluster variance which needs to be minimized in order to obtain compact clusters. The denominators are a function of the distance between the two closest cluster medoids, which represent the minimum cluster separation. To obtain well-separated clusters, the denominators are to be maximized. Therefore, overall both the objectives are to be minimized simultaneously to obtain compact and well-separated clusters from the viewpoint of both gene expression and gene ontology. The objectives f_1 and f_2 are computed as above only if the chromosome is valid, i.e., it does not contain a cluster medoid more than once. For invalid chromosomes, a large value is assigned to the objective functions so that they get out of the competition in subsequent generations.

4.3   Genetic Operators

The *selection* operation is performed using the standard crowded binary tournament selection as in NSGA-II. Subsequently, conventional uniform *crossover* controlled by a crossover probability μ_c is done for yielding the new offspring solutions from the parent chromosomes selected in the mating pool. The *mutation* operation is conducted by substituting the element to be mutated by a randomly chosen different gene indexes from the range $\{1, \ldots, n\}$ such that no element is duplicated in the chromosome. Each position of a chromosome undergoes mutation with mutation probability μ_m. *Elitism* is also incorporated as done in NSGA-II. For details of the different genetic operators, the reader may consult the article in Deb et al. (2002). The algorithm is executed for a fixed number of generations given as input. The near-Pareto-optimal chromosomes of the last generation represent the different solutions to the clustering problem.

4.4   Final Solution from the Non-dominated Front

The multiobjective clustering method generates a non-dominated set of clustering solutions in the final generation. Therefore, it is necessary to obtain a single solution from this non-dominated set. This is done as follows: First, each chromosome in the non-dominated set is updated using the combined distance measure $\mathcal{D}_{exp+GO}$ as discussed before. Thereafter, from each chromosome, two clusterings of the genes are obtained: one based on D_{exp} and another based on D_{GO}. Then, each D_{exp} based clustering is compared with the corresponding D_{GO} based clustering. The clustering pair that matches best is chosen. The idea is to find that clustering balances both expression-based and GO-based dissimilarities. Subsequently, from the chromosome corresponding to the best matching clustering pair, the final clustering solution is obtained based on $\mathcal{D}_{exp+GO}$.

5   Experimental Results and Discussion

In this section, first the yeast cell cycle dataset (Mukhopadhyay et al. 2010) used for experiments is described. Subsequently, the experimental results are discussed.

5.1   Dataset and Preprocessing

Experiments are performed on yeast cell cycle dataset, and this dataset is available publicly at http://genome-www.stanford.edu/cellcycle/. The raw dataset has expression values of 6,178 yeast open reading frames (ORFs) over 18 time points (Fellenberg et al. 2001). The genes without having any GO terms under BP are omitted and thus it produces 4,489 genes. Thereafter, all the genes having missing values are discarded. Finally, this yields 3,354 genes and these are used for the clustering purpose in subsequent phases. Here, normalization is performed over the dataset such that each row has mean 0 and variance 1.

5.2   Experimental Setup

At first, four pairwise distance matrices are constructed based on different distance metrics (Mukhopadhyay et al. 2010). These are expression-based correlation distance, Resnik's GO-based distance, Lin's GO-based distance, and Weighted Jaccard GO-based distance, respectively. The dimension of each matrix is $n \times n$. Here, n denotes the number of genes. The gene expression values have been used for computing the correlation-based distance. On the other hand, the other three distances are measured using the R package csbl.go (Ovaska et al. 2008). This package asks for the gene identifiers along with the GO terms associated with them as the input and returns the pairwise distances for the metric (Resnik, Lin, and Weighted Jaccard) under consideration as output.

The multiobjective clustering method has been executed for the following four distance matrices (Mukhopadhyay et al. 2010):

1. Expression-based distance matrix (D_{exp}).
2. Combined expression-based and Resnik's distance matrices $(\mathcal{D}_{exp+Res})$.
3. Combined expression-based and Lin's distance matrices $(\mathcal{D}_{exp+Lin})$.
4. Combined expression-based and Weighted Jaccard distance matrices $(\mathcal{D}_{exp+WJ})$.

For each case, the algorithm is run to group the genes into 100 clusters for the yeast cell cycle data. A number of generations, population size, crossover probability, and mutation probability are set to 100, 50, 0.8, and 0.1, respectively. The values of the parameters are chosen based on several experiments. For validation of the clusters, functional enrichment of the clusters has been studied using GO and KEGG pathways.

5.3 Study of GO Enrichment

The multiobjective technique yielded 100 gene clusters for each of the four runs as mentioned above for the yeast cell cycle dataset. To investigate the functional enrichment in the clusters, the hypergeometric test is applied to compute the degree of functional enrichment (*p-values*) that measures the probability of finding the number of genes involved in a given GO term within a cluster of genes. If most of the genes in a cluster have the same biological function, then it is likely that this has not taken place by chance and the *p-value* of the category will be very near to 0. The GOstats package from *R* is used to conduct the hypergeometric test. The hypergeometric test has been applied to the gene set of each cluster obtained by executing multiobjective clustering algorithm on the four distance matrices (one expression-based and three combined matrices).

For each cluster, the most significant GO term as well as the corresponding p-value is extracted. After that, the clusters are sorted in descending order of significance (increasing order of the p-value of the most significant GO term) for each distance metric. In Fig. 1, the p-values of the sorted list have been plotted for yeast cell cycle dataset. The plots of −log(p-value) are shown for better readability. Higher value of −log(p-value) implies lower value of p-value and hence, higher significance. The plots for each of the four distance metrics have been shown. The figure clearly shows that irrespective of the GO-based distance measure employed, the clusters resulted using combined distance have smaller p-values (larger −log(p-values)) for the most significant GO terms than that yielded using expression-based distance. This

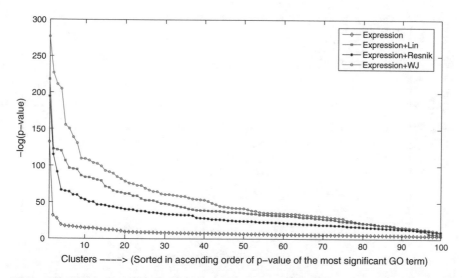

Fig. 1 The p-value plots of the most significant GO terms for all the clusters for each of the four distance metrics for yeast cell cycle data. The clusters are arranged in increasing order of p-value (decreasing order of −log(p-value))

indicates that the combined distance metrics produce more biologically relevant and functionally enriched clusters than the case while only expression-based distance is used. Therefore, it can be said that incorporation of GO information both in distance metric and in the objective function of the multiobjective clustering method yields more biologically enriched clusters.

Furthermore, it is evident from Fig. 1 that here $\mathcal{D}_{exp+WJ}$ based clustering produces more significant clusters than both $\mathcal{D}_{exp+Res}$ and $\mathcal{D}_{exp+Lin}$ based clusterings. It appears that $\mathcal{D}_{exp+Lin}$ based clustering performs better than $\mathcal{D}_{exp+Res}$ based clustering. This signifies the better performance of the $\mathcal{D}_{exp+WJ}$ based clustering for yeast cell cycle data.

Also, the number of clusters that have the most significant p-value (associated with the most significant GO term) less than a particular cut-off p-value is computed. Figure 2 shows the plots of the number of clusters against different cut-off p-values. It is apparent that the curves for the combined distance metrics stay higher compared to the curve corresponding to D_{exp} based clustering. This also demonstrates that the combined distance metrics yield more number of biologically significant clusters. Moreover, it is apparent that the curves corresponding to $\mathcal{D}_{exp+WJ}$ and $\mathcal{D}_{exp+Lin}$ are situated in higher position than that for $\mathcal{D}_{exp+Res}$.

For illustration, Table 1 reports the most significant unique GO terms with their p-values for the five most significant clusters yielded by all the four distance metrics. Evidently, the p-values produced using the combined metrics are much smaller compared to that for D_{exp}. This also demonstrates the better performance of the combined distance metrics.

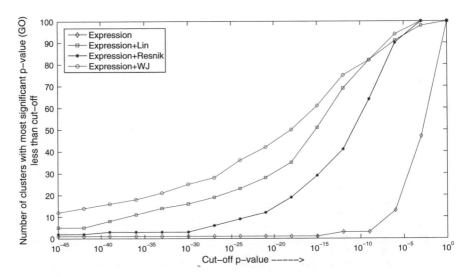

Fig. 2 The plots of number of clusters within certain cut-off p-values corresponding to the most significant GO terms for each of the four distance metrics for yeast cell cycle data

Table 1 The most significant GO terms and corresponding p-values of the top five unique most significant clusters for different distance matrices for yeast cell cycle data

Clusters	Expression		Expression+Lin		Expression+Resnik		Expression+WJ	
	GO term	p-value	GO term	p-value	GO term	p-value	GO term	p-value
Cluster 1	GO:0042254	3.45E−58	GO:0006364	2.27E−95	GO:0006364	3.39E−85	GO:0034470	7.13E−121
Cluster 2	GO:0006412	1.16E−14	GO:0006412	5.42E−54	GO:0009101	1.11E−50	GO:0045449	3.37E−99
Cluster 3	GO:0034605	8.22E−13	GO:0006397	1.82E−53	GO:0006790	1.98E−40	GO:0006486	2.05E−92
Cluster 4	GO:0007109	2.24E−08	GO:0006457	7.14E−53	GO:0016192	1.23E−29	GO:0006508	1.04E−89
Cluster 5	GO:0000722	4.00E−08	GO:0006810	4.04E−47	GO:0046483	4.55E−29	GO:0016481	2.92E−68

For further investigation, the cut-off p-value is set to 0.0001 to get the clusters for which the p-value of the most significant GO term is smaller than 0.0001. A set of 19 unique clusters for D_{exp}, 85 unique clusters for $\mathcal{D}_{exp+Res}$, 57 unique clusters for $\mathcal{D}_{exp+Lin}$, and 86 unique clusters for $\mathcal{D}_{exp+WJ}$ by this filtering. Thus, $\mathcal{D}_{exp+WJ}$ based clustering produces the highest number of clusters with most significant p-values less than the cut-off 0.0001.

In Fig. 3b, the four-way Venn diagram (for the four distance matrices) involving the most significant GO terms of the clusters with p-values lower than 0.0001 is depicted. The Venn diagram shows that for D_{exp}, 15 clusters are there for which the most significant GO terms are not the most significant terms of any other cluster obtained using other distance measures, whereas $\mathcal{D}_{exp+Res}$, $\mathcal{D}_{exp+Lin}$, and $\mathcal{D}_{exp+WJ}$ have produced 48, 24, and 56 such clusters, respectively. Moreover, for each distance metric, only one cluster has the same most significant GO terms. In Table 2, this GO term is reported for the common cluster along with the p-value for each distance metric. This is evident from the table that the p-values associated with the combined distance metrics are much smaller than that for D_{exp}.

Table 3 reports, for each of the four distance metrics, the GO terms and associated p-values for the 5 topmost clusters whose most significant GO terms do not share

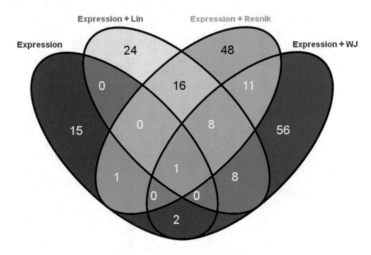

Fig. 3 The four-way Venn diagram (corresponding to the four distance metrics) among the most significant GO terms for yeast cell cycle data (for the clusters having p-values less than 0.0001)

Table 2 The most significant GO terms for the single common cluster along with the p-values for each of the four distance metrics for yeast cell cycle data

GO term	Expression	Expression+Lin	Expression+Resnik	Expression+WJ
GO:0006412 (protein biosynthesis)	3.64E−07	5.42E−54	1.47E−09	7.76E−46

Table 3 The most significant GO terms and corresponding p-values for the top five clusters (for each of the four distance metrics) whose most significant GO terms are not most significant term of any other cluster produced by other distance measures for yeast cell cycle data

Clusters	Expression		Expression+Lin		Expression+Resnik		Expression+WJ	
	GO term	p-value	GO term	p-value	GO term	p-value	GO term	p-value
Cluster 1	GO:0042254	3.45E−58	GO:0007049	2.71E−37	GO:0022613	8.06E−27	GO:0034470	7.13E−121
Cluster 2	GO:0034605	8.22E−13	GO:0051179	1.19E−27	GO:0006413	9.86E−27	GO:0016481	2.92E−68
Cluster 3	GO:0007109	2.24E−08	GO:0008033	1.01E−21	GO:0044283	6.05E−24	GO:0008202	1.90E−57
Cluster 4	GO:0000722	4.00E−08	GO:0006732	2.90E−21	GO:0006913	1.13E−22	GO:0055080	2.55E−48
Cluster 5	GO:0006261	5.37E−08	GO:0009228	9.14E−21	GO:0051649	5.40E−21	GO:0006609	3.29E−48

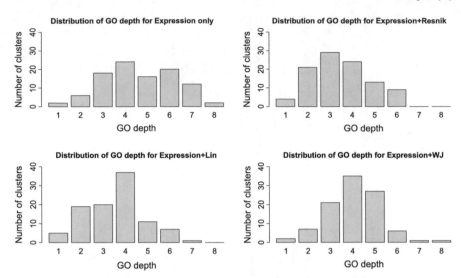

Fig. 4 Distribution of GO depth of the most significant GO terms (Number of clusters for different values of GO depth) for Heidelberg yeast dataset for different distance metrics

the same of any other cluster obtained using other distance measures. The clusters are sorted in increasing order of the p-values. From Table 3, it can be noticed that the p-value of the most significant GO term for the topmost cluster for $\mathcal{D}_{exp+WJ}$ is much lower than that of the topmost clusters for the D_{exp}, $\mathcal{D}_{exp+Res}$, and $\mathcal{D}_{exp+Lin}$. This indicates that the p-values for the combined dissimilarities are much less compared to expression-based clustering only.

To further study the homogeneity of the clusters, the depth of the most significant GO term for each cluster has been computed from the GO DAG (distance from the root—Biological Process), and the distribution has been plotted in Fig. 4. Here, the number of clusters for different values of GO depth of the most significant GO terms is shown. As is apparent from the figure, the clustering methods with combined dissimilarities produce more number of clusters whose most significant GO terms have low depth (nearer to the root), whereas for the expression-based clustering, the depths of the most significant GO terms are more uniformly distributed. This indicates that the combined techniques produce clusters that are biologically very homogeneous, but the price is that the biological process which is characteristic for the cluster, is more general compared to expression-based clustering.

To establish that the clusters are indeed more homogeneous for the combined dissimilarities, the GO compactness of the clusters has been computed. The GO compactness of a cluster is defined as the average of the path lengths between each pair of top five most significant GO terms of that cluster. Lower value of this implies better compactness in terms of GO. The path length between two GO terms has been defined as the sum of their distances from their lowest common ancestor (minimum subsumer) in the GO DAG. Figure 5 shows the distribution of the GO compactness,

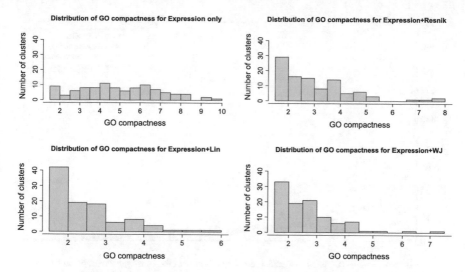

Fig. 5 Distribution of GO compactness (Number of clusters for different ranges of GO compactness) for yeast cell cycle dataset for different distance metrics. GO compactness of a cluster is defined as the average of the path lengths between each pair of top five most significant GO terms

i.e., the number of clusters yielded by the four distance metrics in different ranges of GO compactness. It is evident from the figures that the clustering methods based on combined distance metrics produce more number of clusters with low value of average pairwise distance (high GO compactness), whereas for expression-based clustering, this value is more or less uniformly distributed. This indicates that combined clustering methods are able to provide more compact clusters in terms of GO, i.e., they produce more biologically homogeneous clusters.

5.4   Study of KEGG Pathway Enrichment

As GO information has been incorporated in the proposed multiobjective clustering of gene expression data, validation of clustering results using GO-based analysis may be biased toward the combined distance measures. Therefore, in this section, the GO-based analyses discussed in the previous section have been repeated for KEGG pathways. Hence, the main goal of this study is to find whether the clusters obtained by the different distance measures are enriched with KEGG pathways. The software EXPANDER (Sharan et al. 2003; Shamir et al. 2005) has been utilized to accomplish the KEGG pathway study.

The most important KEGG pathway with its p-value has been extracted for each cluster. The clusters are arranged in decreasing order of significance (increasing order of the p-value) for each distance metric. The p-values from the sorted list have been shown in Fig. 6. The figure clearly shows that no matter which distance GO-based

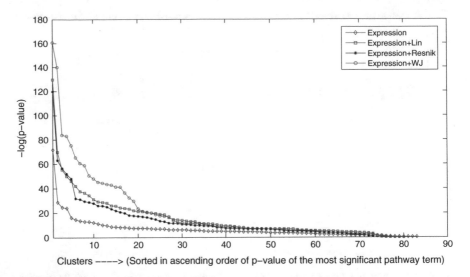

Fig. 6 The p-value plots of the most significant KEGG pathways for all the clusters for each of the four distance metrics. The clusters are arranged in increasing order of p-value (decreasing order of −log(p-value)) for yeast cell cycle data

distance metric is used, the clusters generated by the combined distance metrics are more significant, i.e., they have much smaller p-value (larger −log(p-value)) for the most important KEGG pathways than that obtained using expression-based distance. It is also noticeable that $\mathcal{D}_{exp+WJ}$ outperforms the other combined metrics. These results conform to the findings by GO-based study. It indicates that the clusters obtained using combined distance metrics contain genes which share many KEGG pathways than the clusters obtained using expression-based distance metric. It can be easily verified from Fig. 7, which shows the plotting of number of clusters having most significant p-values within the cut-off p-values. It can also be noted that the plots for the combined distance metrics are situated in the upper side, whereas the same for D_{exp} is mostly situated on the lower region. This suggests that the combined distances yield a larger number of KEGG pathway enriched clusters.

The most significant KEGG pathways with their p-values for the most important top five clusters obtained from the four distance metrics have been reported in Table 4 for illustration. It is clearly visible that p-values obtained for combined metrics are always very lower than the p-values obtained for D_{exp}. This indicates that the combined metrics produce a better result than D_{exp} based clustering.

Subsequently, the cut-off p-value is fixed at 0.001 to obtain the most significant KEGG pathway enriched clusters with p-value less than 0.001. This resulted in the following number of unique clusters for different distance metrics: for D_{exp} it is 15, for $\mathcal{D}_{exp+Res}$ it is 24, for $\mathcal{D}_{exp+Lin}$ it is 30, and for $\mathcal{D}_{exp+WJ}$, it is 31. Hence, $\mathcal{D}_{exp+WJ}$ based clustering gives the maximum number of most significant KEGG pathway

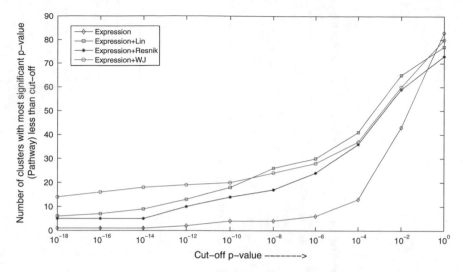

Fig. 7 The plots of number of clusters within certain cut-off p-values corresponding to the most significant KEGG pathways for each of the four distance metrics for yeast cell cycle data

enriched clusters for which the p-value is less than 0.001. Moreover, as expected, D_{exp} based clustering provided the minimum number of such clusters.

Figure 8 shows the four-way Venn diagram (related to the four distance metrics) involving the most important KEGG pathways of the clusters which have p-values lower than 0.001. It comes out from the diagram that for D_{exp}, there are only three such clusters. The most important KEGG pathways of these three clusters do not match with that of any other cluster reported by other distance metrics. On the other hand, $\mathcal{D}_{exp+Res}$, $\mathcal{D}_{exp+Lin}$, and $\mathcal{D}_{exp+WJ}$, respectively, have yielded 5, 10, and 8 such clusters. It is also noticeable that for every distance measure, only eight clusters are present that have the same most important KEGG pathways. The KEGG pathways for these clusters with their corresponding p-values for each distance metric have been reported in Table 5.

For each distance metric, the KEGG pathways and their p-values for the top five clusters that do not share the most important KEGG pathways with any other cluster given by other distance metrics have been reported in Table 6 for illustration. The clusters are arranged in increasing order of their p-values. It is clear that the p-values obtained for the combined distance metrics are smaller than that obtained for D_{exp}.

The results obtained in both GO and KEGG pathway-based studies suggest that the inclusion of the gene ontology knowledge in multiobjective clustering enhances the clustering performance by producing biologically relevant clusters compared to the clusters of genes produced by the expression dataset only. This improvement is evident no matter which GO-based distance measure is chosen. Moreover, the enrichment study also suggests that among the three GO-based dissimilarities,

Table 4 The most significant KEGG pathways and corresponding p-values of the top five unique most significant clusters for different distance metrics for yeast cell cycle data

Clusters	Expression		Expression+Lin		Expression+Resnik		Expression+WJ	
	Pathway	p-value	Pathway	p-value	Pathway	p-value	Pathway	p-value
Cluster 1	Ribosome	7.42E−32	Ribosome	4.97E−57	Ribosome	9.302E−53	Ribosome	1.872E−70
Cluster 2	Ribosome	3.15E−13	Metabolic pathways	4.89E−31	Metabolic pathways	3.967E−28	Aminoacyl-tRNA biosynthesis	1.692E−61
Cluster 3	Ribosome	2.06E−11	Proteasome	8.71E−25	Metabolic pathways	2.818E−25	Glycosylphosphatidylinositol(GPI)-anchor biosynthesis	4.075E−37
Cluster 4	DNA replication	3.81E−11	Ribosome	2.01E−22	Ribosome	2.703E−23	Proteasome	8.94E−37
Cluster 5	Proteasome	9.94E−08	Metabolic pathways	1.06E−20	N-Glycan biosynthesis	1.389E−21	N-Glycan biosynthesis	2.156E−33

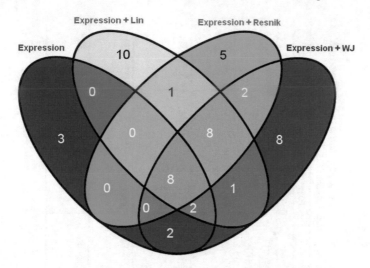

Fig. 8 The four-way Venn diagram (corresponding to the four distance metrics) among the most significant KEGG pathways for yeast cell cycle data (for the clusters having p-values less than 0.001)

Table 5 The most significant KEGG pathways for eight common clusters along with the p-values for each of the four distance metrics for yeast cell cycle data

Pathway term	Expression	Expression+Lin	Expression+Resnik	Expression+WJ
Ribosome	7.42E−32	4.97E−57	9.30E−53	1.872E−70
DNA replication	3.81E−11	4.60E−17	3.01E−14	2.417E−20
Proteasome	9.94E−08	8.71E−25	1.48E−13	8.94E−37
Citrate cycle (TCA cycle)	2.53E−06	1.52E−16	7.24E−13	2.501E−19
Mismatch repair	1.40E−05	4.23E−07	2.38E−06	2.002E−08
N-Glycan biosynthesis	1.89E−04	1.29E−15	1.39E−21	2.156E−33
Metabolic pathways	5.62E−04	4.89E−31	3.97E−28	1.441E−21
RNA degradation	5.67E−04	9.05E−09	1.40E−08	1.187E−18

$\mathcal{D}_{exp+WJ}$ based multiobjective clustering method gives better performance than the other two combined distance measures, viz., $\mathcal{D}_{exp+Res}$ and $\mathcal{D}_{exp+Lin}$.

Table 6 The most significant KEGG pathways and corresponding p-values for the top five clusters (for each of the four distance metrics) whose most significant KEGG pathways are not most significant term of any other cluster produced by other distance measures for yeast cell cycle data

Clusters	Expression		Expression+Lin		Expression+Resnik		Expression+WJ	
	Pathway	p-value	Pathway	p-value	Pathway	p-value	Pathway	p-value
Cluster 1	Sulfur metabolism	4.07E−04	Purine metabolism	3.11E−14	Arginine and proline metabolism	7.96E−11	Glycine, serine and threonine metabolism	6.60E−11
Cluster 2	Selenoamino acid metabolism	5.72E−04	Cysteine and metabolism	5.87E−10	Amino sugar and nucleotide metabolism	1.52E−06	Glycerolipid metabolism	3.79E−08
Cluster 3	Nitrogen metabolism	8.15E−04	Metabolism of xenobiotics by cytochrome P450	1.91E−09	Folate biosynthesis	9.70E−06	Glycerophospholipid metabolism	4.33E−06
Cluster 4	–		Porphyrin and chlorophyll metabolism	5.75E−07	Pyruvate metabolism	3.00E−05	Protein export	8.05E−06
Cluster 5	–		One carbon pool by folate	7.92E−07	Biosynthesis of unsaturated fatty acids	7.50E−05	Endocytosis	4.43E−05

6 Conclusion

In this chapter, a multiobjective clustering technique has been presented that finds the clusters of genes in a gene expression dataset by considering the GO information. The gene-to-gene distance metrics based on expression values and GO knowledge have been merged. Furthermore, two objective functions, one from expression point of view and another from GO point of view, have been optimized simultaneously to obtain the right balance between experimental and ontological information. Pearson correlation-based distance has been employed for computing the distances between the genes from experimental gene expression values. On the other hand, to compute distances between genes from biological point of view, three GO-based semantic similarity metrics, viz., Resnik's, Lin's, and Weighted Jaccard similarity measures, have been utilized. Experimental results based on GO and KEGG pathway enrichment analysis on time series gene expression data of yeast cell cycle have been described. The results suggest that the combined metrics of expression-based and GO-based distances yield improved clustering performance in comparison with only expression-based clustering. Furthermore, weighted Jaccard similarity measure has been found to outperform the other two GO-based semantic similarity metrics.

As a future scope of work, other GO-based semantic similarity measures can be tested. Moreover, other kind of biological knowledge, such as pathway information, can directly be incorporated in the clustering process as well. From the optimization point of view, other multiobjective metaheuristic techniques can be utilized for the clustering purpose. The author is working in these directions.

References

A.A. Alizadeh, M.B. Eisen, R. Davis, C. Ma, I. Lossos, A. Rosenwald, J. Boldrick, R. Warnke, R. Levy, W. Wilson, M. Grever, J. Byrd, D. Botstein, P.O. Brown, L.M. Straudt, Distinct types of diffuse large B-cell lymphomas identified by gene expression profiling. Nature **403**, 503–511 (2000)

M. Ashburner, C.A. Ball, J.A. Blake, D. Botstein, H. Butler, J.M. Cherry, A.P. Davis, K. Dolinski, S.S. Dwight, J.T. Eppig, M.A. Harris, D.P. Hill, L. Issel-Tarver, A. Kasarskis, S. Lewis, J.C. Matese, J.E. Richardson, M. Ringwald, G.M. Rubin, G. Sherlock, Gene Ontology: tool for the unification of biology. The gene ontology consortium. Na. Genet. **25**, 25–29 (2000)

S. Bandyopadhyay, U. Maulik, J.T.L. Wang, *Analysis of Biological Data: A Soft Computing Approach* (World Scientific, 2007)

S. Bandyopadhyay, A. Mukhopadhyay, U. Maulik, An improved algorithm for clustering gene expression data. Bioinformatics **23**(21), 2859–2865 (2007)

C.A. Coello Coello, A comprehensive survey of evolutionary-based multiobjective optimization techniques. Knowl. Inf. Syst. **1**(3), 129–156 (1999)

C. Coello Coello, Evolutionary multiobjective optimization: a historical view of the field. IEEE Comput. Intell. Mag. **1**(1), 28–36 (2006)

K. Deb, S. Agrawal, A. Pratap, T. Meyarivan, A fast elitist non-dominated sorting genetic algorithm for multi-objective optimization: NSGA-II, in *Proceedings of the Parallel Problem Solving from Nature VI Conference*. Lecture Notes in Computer Science No. 1917 (Springer, 2000), pp. 849–858

K. Deb, *Multi-Objective Optimization Using Evolutionary Algorithms* (Wiley, England, 2001)

K. Deb, A. Pratap, S. Agrawal, T. Meyarivan, A fast and elitist multiobjective genetic algorithm: NSGA-II. IEEE Trans. Evol. Comput. **6**, 182–197 (2002)

M.B. Eisen, P.T. Spellman, P.O. Brown, D. Botstein, Cluster analysis and display of genome-wide expression patterns, in *Proceedings of the National Academy of Sciences, (USA)* (1998), pp. 14863–14868

K. Fellenberg, N.C. Hauser, B. Brors, A. Neutzner, J.D. Hoheisel, M. Vingron, Correspondence analysis applied to microarray data. Proc. Natl. Acad. Sci. **98**(19), 10781–10786 (2001)

A.K. Jain, R.C. Dubes, *Algorithms for Clustering Data* (Prentice-Hall, Englewood Cliffs, NJ, 1988)

M. Kanehisa, S. Goto, KEGG: kyoto encyclopedia of genes and genomes. Nucleic Acids Res. **28**, 27–30 (2000)

D. Lin, An information-theoretic definition of similarity, in *Proceedings of the 15th International Conference on Machine Learning (ICML-98)* (Morgan Kaufmann Publishers Inc., San Francisco, CA, USA, 1998), pp. 296–304

D.J. Lockhart, E.A. Winzeler, Genomics, gene expreesion and DNA arrays. Nature **405**, 827–836 (2000)

U. Maulik, A. Mukhopadhyay, S. Bandyopadhyay, Combining pareto-optimal clusters using supervised learning for identifying co-expressed genes. BMC Bioinform. **10**(27) (2009)

U. Maulik, S. Bandyopadhyay, Performance evaluation of some clustering algorithms and validity indices. IEEE Trans. Pattern Anal. Mach. Intell. **24**(12), 1650–1654 (2002)

A. Mukhopadhyay, U. Maulik, Unsupervised pixel classification in satellite imagery using multi-objective fuzzy clustering combined with SVM classifier. IEEE Trans. Geosci. Remote Sens. (in press)

A. Mukhopadhyay, U. Maulik, S. Bandyopadhyay, A survey of multiobjective evolutionary clustering. ACM Comput. Surv. **47**, 61:1–61:46 (2015)

A. Mukhopadhyay, U. Maulik, S. Bandyopadhyay, Multi-objective genetic algorithm based fuzzy clustering of categorical attributes. IEEE Trans. Evol. Comput. (in press)

A. Mukhopadhyay, U. Maulik, S. Bandyopadhyay, B. Brors, GOGA: GO-driven genetic algorithm-based fuzzy clustering of gene expression data, in *2010 International Conference on Systems in Medicine and Biology (ICSMB)* (IEEE, 2010), pp. 221–226

A. Mukhopadhyay, U. Maulik, S. Bandyopadhyay, C.A.C. Coello, A survey of multiobjective evolutionary algorithms for data mining: part I. IEEE Trans. Evol. Comput. **18**(1), 4–19 (2014a)

A. Mukhopadhyay, U. Maulik, S. Bandyopadhyay, C.A.C. Coello, Survey of multiobjective evolutionary algorithms for data mining: part II. IEEE Trans. Evol. Comput. **18**(1), 20–35 (2014b)

K. Ovaska, M. Laakso, S. Hautaniemi, Fast gene ontology based clustering for microarray experiments. BioData Min. **1**(11) (2008)

C. Pesquita, D. Faria, H.B.A.O. Falco, F.M. Couto, An information-theoretic definition of similarity, in *Proceedings of the 10th Annual Bio-Ontologies Meeting (Bio-Ontologies-07)* (2007), pp. 37–40

P. Resnik, Using information content to evaluate semantic similarity in a taxonomy, in *Proceedings of the 14th International Joint Conference on Artificial Intelligence (IJCAI-95)* (Morgan Kaufmann Publishers Inc., San Francisco, CA, USA, 1995), pp. 448–453

R. Shamir, A. Maron-Katz, A. Tanay, C. Linhart, I. Steinfeld, R. Sharan, Y. Shiloh, R. Elkon, EXPANDER—an integrative program suite for microarray data analysis. BMC Bioinform. **6**(232) (2005)

R. Sharan, M.-K. Adi, R. Shamir, CLICK and EXPANDER: a system for clustering and visualizing gene expression data. Bioinformatics **19**, 1787–1799 (2003)

E. Zitzler, M. Laumanns, L. Thiele, SPEA2: Improving the Strength Pareto Evolutionary Algorithm, Technical Report 103, Gloriastrasse 35, CH-8092 Zurich, Switzerland (2001)

E. Zitzler, L. Thiele, An Evolutionary Algorithm for Multiobjective Optimization: The Strength Pareto Approach, Technical Report 43, Gloriastrasse 35, CH-8092 Zurich, Switzerland (1998)

Interval-Valued Goal Programming Method to Solve Patrol Manpower Planning Problem for Road Traffic Management Using Genetic Algorithm

Bijay Baran Pal

1 Introduction

The significant improvement of socio-economic conditions along with growth of population and Cultural Revolution across the countries, different types of motor vehicles have increase enormously in major towns and cities in recent years. It is worth mentioning here that metropolitan cities, or more popularly, metro cities have been modernized extensively as the epicentres of different countries, and they have become densely populated ones and vehicular traffic has increased a lot over the last few years. As a consequence, traffic congestion owing to increase in traffic density has become a regular feature on metro roadways.

It is to be noted, however, that congested traffic can cause considerable delay on roads for motorists, and inconsiderate driving by violating traffic rules frequently takes place on roadways of densely populated cities. Eventually, the occurrence of accidents resulting in injuries and death cases has become common phenomenon in most of the metro cities. To cope with such a situation, although automatic traffic signal system and other technological devices have been introduced to control flow of traffic, car crash and other untoward road incidents are found common phenomena owing to inexact behaviour of motorists, like hurried driving for overtaking other ones by overlooking overhead signals, mainly at junctions of approach roads. As such, automatic traffic control system in isolation cannot be accounted for best possible elimination of untoward road incidents on city roadways.

In context to the above, it may be mentioned that manual operations on traffic by deploying traffic patrol personnel on roadways, particularly in road segment areas, would give a comprehensive effect in the environment of controlling road traffic. As a matter of fact, mathematical modelling for quantitative measuring of road safety is

B. B. Pal (✉)
Department of Mathematics, University of Kalyani, Kalyani 741235, West Bengal, India
e-mail: bbpal18@hotmail.com

© Springer Nature Singapore Pte Ltd. 2018
J. K. Mandal et al. (eds.), *Multi-Objective Optimization*,
https://doi.org/10.1007/978-981-13-1471-1_4

essentially needed and thereby taking managerial decision for deployment of patrol-men in traffic management.

The mathematical theory of traffic flow was originated as early as 1921 when Knight (1921) produced an analysis of traffic equilibrium in the context of predic-tion of traffic patterns in transportation networks that are subject to congestion. A comprehensive view of the anatomy of urban road traffic control was introduced by Buchanan (1964) and widely circulated in the literature of traffic management.

The effective use of analytical process of operations research (OR) to solve traffic management problems was first well discussed by Larson (1972). The primary aim of using OR technique for physical control of traffic is to optimize various perfor-mance criteria that are concerned with enforcement of traffic rules and regulations to deterring traffic law violations and untoward road incidents. Here, it is worthy to note that the traffic management problems involve a number of incommensurable and conflicting objectives, because a number of performance measures are to be considered there with regard to optimal control of traffic in metro roadways.

To cope with the above situation, GP approach (Ignizio 1976), an efficient tool for solving problems with multiplicity of objectives in precise decision-making premises to traffic management problems was initially studied by Lee et al. (1979). A study on the approach was further extended by Taylor et al.(1985) to overcome the com-putational difficulty occurs for hidden nonlinearity (Hannan 1977) in model goals defined for measuring various performances against the deployment of patrol per-sonnel to control vehicular traffic on city roadways. The extensive literature on road traffic management has been well documented (Kerner 2009) in a previous study. However, it is worth mentioning that the target levels of goals in GP method are introduced by decision-maker (DM) in decision-making context. But, there are var-ious uncertain factors which come into play the active roles to many managerial problems, where significant computational difficulties arise in optimizing objectives by assigning precise target levels to them in the choice set.

To avoid the above shortcoming, fuzzy goal programming (FGP) in GP config-uration, a new version of fuzzy programming (FP) (Zimmermann 1978) in the area fuzzy sets (Zadeh 1965) was studied (Pal and Moitra 2003) towards measuring goal-oriented solution in imprecise environment. Then, application of FGP method to different kinds of decision-making (MODM) problems has been discussed (Biswas and Pal 2005; Pal et al. 2012; Slowinski 1986) previously. In fuzzy environment, however, goals are viewed as fuzzy numbers with known membership functions in contrast to known probability distributions of random numbers in stochastic pro-gramming (SP) (Liu 2009). But, in case of some real-world problems, it may not be possible to specify membership functions of fuzzily described goals appropriately or accurate probability distributions of random numbers to establish analytical model owing to lack of obtaining complete information in uncertain environment.

However, to model problems in some practical cases, data are found to be inexact in nature, where a set of real numbers in a bounded interval (Jiang et al. 2008), instead of fuzzy/random numbers, are involved with regard to setting parameter values to problems. An interval is potentially an extension of conceptual frame of real numbers and it is considered a real subset of the real line $\Re$ (Moore et al. 2009).

Fig. 1 Graphical
representation of closed
interval

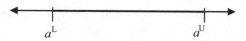

a^L a^U

In reality, the notion 'about' is usually used to set parameter values of a problem and certain tolerance ranges are considered to take their possible values in an imprecise environment, whereas in the situation of applying interval parameter sets, the notion 'region' is used to take possible parameter values bounded by certain lower and upper limits, because not all subsets of numbers on the real line are interesting in inexact environment.

More to say about interval parameters, it may be noted that interval can be viewed degenerated fuzzy numbers by applying the notion of α- cuts in fuzzy sets (Zimmermann 1991). Such an aspect to model optimization problems along with numerical illustrations has been well documented (Tong 1994) in a previous study.

Mathematically, an interval is defined by an ordered pair as

$$A = [a^L, a^U] = \{a : a^L \le a \le a^U; a \in \Re\},$$

where a^L and a^U are lower and upper bounds, respectively, of interval A on the real line $\Re$, and where L and U are used to indicate lower and upper, respectively.

The notion of a *closed interval* is depicted in Fig. 1.

Again, the *width* $w[A]$ and *midpoint* $m[A]$ of interval A can be defined as

$$w[A] = (a^U - a^L) \text{ and } m[A] = \tfrac{1}{2}(a^L + a^U), \text{ respectively.}$$

Here, in particular, if $a^L = a^U = a$ is considered, then $A = [a, a]$ indicates a real number a and it is said to degenerate interval $[a, a]$. In such a sense, $1 = [1, 1]$ is accounted for representing an interval.

To solve problems with interval data, interval programming method, initially introduced by Bitran (1980), has appeared as a prominent tool in the literature of optimization methods. In interval programming, interval arithmetic rules (Moore et al. 2009) are used for analysing problems with uncertain data. Historically, the conceptual frame of interval programming could be traced in the third century BC, when the famous mathematician Archimedes defined that the value of π (pi) practically lies in the range $(223/71 < \pi < 22/7)$. In real practice, interval computation is a numerical method of putting bounds on rounding errors to yield a reliable result of an entity, whose value is conceptually put as a range of possibilities. The study on interval analysis and crucial role of interval arithmetic for optimization has been discussed extensively in a book prepared by Hansen (2003).

Now, in traffic control system, since road network is a very complicated system and traffic phenomena are nonlinear in nature, it is nearly impossible to predict the characteristics of moving vehicles either exactly or imprecisely to design model for quantifying performances against the deployment of patrolmen to control vehicular traffic. In case of such a real-world problem, interval data instead of fuzzy/probabilistic one can reasonably be introduced there to model the problem.

It is worth noting here that the potential use of such a method to different kinds of practical problems along with diagrammatic representations of them has been well discussed (Kulpa 2001) previously.

The IVGP method for solving MODM problems was first presented by Inuiguchi and Kume (1991). The effectiveness of using IVGP approach to practical problems has also been discussed (Ida 2003) in a previous study. The methodologies concerned with IVGP discussed in previous studies have been extensively surveyed by Olivera and Antunes (2007). Again, IVGP method based on GA (Michalewicz 1996) to patrolmen deployment problem has also been discussed by Pal et al. (2009) in a previous study. But, the notion of using interval arithmetic to develop model goals was not fully explored therein and potential incorporation of interval coefficients was not considered in the formulated model. Furthermore, the study on modelling traffic control problems with interval data is very thin in literature.

In this chapter, a *priority-based* IVGP formulation of patrol manpower alloca-tion problem with interval parameter sets is considered for optimal allocation of patrolmen to various road segment areas at different shifts of a time period on metro roadways to deterring violation traffic rules and reduction of road incidents in traffic management system. In the proposed approach, the objective parameters concerned with utilization of resources as well as measuring performances of various traffic control activities are considered intervals to make a proper solution for patrolmen deployment in the inexact environment. Then, in model formulation, the functional expressions of objectives, which measure the performances of various activities and inherently fractional in nature, are transformed into their linear equivalents to over-come computational burden occurs for nonlinearity in model goals. Then, from the optimistic point of view, IVGP model of the problem under a *pre-emptive priority* structure (Romero 2004) as in conventional GP approach is designed to minimize regrets arising out of possible deviations from target values of model goals of the problem. To execute the model, a GA approach (Michalewicz 1996) is adopted to achieve target levels of defined goal on the basis of priorities introduced to them. The sensitivity analysis is also made by rearranging priorities of model goals to present how solution changes with changes of priorities, and then *Euclidean distance function* is applied to reach ideal point dependent decision regarding deployment of patrolmen in traffic control horizon.

A case example of the metro-city Kolkata of West Bengal in India is examined to present the effective use of the method. The successful implementation of the model is also highlighted by comparing the resulting decision with *minsum* IVGP approach studied by Inuiguchi and Kume (1991) previously.

This chapter is further arranged in the following order. Section 2 allocates devel-opment of the framework of model, and various aspects of executing model by applying GA along with the use of *Euclidean distance function* on sensitized solu-tions for decision identification. Section 3 introduces the notational description of variables and parameters associated with modelling of traffic control problem. In Sect. 4, development of objective goals and constraints of the proposed problem as well as evolutionary function associated with GA random search process are pre-sented. Section 5 contains an illustrative case example, along with comparison of solutions of present IVGP approach and a goal-oriented method of the previous study to present the effective use of the model. Finally, conclusions and future scope are discussed in Sect. 6.

Now, formulation of general IVGP model is discussed in Sect. 2.

2 IVGP Formulation

In IVGP approach, model parameters of an optimization problem are considered intervals, instead of considering their precise values as assumed in traditional optimization methods, and thereby analysing the perspective of taking decision in uncertain environment. Conceptually, IVGP can be viewed as an extension of conventional GP with a bounded set of parameter values of objective goals in the form of intervals to measure possible attainment of objective values in the ranges specified in decision situation.

Now, chronological development of IVGP model through conventional GP formulation is discussed as follows.

Let $F_k(\mathbf{x}), k = 1, 2, \ldots, K$ be the objectives of DM that are to be optimized in MODM situation, where $\mathbf{x}$ denotes a vector of decision variables.

Then, let t_k be the target level introduced to kth objective in the sequel of representing objectives as standard goals in GP methodology.

Then, kth objective with target value t_k, termed objective goal, appears as:

$$F_k(\mathbf{x}) = t_k, k = 1, 2, \ldots, K \tag{1}$$

In conventional GP approach, the rigid form goals in (1) are made flexible by incorporating under- and over-deviational variables to each of them with regard to achievement of their aspired goal levels to the extent possible, because achievements of target levels of all goals rarely take place in practical cases due to limitations on utilization of resources, which are scarce in nature in most of decision-making contexts.

The flexible goals take the form (Ignizio 1976):

$$F_k(\mathbf{x}) + d_k^- - d_k^+ = t_k, k = 1, 2, \ldots, K \tag{2}$$

where d_k^- and d_k^+ (≥ 0) represent under- and over-deviational variables, respectively.

Now, since objectives are in general incommensurable and frequently conflict each other to reach the best of their respective aspired levels, minimization of deviational variables (that are unwanted) in executable objective function (called goal achievement function) is taken into account according to the importance of achieving target levels of goals in decision situation.

However, in an inexact environment, instead of introducing the coefficients precisely and target level exactly to kth objective, the vector of interval coefficients $[\mathbf{c}_k^L, \mathbf{c}_k^U]$ and target interval $[t_k^L, t_k^U]$ are incorporated to analyse the region for possible achievement of defined objective goal, where c_k^L and c_k^U represent the vector of lower and upper limits of interval coefficient vectors, and t_k^L and t_k^U denote lower and upper bounds of target interval, respectively, associated with kth objective $F_k(\mathbf{x})$, $k = 1, 2, \ldots, K$.

Then, basic mathematical structure of the IVGP formulation described in this chapter is presented as follows.

Find $\mathbf{x}$ so as to

satisfy $[\mathbf{c}_k^L, \mathbf{c}_k^U]\mathbf{X} = [t_k^L, t_k^U]$, $k = 1, 2, \ldots, K$

subject to $\mathbf{X} \in S = \{\mathbf{X} | \mathbf{CX} \begin{pmatrix} \leq \\ \geq \\ = \end{pmatrix} \mathbf{b}, \mathbf{X} \geq 0; \mathbf{X} \in \mathbf{R}^n, \mathbf{b} \in \mathbf{R}^m\}$

$$\mathbf{x}^L \leq \mathbf{x} \leq \mathbf{x}^U, \tag{3}$$

where $\mathbf{C}$ is a real matrix and $\mathbf{b}$ is a constant vector, $\mathbf{X}^L$ and $\mathbf{X}^U$ indicate vectors of lower and upper limits, respectively, associated with vector $\mathbf{x}$. Also, it is considered that the feasible region $S(\neq \phi)$ is bounded.

The objectives in (3) are termed interval-valued goals and the associated intervals define bounded regions within which goals possibly take their values with regard to arriving at optimal solution in the decision-making horizon.

Now, basic arithmetic operations on intervals have been well documented (Moore et al. 2009) previously. It is to be mentioned here that 'interval arithmetic' describes a set of operations on intervals, whereas 'classical arithmetic' settles operations on individual numbers.

The extended sum operation which is primarily concerned with IVGP formulation is presented as follows.

If $\{A_j = [a_j^L, a_j^U], j = 1, 2, \ldots, n\}$ be a set of n intervals and $\{x_j(\geq 0), j = 1, 2, \ldots, n\}$ be a collection of decision variables, then the possible extended sum of the intervals can be obtained as:

$$\left(\overset{n}{\underset{j=1}{+}} \right) A_j x_j = [\sum_{j=1}^{n} a_j^L x_j, \sum_{j=1}^{n} a_j^U x_j] \tag{4}$$

Now, following interval arithmetic rules, the objective goals in (3) can be designated by planned interval goals in the process of designing the model, where the values possibly taken by objective goals in their respective planned intervals are called planned values of them. Then, performing interval arithmetic operation, the defined planned interval goals can be converted into two deterministic goals that are similar to the form of goal defined in (1) to make flexible goal as defined in (2).

Now, deterministic equivalents of planned interval goals are discussed in Sect. 2.1.

2.1 Deterministic Flexible Goals

The interval, say D_k, occurs for the question of possible deviations in connection to achievement of a value of kth objective goal in (3) within its specified range can be determined as (Inuiguchi and Kume 1991):

$$D_k = [t_k^L - \sum_{j=1}^{n} c_{kj}^U x_j, t_k^U - \sum_{j=1}^{n} c_{kj}^L x_j] \tag{5}$$

Here, it can easily be followed that the exact upper and lower values of D_k cannot be easily determined owing to inexactness of achieving the functional values of objectives as well as incommensurability and conflict nature of goals for achieving their values in decision situation. However, to cope with the situation of measuring possible deviations, each of the defined planned interval goals (called interval-valued goal) is converted into its equivalent to two flexible goals by incorporating individual lower and upper bounds of target interval as lower and upper target levels, termed the least and best planned values, respectively, and introducing both under- and over-deviational variables as defined in (2).

The flexible goals corresponding to kth goal in (3) appear as (Olivera and Antunes 2007):

$$\sum_{j=1}^{n} c_{kj}^U x_j + d_{kL}^- - d_{kL}^+ = t_k^L, \sum_{j=1}^{n} c_{kj}^L x_j + d_{kU}^- - d_{kU}^+ = t_k^U; k = 1, 2, \ldots, K \tag{6}$$

where $(d_{kL}^-, d_{kU}^-), (d_{kL}^+, d_{kU}^+) (\geq 0)$ denote the sets of under- and over-deviational variables, respectively, introduced to goal expression in (6), and where $d_{kL}^- \cdot d_{kL}^+ = 0$ and $d_{kU}^- \cdot d_{kU}^+ = 0, k = 1, 2, \ldots, K$.

Considering the goal expressions in (6), the expression of D_k in (5), which is considered possible regret interval, can be determined in the following three possible ways:

(i) If $d_{kL}^- = 0$ and $d_{kU}^- = 0$, then $D_k = [d_{kU}^+, d_{kL}^+]$, since $t_k^L < t_k^U$.
(ii) If $d_{kL}^- = 0$ and $d_{kU}^+ = 0$, then $D_k = [0, (d_{kL}^+ \vee d_{kU}^-)]$, where '$\vee$' indicates *max* operator.
(iii) If $d_{kL}^+ = 0$ and $d_{kU}^+ = 0$, then $D_k = [d_{kL}^-, d_{kU}^-]$.

Here, it is to be noted that $d_{kL}^- \cdot d_{kU}^+ = 0$ always holds, because $d_{kL}^- > 0$ and $d_{kU}^+ > 0$, i.e. simultaneous occurrence of $\sum_{j=1}^{n} c_{kj}^U < t_k^L$ and $\sum_{j=1}^{n} c_{kj}^L > t_k^U$ contradicts the conditions that $\mathbf{x} \geq 0$ as well as $c_{kj}^U > c_{kj}^L, t_k^U > t_k^L$.

Therefore, $(d_{kL}^- + d_{kU}^+) < max(d_{kL}^+, d_{kU}^-)$, i.e. $(d_{kL}^- + d_{kU}^+) < (d_{kL}^+ \vee d_{kU}^-)$ always holds true, which is to be accounted for modelling a problem in inexact environment.

However, the above three possible cases lead to take the expression of D_k as

$$D_k = [d_{kL}^- + d_{kU}^+, d_{kL}^+ \vee d_{kU}^-] \tag{7}$$

In context to the above, it is to be followed that attainments of goal values within their specified ranges indicate minimization of the deviational variables attached to goal equations in (6) to the possible extent in decision environment.

Now, IVGP model is presented in Sect. 2.2.

2.2 IVGP Model

The main activity concerning traffic management is to implement an enforcement program which applies selective enforcement pressure to critical road intersection areas at the hours of greatest accidents or rule violation expectancy. Here, it is worthy to mention that traffic movement is inherently inexact in nature and human behaviour is always unpredictable concerning violation of traffic rules in driving mode, and that increases with the decrease of traffic movement. In such a situation, modelling of the problem in the framework of IVGP would be an effective one to express uncertainties in making decision. The advantage of applying such a method is that interval characteristics regarding attainment of objective values are preserved there during execution of the model and interval data information are communicated directly to the optimization process.

Now, to formulate IVGP model, the main concern is with designing of goal achievement function (called regret function) (Inuiguchi and Kume 1991) for minimizing deviational variables attached to objective goals. Here, from the optimistic point of view of DM, minimization of possible regrets is considered as a promising one in the decision situation. In course of model formulation, however, since goals are generally incommensurable and often conflict each other concerning achievements of goals, a *priority-based* IVGP model, which is similar to that of modelling MODM problems in conventional GP (Romero 2004) is considered to arrive at optimal decision.

In *priority-based* IVGP method, the achievements of model goals are considered according to priorities assigned to them. Here, if some of the goals seem to be equally important to assign priorities, they are included at same priority level and relative numerical weights of importance are given there for sequential achievements of them in decision-making context.

Now, the general *priority-based* IVGP model of a MODM problem can be presented as:

Find **x** so as to

Minimize $Z = \left[P_1(\overline{\mathbf{d}}), P_2(\overline{\mathbf{d}}), \ldots, P_r(\overline{\mathbf{d}}), \ldots, P_R(\overline{\mathbf{d}}) \right]$
and satisfy the goal expressions in (6)
subject to

$$d_{r,kL}^- + d_{r,kU}^+ - V_r \leq 0, \quad k \in I_r, r \in \{1, 2, \ldots, R\}, \tag{8}$$

and the system constraints in (3),

where $\mathbf{Z}$ denotes a vector of achievement functions which is composed of R priorities, and where $V_r = \left[\max_{k \in I_r}(d^-_{r,kL} + d^+_{r,kU}) \right]$.

In the above IVGP model, $\mathrm{P}_r(\overline{\mathbf{d}})$ designates the regret function of vector $\overline{\mathbf{d}}$ that consists of under- and over-deviational variables attached to goals included at rth priority level, and where $\mathrm{P}_r(\overline{\mathbf{d}})$ is of the form:

$$\mathrm{P}_r(\overline{\mathbf{d}}) = \{\sum_{k \in I_r} \lambda_r w_{r,k}(d^-_{r,kL} + d^+_{r,kU}) + (1 - \lambda_r)V_r\}, 0 < \lambda_r < 1 \qquad (9)$$

where I_r denotes the set of model goals included at rth priority level P_r, and where $I_r = \{1, 2, \ldots, K\}; 0 < \lambda_r < 1$ with $\sum_{r \in I_r} \lambda_r = 1$, and $w_{r,k}(>0)$ with $\sum_{k \in I_r} w_{r,k} = 1, k \in I_r, r = 1, 2, \ldots, R$, are numerical weights given to goals for their relative achievements at priority level P_r, and $d^-_{r,kL}, d^+_{r,kU}$ are actually renamed for d^-_{kL}, d^+_{kU}, respectively, to represent them at rth priority level.

Here, the priority factors have the relationship:

$P_1 \ggg P_2 \ggg \ldots \ggg P_r \ggg \ldots \ggg P_R$, which means that goal achievement under the priority factor P_r is preferred most to the next priority factor $P_{r+1}, r = 1, 2,\ldots, R–1;$ '$\ggg$' is used to indicate much greater than.

Now, in solution process, execution of problem step-by-step for achievement of model goals according to priorities is summarized in the following algorithmic steps.

2.3   The IVGP Algorithm

In context to the above, however, it is worth mentioning that an optimal solution for achievements of all goal values to their aspired levels exactly that corresponds to ideal point (called utopia point), is a non-attainable case due to limitations on the availability of system resources. Further, DM is often confused with that of assigning proper priorities to goals due to the conflict in nature of them with regard to attaining their aspired levels in a decision situation. To avoid such a difficulty, *Euclidean distance function* (Yu 1973) can be effectively employed to determine appropriate priority structure under which an ideal point oriented solution (nearest to ideal solution point) as best one can be obtained in decision premises.

Now, determination of the priority structure of model goals properly by performing sensitivity analysis on alternative solutions that are obtained by rearranging the goal priorities is discussed in the following section.

2.3.1   Priority Structure Determination

In the situation of modelling a problem, it is mentionable that assignments of priorities to model goal equations for proper solution achievement, which is generally assigned by DM with usual perception, may not always be a satisfactory one, because a confusion in most of the times is created there for assigning priorities according to importance of meeting the target levels of model goals.

The IVGP Algorithm

Step 1.	Convert the interval-valued goals into flexible goals (as given in (6))
Step 2.	Construct the goal achievement function $\mathbf{Z}$ of the problem (as given in (8))
Step 3.	Rename $\mathbf{Z}$ as $(Z)_r$ where r is used to represent the component of $\mathbf{Z}$ when goal achievement at rth priority level P_r is considered (as defined in (9)), and where R is the total priority levels
Step 4.	Set $r = 1$
Step 5.	Determine the value of achievement function $(Z)_r$
Step 6.	Let $(Z^*)_r$ be the optimal value achieved in Step 5
Step 7.	Set $r = r + 1$
Step 8.	If $r \leq R$, go to Step 9; otherwise, go to Step 12
Step 9.	Introduce the constraint $\{\sum_{k \in I_{r-1}} \lambda_{r-1} w_{r-1,k}(d^-_{r-1,kL} + d^+_{r-1,kU}) + (1 - \lambda_{r-1})V_{r-1} \leq (Z^*)_{r-1}\}$ as system constraints set in (8) Here, it may be noted that the additional constraint defined in Step 9 actually acts as an insurance against any deterioration of the achieved values of goals at the higher priority level P_{r-1} for further evaluation of the problem for attainments of goals at lower priority level P_{r-2}
Step 10.	Return to Step 5
Step 11.	Test the convergence: If $r > R$, STOP; the optimality is reached
Step 12.	Identify the solution $\mathbf{x}^* = (x_1^*, x_2^*, \ldots, x_n^*)$

Note 1 The main merit of using such solution approach is that the achievements of goals step-by-step according to assigned priorities and relative numerical weights can be obtained in the process of executing the problem. Here, at the first step, minimization of the regret function $(Z)_1$ associated with goals at first priority level (P_1) is considered. When the value $(Z^*)_1$ is reached, the execution to the next step is made to evaluate $(Z)_2$ for possible achievements of goals at second priority level (P_2). Here, in no situation achievement of goals at P_1 can be sacrificed for achievement of any goal at P_2. The execution process is continued with sequential selection of priorities until $(Z)_R$ is evaluated with regard to taking final decision

Note 2 To employ the above-mentioned solution search process, it is worth noting that since the solution space is bounded and decision variables are with their upper and lower limits, a finite number of feasible solutions would have to be searched there to reach optimality. Therefore, the algorithm always terminates after a finite number of iterative steps and the method can be effectively used to obtain optimal solution

However, the process of selecting an appropriate priority structure to arrive at best solution is discussed as follows.

Let Q be the total number of different priority structures raised in the planning horizon. As such, a set of Q different solutions can be obtained in the decision situation.

Then, let $\{x_j^q, j = 1, 2, \ldots, n\}$ be the solution set obtained under qth priority structure, $q = 1, 2, \ldots, Q$.

Here, it is to be followed that the maximum of jth decision x_j^q, $\forall q$, would be an ideal one to achieve optimal solution in the decision-making horizon.

Let $x_j^* = \max\limits_{q=1}^{Q}(x_j^q), j = 1, 2, \ldots, n$.

Then, the set $\{x_j^*; j = 1, 2, \ldots, n\}$ constitutes ideal solution point.

Now, since the achievement of ideal solution is generally a nontrivial case in real decision situation, the solution set that is closest to ideal solution set might be considered best one and the associated priority structure would be the appropriate one in the context of making decision. Again, to find the closeness of Q different solution sets from the ideal one, the conventional *Euclidean distance function* can be applied to identify appropriate priority structure to find optimal solution.

The *Euclidean distance*, D^q (say), between the ideal solution point and solution point obtained under qth priority structure is defined as

$$D^{(q)} = \left[\sum_{j=1}^{n} (x_j^* - x_j^q)^2 \right]^{1/2}, q = 1, 2, \ldots, Q \tag{10}$$

Here, it is to be realized that the solution set which is closest to ideal set must correspond to the minimum of all the distances defined in the evaluation process.

Let $D^s = \min\limits_{q=1}^{Q}\{D^{(q)}\}$, where *min* stands for minimum.

Then, sth priority structure would be the appropriate one to obtain solution of the problem.

Now, GA computational process employed to the problem is presented in Sect. 2.4.

2.4   GA Computational Scheme for IVGP Model

In GA random solution search approach (Goldberg 1989; Deb 2009), new population (i.e. new solution set) is generated through execution of problem by employing probabilistically stipulated operators: *selection, crossover* and *mutation*. Here, real-value coded chromosomes are adopted to perform GA computation towards searching solution in random manner. The fitness of a chromosome is evaluated in the premise of feasible solutions set towards optimizing objectives of the problem. In the present IVGP model, since evaluation function Z is single-objective in nature with the characteristic of linear program, instead of using multiobjective GA scheme (Deb 1999), roulette-wheel selection (Goldberg 1989), arithmetic crossover (Hasan and Saleh 2011) and uniform mutation (Craenen et al. 2001) are addressed to find solution in the domain of interest.

The GA computational scheme with main functions adopted in the environment of making decision is described in algorithmic steps as follows.

The GA Algorithm

Step 1.	*Representation and initialization* Let E denote the real-valued representation of a chromosome in a population as $E = \{x_1, x_2, \ldots, x_n\}$. The population size is denoted by P, and P size chromosomes are randomly initialized in the search domain
Step 2.	*Fitness function* The evaluation function to determine the fitness of a chromosome is of the form: $$eval(E_v)_r = (Z_v)_r = \left\{ \sum_{k=1}^{K} w_{rk}^- d_{rk}^- \right\}_v, v = 1, 2, \ldots, P, \qquad (11)$$ where the function Z in (8) is actually renamed $(Z_v)_r$ to measure fitness of vth chromosome at a step when evaluation for goal achievement at rth priority level (P_r) is considered The optimal value Z_k^* of kth objective for the fittest chromosome at a generation is computed as: $$Z_k^* = min\{eval(Z_v)_r | v = 1, 2, \ldots, P\}, k = 1, 2, \ldots, K \qquad (12)$$
Step 3.	*Selection* The simple roulette-wheel selection (Goldberg 1989) is applied to select parents in a mating pool for production of offspring in the evolutionary system. The fitness proportionate strategy is made here by using probability distribution rule to select fitter chromosomes in new generation. Eventually, an individual with higher fitness probability is copied to place in a mating pool
Step 4.	*Crossover* The probability of crossover is defined by P_c. The arithmetic crossover (single-point crossover) (Michalewicz 1996) is implemented to explore promising regions of search space, where resulting offspring always satisfies system constraints set $S(\neq \phi)$. Here, a chromosome is selected as a parent for a defined random number $r \in [0,1]$, if $r < P_c$ is satisfied For example, arithmetic crossover over the selected parents $C_1, C_2 \in S$ is obtained as $C_1^1 = \alpha_1 C_1 + \alpha_2 C_2, \quad C_1^2 = \alpha_2 C_1 + \alpha_1 C_2$ in course of generating two offspring C_1^1 and $C_1^2,(C_1^1, C_1^2 \in S)$, where $\alpha_1, \alpha_2 \geq 0$, with $\alpha_1 + \alpha_2 = 1$
Step 5.	*Mutation* Mutation operation is accomplished on a population after conducting crossover operation. It recasts position of one or more genes of a chromosome to gain extra variability of fitness strength. Here, parameter P_m as probability of mutation is defined in the genetic system. The uniform mutation (Craenen et al. 2001) is conducted to exploit best solution that lies in the range specified for evaluation of the problem. A chromosome is selected here for mutation when a defined random number $r < P_m$ is satisfied for $r \in [0,1]$
Step 6.	*Termination* The termination of genetic search process arises when fittest chromosome is found at a certain generation in course of evaluating the problem

Henceforth, different perspectives of designing IVGP model of traffic management problem are described sequentially in the chapter. The decision variables and various interval parameter sets affiliated to the problem are now introduced in Sect. 3.

3   Definitions of Variables and Parameters

(a)   Decision variables

Two kinds of variables, independent variables and dependent variables which are inherent to problem are defined below.

(i)   *Independent variables*

p_{ij} = Allocation of patrolmen to road segment area i, shift j at a time period under consideration, i = 1, 2, …, m; j = 1, 2, …, n.

(ii)   *Dependent variables*

In the environment of physical control of traffic, it is to be noted that the key aim of deploying patrol units on city roadways is to reduce rate of accident, i.e. accident rate reduction by performing two kinds of operations, physical contact (PC) and site contact (SC), which are in essence separately involved with the problem to smoothing the flow of traffic. Here, it is to be observed that, although the primary aim of traffic operation is to deter rule violations and accidents, the operational modes of PC and SC are different in the sense that PC means physical contact with motorists to issuing warnings, citations, etc., against errant driving, personal intervention to check up valid driving license and related documents, investigations and legal actions against accidents and crime committed on the roadways, and others. Further, it is reactive in nature and has a greater impact on motorist behaviour as well as maintenance of peace and harmony on roadways. On the other hand, SC means constant vigilance to assist motorist, provide directions to a variety of traffic flows to most effective routes as well as to maintain order and control of pedestrian flow. As a matter of fact, both kind operations provide the underlying basis for intersectional traffic control mechanism, and they would have to be considered individually for quantitative measures for reduction of untoward road incidents.

Now, a set of variables that are concerned with various performance measuring criteria and which is defined in terms of p_{ij} is presented below.

AR_{ij}	Accident rate reduction (AR) contributed to road segment area i, shift j at a time period
PC_{ij}	Number of PCs attained by a patrolman in road segment area i, shift j at a time period
SC_{ij}	Number of SCs made by a patrolman in road segment area i, shift j at a time period
C_{ij}	Cost of deploying a patrolman to road segment area i, shift j at a time period
EB	Estimated budget to meet expenditure incurred for deployment of patrolmen at a time period

(b) **Definition of intervals**

The intervals that are inherently associated with objectives of the problem are defined as follows.

$[AR_{ij}^{\mathrm{L}}, AR_{ij}^{\mathrm{U}}]$	Target interval for AR at road segment area i in all shifts of a time period
$[PC_{ij}^{\mathrm{L}}, PC_{ij}^{\mathrm{U}}]$	Target interval of number of PCs required at road- segment area i, shift j at a time period
$[SC_{ij}^{\mathrm{L}}, SC_{ij}^{\mathrm{U}}]$	Target interval of the number of SCs required at road segment area i, shift j at a time period

Now, interval-valued goals and system constraints are discussed in Sect. 4.

4   Descriptions of Goals and Constraints

The different kinds of interval-valued goals, which measure performances against deployment of patrol personnel, are presented in Sect. 4.1.

4.1   Performance Measure Goals

The primary concern for effective management of traffic is to optimize law enforcement criteria against deployment of police patrol personnel, which is a macro-level study concerning quantitative measures for maintenance of traffic rules and regulations. It is to be noted here that various operational objectives are involved therein concerning enforcement of traffic laws and thereby controlling traffic during a time period under consideration. Again, in a patrolmen deployment decision situation, the three main objectives concerned with optimization problem are: increases in AR, PC and SC, which are typically represented by performance measure functions towards patrolmen deployment policy making and thereby deterring traffic rule violations and untoward road incidents.

Here, it is worth mentioning that total elimination of untoward road incidents is almost impossible in actual practice and often uncertain in nature owing to inexact nature of traffic environment, particularly in rush hours traffic. Therefore, it can be recognized that the managerial policy of accelerating the reduction of accident rate, instead of minimizing accident rate directly, by enhancing the number of PCs and SCs with the increase in number of patrolmen could be effective one in traffic management horizon.

Fig. 2 Graphical representation of performance measure function

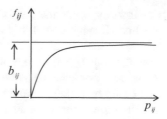

In context to the above, a multiplicative inverse relationship in parametric functional form can be established as a straightforward and logical one to define various criteria with regard to measuring performances against the deployment of patrol units in course of modelling the problem.

Now, the general mathematical expression to represent performance measure function is described in the following section.

4.1.1 Description of Performance Measure Function

The performance measure function which defines the relationship between p_{ij} and each of AR_{ij}, PC_{ij} and SC_{ij} for controlling traffic can be algebraically presented as

$$f_{ij} = b_{ij} - \frac{c_{ij}}{p_{ij}}, \; i = 1, 2, \ldots, m; j = 1, 2, \ldots, n, \tag{13}$$

where $f_{ij}(\geq 0)$ represents performance measure function, that is fitted against $p_{ij}(\geq 0)$, $\forall i, j$; b_{ij} and c_{ij} denote values of parameters.

It is to be followed that p_{ij} is always bounded with integer restrictions. Let it be assumed that $l_{ij} \leq p_{ij} \leq L_{ij}$, where l_{ij} and L_{ij} designate lower and upper bounds, respectively.

The graphical representation of performance measure function is depicted in Fig. 2.

Now, the mathematical expression of f_{ij} can be analytically explained as follows.

The graph shows that the given function is a strictly convex function, and f_{ij} is infeasible when $p_{ij} = 0$. Therefore, $f_{ij} = 0$ is taken into account as and when $p_{ij} = 0$ to avoid any situation of undefined functional value. Here, the least value of f_{ij} is obtained for the lowest value of $p_{ij}(>0)$. Again, it is to be observed that when p_{ij} increases, f_{ij} also increases monotonically at a reduced rate.

As a matter of consequence, the functional expression would be parabolic in form and that becomes asymptotic after a certain limiting value of p_{ij}, i.e. beyond the specified upper bound L_{ij}. It may be mentioned that induction of both the limits on p_{ij} depends on physical characteristics of the problem in decision environment. Therefore, it may be said that the defined functional relationship in (13) is both logical and anticipatory to measure the performances of various criteria against deployment of patrolmen to control traffic.

However, to define the functional relationship, two-point parameter estimation method (Rosenblueth 1981) can be used to estimate the values of parameters b_{ij} and c_{ij}, where some observational numeric data can be used to establish a set of simultaneous equations by following the relationship in (13) and thereby fitting the parametric nonlinear curve represented by f_{ij}. The effective use of the defined curve can easily be followed from the subsequent presentation of criteria as performance measure functions associated with the proposed traffic control problem.

Now, to avoid operational error with fractional function, the linearization technique (Kornbluth and Steuer 1981) is addressed here to take linear form of a fractional function.

4.1.2 Linearization of Performance Measure Function

In course of formulating IVGP model, since the mathematical expression in (13) represents an objective function of the problem in inexact environment, it would be considered as an interval-valued goal by introducing interval coefficient sets $[b_{ij}^{L}, b_{ij}^{U}]$ and $[c_{ij}^{L}, c_{ij}^{U}]$ and assigning target interval $[T_{ij}^{L}, T_{ij}^{U}]$.

Then, the goal expression in (13) appears as

$$([b_{ij}^{L}, b_{ij}^{U}] - \frac{[c_{ij}^{L}, c_{ij}^{U}]}{p_{ij}}) = [T_{ij}^{L}, T_{ij}^{U}] \tag{14}$$

Now, in the process of linearization of the fractional expression in (14), it may apparently seem that the traditional subtraction rule in transformation technique can be used. But, it is not an effective one, because subtraction is not the reverse of addition for operations on intervals. Here, interval arithmetic rules for subtraction (Jiang et al. 2008) can be used to determine the equivalent linear planned interval goal of the fractional expression in (14). The linearization process is described below.

• *Linear transformation of fractional interval-valued function*

For simplicity, let it be considered $t_{ij} = 1/p_{ij}$. Then, it is to be followed that $[c_{ij}^{L}, c_{ij}^{U}]t_{ij} \neq [b_{ij}^{L} - T_{ij}^{U}, b_{ij}^{U} - T_{ij}^{L}]$, because $[b_{ij}^{L}, b_{ij}^{U}] - [b_{ij}^{L} - T_{ij}^{U}, b_{ij}^{U} - T_{ij}^{L}] = [b_{ij}^{L} - b_{ij}^{U} + T_{ij}^{L}, b_{ij}^{U} - b_{ij}^{L} + T_{ij}^{U}] \neq [T_{ij}^{L}, T_{ij}^{U}]$.

Here, the real perspective is such that the necessary condition for the existence of solution would be $w[B] \geq w[T]t_{ij}$, since $w[B] - w[C]t_{ij} = w[T]$ with $t_{ij} \geq 0$, where B, C and T are used to represent the successive intervals.

With regard to the above condition, the simultaneous linear equations are obtained as

$$b_{ij}^{L} - c_{ij}^{U}t_{ij} = T_{ij}^{L} \text{ and } b_{ij}^{U} - c_{ij}^{L}t_{ij} = T_{ij}^{U} \tag{15}$$

Then, the planned interval goal to represent the solution space is obtained as $[c_{ij}^{L}, c_{ij}^{U}] t_{ij} = [b_{ij}^{U} - T_{ij}^{U}, b_{ij}^{L} - T_{ij}^{L}]$. As such, the specified target interval of the defined goal in (14) is satisfied.

Now, the equations in (15) in terms of p_{ij} and in linear form are successively derived as

$$(b_{ij}^{L} - T_{ij}^{L})p_{ij} = C_{ij}^{U} \text{ and } (b_{ij}^{U} - T_{ij}^{U})p_{ij} = C_{ij}^{L} \tag{16}$$

Following the equations in (16), the equivalent linear planned interval goal of the expressions in (14) is obtained as

$$[(b_{ij}^{U} - T_{ij}^{U})p_{ij}, \quad (b_{ij}^{L} - T_{ij}^{L})p_{ij}] = [c_{ij}^{L}, c_{ij}^{U}] \tag{17}$$

The effective use of the expression in (17) in interval form to represent performance measure goals in traffic control system is discussed as follows.

(a) **AR goals**

In the horizon of deterring traffic accidents, it may be mentioned that total elimination of accidents on roadways is almost impossible owing to inexactness of different human factors that actively play roles at turning points of flow of traffic, particularly in a congested traffic situation. Therefore, the notion of accelerating AR, that is, decrease in accident rate to the extent possible would be the main concern in traffic management system.

Following the expression in (17), the algebraic structure of AR function takes the form:

$$AR_{ij} = R_{ij} - \frac{r_{ij}}{p_{ij}}, i = 1, 2, \ldots, m; j = 1, 2, \ldots, n \tag{18}$$

where R_{ij} and $r_{ij}(R_{ij} > r_{ij})$ denote estimated parameters in precise sense.

Now, it is to be followed that the accurate functional relationship of accident rate reduction versus patrolmen deployment cannot be represented by well-behaved function owing to inexactness of measuring the characteristics of visual traffic, like traffic volume, accident frequency, etc., and a change of prediction can be realized over time. To overcome the difficulty, the interval parameter sets $[R_{ij}^{L}, R_{ij}^{U}]$ and $[r_{ij}^{L}, r_{ij}^{U}]$ need be taken into account here, and conventional function approximation method with the use of data fitting technique can be used to determine performance measure function, which represents a close relationship between AR_{ij} and p_{ij}. In the approximation process, certain pairwise data sets each consisting of accident rate (Accident rate is quoted in terms of periodic traffic volumes with the numbers of accidents) and patrol unit allocation obtained by going through certain predictable mechanism/long-term actual data recorded through day-to-day observations within existing structure are fitted there to estimate the parameter values of intervals numerically and thereby obtaining approximate function to represent the system. In actual practice, the approximated function assures smoothness of the corresponding curve to fit data in the sense that the curve passes as closely as possible to data points arising out of the reduction of accident rate in decision horizon.

Fig. 3 An exemplar of patrol personnel on duty on a city road

The above observations show that the use of the notion of increasing reduction of accident rates in road segment areas by deploying patrol units adequately is analytically more effective in the context of deterring untoward road incidents.

Then, following the expression in (14), accident rate reduction goal with interval data can be presented as

$$[R_{ij}^{L}, R_{ij}^{U}] - \frac{[r_{ij}^{L}, r_{ij}^{U}]}{p_{ij}} = [AR_{ij}^{L}, AR_{ij}^{U}], i = 1, 2, \ldots, m; j = 1, 2, \ldots, n \quad (19)$$

where $R_{ij}^{U} > r_{ij}^{U}$ and $R_{ij}^{L} > r_{ij}^{L}$ are estimated parameters.

It is to be noted that the increase in number of patrolmen over a certain limit could not provide any extra effect to determine AR, because there is a limitation on vehicular capacity of each of the approach roads to a road segment area in city road network.

An instant of Patrol personnel on duty is depicted in Fig. 3.

Following the expression in (17), the linear planned interval goal corresponding to the interval goal expression in (19) can be obtained as

$$[(R_{ij}^{U} - AR_{ij}^{U})p_{ij}, (R_{ij}^{L} - AR_{ij}^{L})p_{ij}] = [r_{ij}^{L}, r_{ij}^{U}], i = 1, 2, \ldots, m; j = 1, 2, \ldots, n \quad (20)$$

Now, different planned interval goals associated with AR are discussed as follows.

(i) *Segmentwise AR goal*

The AR at ith road segment area in all the shifts at a time period appears as

Fig. 4 A sample of physical contact in a road segment area on a city road

$$\left[\sum_{j=1}^{n}\{(R_{ij}^{U} - AR_{ij}^{U})p_{ij}\}, \sum_{j=1}^{n}\{(R_{ij}^{L} - AR_{ij}^{L})p_{ij}\}\right] = \sum_{j=1}^{n}[r_{ij}^{L}, r_{ij}^{U}], i = 1, 2, \ldots, m$$

(21)

(ii) *Shiftwise AR goal*

Similar to the expression in (21), AR of all the road segment areas during *j*th shift takes the form:

$$\left[\sum_{i=1}^{m}\{(R_{ij}^{U} - AR_{ij}^{U})p_{ij}\}, \sum_{i=1}^{m}\{(R_{ij}^{L} - AR_{ij}^{L})p_{ij}\}\right] = \sum_{i=1}^{m}[r_{ij}^{L}, r_{ij}^{U}], j = 1, 2, \ldots, n$$

(22)

Similarly, other goal equations can be mathematically presented as follows.

(b) **PC goals**

It is the most important part of measuring performances against deployment of patrolmen to deterring violations of traffic rules and accidents.

A sample of physical contact is displayed in Fig. 4.

Here, in an analogous to the expression in (19), the expression of PC_{ij} can be defined as

$$[P_{ij}^L, P_{ij}^U] - \frac{[q_{ij}^L, q_{ij}^U]}{p_{ij}} = [PC_{ij}^L, PC_{ij}^U], i = 1, 2, \ldots, m; j = 1, 2, \ldots, n \quad (23)$$

where $P_{ij}^U > q_{ij}^U$ and $P_{ij}^L > q_{ij}^L$ are associated parameters.

Then, goal equations can be described as follows.

(i) *Segmentwise PC goal*

The frequency of conducting the number of PCs actually depends on traffic density at a road segment area.

Therefore, segmentwise PC goal equation takes the form:

$$\left[\sum_{j=1}^n \{(P_{ij}^U - PC_{ij}^U)p_{ij}\}, \sum_{j=1}^n \{(P_{ij}^L - PC_{ij}^L)p_{ij}\} \right] = \sum_{j=1}^n [q_{ij}^L, q_{ij}^U], i = 1, 2, \ldots, m$$

$$(24)$$

(ii) *Shiftwise PC goal*

In an analogous to the expression in (22), shiftwise PC goal equation can be expressed as

$$\left[\sum_{i=1}^m \{(P_{ij}^U - PC_{ij}^U)p_{ij}\}, \sum_{i=1}^m \{(P_{ij}^L - PC_{ij}^L)p_{ij}\} \right] = \sum_{i=1}^m [q_{ij}^L, q_{ij}^U], j = 1, 2, \ldots, n$$

$$(25)$$

(c) **SC goals**

Similar to performing PC operation, SC operation is necessarily an integral part of controlling traffic, which is mostly concerned with controlling vehicle speed, diverging vehicles to different link roads to avoid traffic congestion and preventing violations of general traffic rules by both motorists and pedestrians when crossover takes place at a road segment area.

An example of sight-contact is depicted in Fig. 5.

In an analogous to PC goal expression in (23), the SC goal expression can be defined as

$$[S_{ij}^L, S_{ij}^U] - \frac{[s_{ij}^L, s_{ij}^U]}{p_{ij}} = [SC_{ij}^L, SC_{ij}^U], i = 1, 2, \ldots, m; j = 1, 2, \ldots, n \quad (26)$$

where $S_{ij}^U > s_{ij}^U$ and $S_{ij}^L > s_{ij}^L$ designate associated parameters.

Fig. 5 An example of sight-contact at a road segment on a city roadway

The goal equations associated with SC are discussed as follows.

(i) *Segmentwise SC goal*

To provide congenial traffic environment to road users, similar to the expression in (24), segmentwise SC goal appears as

$$\left[\sum_{j=1}^{n} \{(S_{ij}^{U} - SC_{ij}^{U}) p_{ij}\}, \sum_{j=1}^{n} \{(S_{ij}^{L} - SC_{ij}^{L}) p_{ij}\} \right] = \sum_{j=1}^{n} [s_{ij}^{L}, s_{ij}^{U}], i = 1, 2, \ldots, m$$

(27)

(ii) *Shiftwise SC goal*

Similar to the shiftwise PC goal presented in (25), here the goal equation appears as

$$\left[\sum_{i=1}^{m} \{(S_{ij}^{U} - SC_{ij}^{U}) p_{ij}\}, \sum_{i=1}^{m} \{(S_{ij}^{L} - SC_{ij}^{L}) p_{ij}\} \right] = \sum_{i=1}^{m} [s_{ij}^{L}, s_{ij}^{U}], j = 1, 2, \ldots, n$$

(28)

Now, system constraints set associated with proposed model is defined in Sect. 4.2.

4.2 System Constraints

(a) Budget utilization constraint

An estimated budget would always be provided to meet day-to-day incurring cost for deployment of patrol units in different time periods. It may be mentioned here that cash expenditure is mainly involved with fuel charges and miscellaneous ones for the use of patrol vehicles for frequent vigilance to nearby areas of road segments to keep harmony on roadways. However, different cost factors C_{ij} per patrol unit at different time period may arise and that depend upon accident frequency and traffic load at various road segment areas. Since allocation of budget is always restricted in nature, budget utilization constraint takes the form:

$$\sum_{i=1}^{m}\sum_{j=1}^{n} C_{ij} p_{ij} \leq EB \tag{29}$$

(b) Patrolmen deployment constraint

As discussed previously, it may further be mentioned that certain limitations on deployment of patrolmen are considered for any road segment area during any shift of the time period, and that depend on traffic load on roadways.

The associated constraint is of the form:

$$l_{ij} \leq p_{ij} \leq L_{ij}, \tag{30}$$

where l_{ij} and L_{ij} are lower and upper limits, respectively.

Now, construction of flexible goal equations associated with the linear planned interval goals defined above and thereby designing IVGP model is illustrated through an example presented in Sect. 5.

5 An Illustrative Example

The traffic management problem of Traffic Police department of metro-city Kolkata in India is selected. It may be noted that Kolkata is a densely populated city with population density is approximately 24,500 per square kilometre. The total road network is nearly 1850 km with a large number of road segments, and vehicle density is about 814.80 per kilometre. It may be mentioned here that ever more increase in number of motor vehicles without a scope of increase in road surface area has become huge every day's challenge to the road traffic control department of city Kolkata.

However, the main activity concerning management of traffic is to implement an enforcement program which applies selective enforcement pressure to critical road intersection areas at highest accident hours or rule violation expectancy. It may be pointed out that traffic movement is inherently inexact in nature and human behaviour is always unpredictable concerning violation of traffic rules in driving mode, and that increases with the decrease in traffic movement. In such a situation, modelling of the problem in the framework of IVGP would be an effective one to express uncertainty in making patrol units deployment decision.

The data of the proposed model were collected from the Annual Review Bulletin (Annual Review Bulletin 2012–2016), Traffic Police Department of Kolkata, India. The relevant information documented in the Bulletins of proceeding some years were also collected for numerical estimation of model parameters of the problem. Here, the yearly overall performance is estimated by considering day-to-day performance. The records show that most accident-prone road segment areas are: *BT road*, *AJC road*, *Strand road* and *EM Bypass*. The road segment areas are successively numbered: 1, 2, 3 and 4. Again, most of the road accident cases take place during the time interval 6.00 am–10.00 pm. Generally, the four shifts: 6.00 am–10.00 am, 10.00 am–2.00 pm, 2.00 pm–6.00 pm and 6.00 pm–10.00 pm can be sequentially arranged to deploy patrolmen to road segment areas according to traffic density and accident frequency. The shifts are sequentially numbered as 1, 2, 3 and 4, respectively.

The geographical representation of selected road segment areas of Kolkata is shown in Fig. 6.

Then, in course of formulating model goals, conventional two-point parameter estimation method (Rosenblueth 1981) is used and Microsoft EXCEL2007-Window is employed to estimate numerically the interval parameter sets associated with the defined performance measure goals. The estimated values of the limits on intervals are presented in Table 1.

Now, using the data presented in Table 1, the interval-valued goals can easily be built by following the equations in (21), (22), (24), (25), (27) and (28).

Then, following the goal equations defined in (6), executable goal equations are structured and presented in Sect. 5.1.

In the context of constructing goal equations, it is worthy to present here that when segmentwise AR goal for the *BT road* (for $i = 1$) during all the four shifts $(j = 1, 2, \ldots, 4)$ of the time period is considered, then interval-valued goal can be algebraically presented in the form:

$$\left[\sum_{j=1}^{4} \{(R_{ij}^{U} - AR_{ij}^{U})p_{ij}\}, \ \sum_{j=1}^{4} \{(R_{ij}^{L} - AR_{ij}^{L})p_{ij}\} \right] = \sum_{j=1}^{4} [r_{ij}^{L}, r_{ij}^{U}], \ \text{for } i = 1$$

Here, two successive goal equations corresponding to the upper and lower values of planned interval can be obtained by associating the lower and upper bounds, respectively, of the defined target interval.

The goal equations appear as

$$(R_{11}^{L} - AR_{11}^{L})p_{11} + (R_{12}^{L} - AR_{12}^{L})p_{12} + (R_{13}^{L} - AR_{13}^{L})p_{13} + (R_{14}^{L} - AR_{14}^{L})p_{14} = \sum_{j=1}^{4} r_{1j}^{L}$$

and

Table 1 Estimation of interval parameters associated with AR_{ij}, PC_{ij} and SC_{ij} functions

Road segment i	Shift j	AR_{ij}			PC_{ij}			SC_{ij}		
		$[R^L_{ij}, R^U_{ij}]$	$[r^L_{ij}, r^U_{ij}]$	$[AR^L_{ij}, AR^U_{ij}]$	$[P^L_{ij}, P^U_{ij}]$	$[q^L_{ij}, q^U_{ij}]$	$[PC^L_{ij}, PC^U_{ij}]$	$[S^L_{ij}, S^U_{ij}]$	$[s^L_{ij}, s^U_{ij}]$	$[SC^L_{ij}, SC^U_{ij}]$
1	1	[0.78, 0.85]	[0.18, 0.49]	[0.68, 0.77]	[120, 132]	[95, 110]	[98, 113]	[200, 220]	[165, 175]	[167, 185]
1	2	[0.76, 0.80]	[0.17, 0.47]	[0.60, 0.67]	[125, 128]	[89, 100]	[92, 99]	[235, 250]	[195, 225]	[171, 176]
1	3	[0.77, 0.82]	[0.17, 0.47]	[0.61, 0.70]	[135, 150]	[125, 130]	[92, 107]	[195, 210]	[137, 154]	[150, 159]
1	4	[0.81, 0.93]	[0.29, 0.63]	[0.65, 0.79]	[130, 140]	[117, 125]	[99, 109]	[257, 265]	[185, 206]	[212, 214]
2	1	[0.77, 0.84]	[0.27, 0.63]	[0.64, 0.71]	[117, 120]	[83, 95]	[98, 104]	[197, 217]	[165, 174]	[162, 184]
2	2	[0.75, 0.83]	[0.24, 0.55]	[0.55, 0.64]	[110, 119]	[87, 105]	[76, 91]	[215, 235]	[185, 210]	[146, 174]
2	3	[0.75, 0.80]	[0.13, 0.56]	[0.26, 0.70]	[127, 134]	[107, 115]	[104, 113]	[185, 197]	[167, 195]	[146, 164]
2	4	[0.83, 0.97]	[0.16, 0.49]	[0.34, 0.56]	[129, 135]	[118, 127]	[19, 30]	[223, 230]	[189, 205]	[18, 41]
3	1	[0.75, 0.68]	[0.19, 0.51]	[0.24, 0.26]	[135, 147]	[135, 143]	[22, 42]	[183, 195]	[155, 170]	[13, 40]
3	2	[0.74, 0.78]	[0.15, 0.56]	[0.63, 0.69]	[139, 145]	[87, 98]	[120, 128]	[197, 220]	[179, 190]	[159, 182]
3	3	[0.76, 0.83]	[0.21, 0.58]	[0.18, 0.33]	[127, 134]	[107, 114]	[13, 27]	[185, 197]	[117, 127]	[58, 80]
3	4	[0.82, 0.91]	[0.25, 0.64]	[0.69, 0.80]	[147, 156]	[117, 124]	[22, 32]	[195, 215]	[175, 187]	[158, 180]
4	1	[0.69, 0.75]	[0.11, 0.37]	[0.60, 0.68]	[98, 105]	[78, 90]	[80, 90]	[178, 190]	[147, 165]	[145, 161]
4	2	[0.67, 0.73]	[0.18, 0.53]	[0.41, 0.51]	[100, 110]	[87, 98]	[51, 67]	[200, 215]	[172, 180]	[110, 129]
4	3	[0.68, 0.71]	[0.21, 0.50]	[0.58, 0.63]	[108, 112]	[97, 112]	[86, 93]	[183, 195]	[97, 116]	[160, 176]
4	4	[0.74, 0.89]	[0.23, 0.58]	[0.16, 0.43]	[95, 102]	[69, 81]	[14, 33]	[205, 225]	[163, 175]	[30, 62]

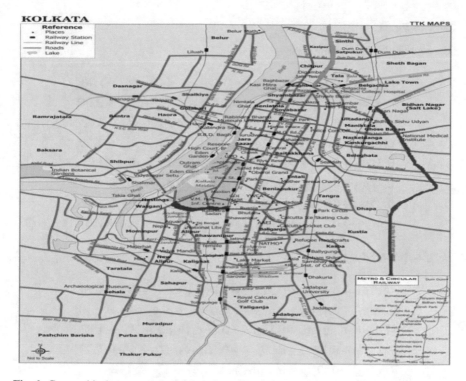

Fig. 6 Geographical representation of the selected road segment areas. *Note* The green line, yellow line, red line and blue line indicate *BT road*, *Stand road*, *AJC road* and *EM Bypass*, respectively, of the road network in Kolkata

$$(R_{11}^{U} - AR_{11}^{U})p_{11} + (R_{12}^{U} - AR_{12}^{U})p_{12} + (R_{13}^{U} - AR_{13}^{U})p_{13} + (R_{14}^{U} - AR_{14}^{U})p_{14} = \sum_{j=1}^{4} r_{1j}^{U}$$

Then, using the data in Table 1, the first two goal equations presented under '(*i*) *Segment-wise AR goals*' in Sect. 5.1 can be obtained.

In a same way, the other goal equations associated with their linear interval-valued goals can easily be constructed.

5.1 Construction of Model Goals

The three kinds of model goal equations that are concerned with AR, PC and SC for measuring performances against deployment of patrol units can be obtained as follows.

(a) **AR goals**

(i) *Segmentwise AR goals*

$$0.10p_{11} + 0.16p_{12} + 0.16p_{13} + 0.16p_{14} + d_{1L}^- - d_{1L}^+ = 1.70,$$
$$0.08p_{11} + 0.13p_{12} + 0.12p_{13} + 0.14p_{14} + d_{1U}^- - d_{1U}^+ = 2.06;$$
$$0.13p_{21} + 0.20p_{22} + 0.12p_{23} + 0.49p_{24} + d_{2L}^- - d_{2L}^+ = 2.02,$$
$$0.13p_{21} + 0.19p_{22} + 0.10p_{23} + 0.38p_{24} + d_{2U}^- - d_{2U}^+ = 2.29;$$
$$0.51p_{31} + 0.11p_{32} + 0.58p_{33} + 0.13p_{34} + d_{3L}^- - d_{3L}^+ = 1.93,$$
$$0.42p_{31} + 0.09p_{32} + 0.50p_{33} + 0.11p_{34} + d_{3U}^- - d_{3U}^+ = 2.29;$$
$$0.09p_{41} + 0.26p_{42} + 0.10p_{43} + 0.58p_{44} + d_{4L}^- - d_{4L}^+ = 1.67,$$
$$0.07p_{41} + 0.22p_{42} + 0.08p_{43} + 0.46p_{44} + d_{4U}^- - d_{4U}^+ = 2.08 \qquad (31)$$

(ii) *Shiftwise AR goals*

$$0.10p_{11} + 0.13p_{21} + 0.51p_{31} + 0.09p_{41} + d_{5L}^- - d_{5L}^+ = 1.76,$$
$$0.08p_{11} + 0.13p_{21} + 0.42p_{31} + 0.07p_{41} + d_{5U}^- - d_{5U}^+ = 2.10;$$
$$0.16p_{12} + 0.21p_{22} + 0.11p_{32} + 0.26p_{42} + d_{6L}^- - d_{6L}^+ = 1.87,$$
$$0.13p_{12} + 0.19p_{22} + 0.09p_{32} + 0.22p_{42} + d_{6U}^- - d_{6U}^+ = 2.17;$$
$$0.16p_{13} + 0.12p_{23} + 0.56p_{33} + 0.10p_{43} + d_{7L}^- - d_{7L}^+ = 1.77,$$
$$0.12p_{13} + 0.10p_{23} + 0.50p_{33} + 0.08p_{43} + d_{7U}^- - d_{7U}^+ = 2.11;$$
$$0.16p_{14} + 0.49p_{24} + 0.13p_{34} + 0.58p_{44} + d_{8L}^- - d_{8L}^+ = 1.92,$$
$$0.14p_{14} + 0.38p_{24} + 0.11p_{34} + 0.46p_{44} + d_{8U}^- - d_{8U}^+ = 2.34 \qquad (32)$$

(b) **PC goals**

(i) *Segmentwise PC goals*

$$22p_{11} + 33p_{12} + 43p_{13} + 31p_{14} + d_{9L}^- - d_{9L}^+ = 395,$$
$$19p_{11} + 29p_{12} + 41p_{13} + 29p_{14} + d_{9U}^- - d_{9U}^+ = 442;$$
$$19p_{21} + 34p_{22} + 23p_{23} + 110p_{24} + d_{10L}^- - d_{10L}^+ = 372,$$
$$16p_{21} + 28p_{22} + 21p_{23} + 105p_{24} + d_{10U}^- - d_{10U}^+ = 425;$$
$$113p_{31} + 19p_{32} + 114p_{33} + 125p_{34} + d_{11L}^- - d_{11L}^+ = 446,$$
$$105p_{31} + 17p_{32} + 107p_{33} + 124p_{34} + d_{11U}^- - d_{11U}^+ = 479;$$
$$18p_{41} + 49p_{42} + 22p_{43} + 81p_{44} + d_{12L}^- - d_{12L}^+ = 331,$$
$$15p_{41} + 43p_{42} + 19p_{43} + 69p_{44} + d_{12U}^- - d_{12U}^+ = 381 \qquad (33)$$

(ii) *Shiftwise PC goals*

$$22p_{11} + 19p_{21} + 133p_{31} + 18p_{41} + d_{13L}^- - d_{13L}^+ = 331,$$
$$19p_{11} + 16p_{21} + 105p_{31} + 15p_{41} + d_{13U}^- - d_{13U}^+ = 408;$$
$$33p_{12} + 34p_{22} + 19p_{32} + 49p_{42} + d_{14L}^- - d_{14L}^+ = 350,$$
$$29p_{12} + 28p_{22} + 17p_{32} + 43p_{42} + d_{14U}^- - d_{14U}^+ = 401;$$
$$43p_{13} + 23p_{23} + 114p_{33} + 22p_{43} + d_{15L}^- - d_{15L}^+ = 436,$$
$$41p_{13} + 21p_{23} + 107p_{33} + 19p_{43} + d_{15U}^- - d_{15U}^+ = 471;$$
$$31p_{14} + 110p_{24} + 125p_{34} + 81p_{44} + d_{16L}^- - d_{16L}^+ = 398,$$
$$24p_{14} + 105p_{24} + 124p_{34} + 69p_{44} + d_{16U}^- - d_{16U}^+ = 440 \qquad (34)$$

(c) **SC Goals**

(i) *Segmentwise SC goals*

$$35p_{11} + 74p_{12} + 51p_{13} + 51p_{14} + d_{17L}^- - d_{17L}^+ = 686,$$
$$33p_{11} + 64p_{12} + 45p_{13} + 46p_{14} + d_{17U}^- - d_{17U}^+ = 760;$$
$$35p_{21} + 69p_{22} + 39p_{23} + 205p_{24} + d_{18L}^- - d_{18L}^+ = 706,$$
$$33p_{21} + 61p_{22} + 33p_{23} + 189p_{24} + d_{18U}^- - d_{18U}^+ = 784;$$
$$170p_{31} + 38p_{32} + 127p_{33} + 37p_{34} + d_{19L}^- - d_{19L}^+ = 626,$$
$$155p_{31} + 38p_{32} + 117p_{33} + 35p_{34} + d_{19U}^- - d_{19U}^+ = 674;$$
$$33p_{41} + 90p_{42} + 23p_{43} + 175p_{44} + d_{20L}^- - d_{20L}^+ = 579,$$
$$29p_{41} + 86p_{42} + 19p_{43} + 163p_{44} + d_{20U}^- - d_{20U}^+ = 636 \qquad (35)$$

(ii) *Shiftwise SC goals*

$$35p_{11} + 35p_{21} + 170p_{31} + 33p_{41} + d_{21L}^- - d_{21L}^+ = 632,$$
$$33p_{11} + 33p_{21} + 155p_{31} + 29p_{41} + d_{21U}^- - d_{21U}^+ = 684;$$
$$74p_{12} + 69p_{22} + 38p_{32} + 90p_{42} + d_{22L}^- - d_{22L}^+ = 731,$$
$$64p_{12} + 61p_{22} + 38p_{32} + 86p_{42} + d_{22U}^- - d_{22U}^+ = 805;$$
$$51p_{13} + 39p_{23} + 127p_{33} + 23p_{43} + d_{23L}^- - d_{23L}^+ = 518,$$
$$45p_{13} + 33p_{23} + 117p_{33} + 19p_{43} + d_{23U}^- - d_{23U}^+ = 592;$$
$$51p_{14} + 205p_{24} + 37p_{34} + 175p_{44} + d_{24L}^- - d_{24L}^+ = 712,$$
$$46p_{14} + 189p_{24} + 35p_{34} + 163p_{44} + d_{24U}^- - d_{24U}^+ = 773 \qquad (36)$$

Now, two kinds of model constraints are discussed in Sect. 5.2.

5.2   *Description of Constraints*

(a)   **Budget utilization**

In the environment of utilization of budget, it may be mentioned that the cash expenditure associated with deployment per patrol unit per road segment area is approximately the same in all the shifts of the time period.

The data associated with utilization of budget are $(C_{1j}, C_{2j}, C_{3j}, C_{4j}) = $ Rs. (220, 245, 200, 200), $(j = 1, 2, 3, 4)$, and the allocated budget $=$ Rs. 13500.

The constraint appears as

$$220 \sum_{j=1}^{4} p_{1j} + 245 \sum_{j=1}^{4} p_{2j} + 200 \sum_{j=1}^{4} p_{3j} + 200 \sum_{j=1}^{4} p_{4j} \leq 13500 \qquad (37)$$

(b)   **Variable limitation**

In the context of smoothing traffic operation, it may be mentioned that there is no extra effect when more than five patrolmen are deployed to a road segment area during any shift of a time period.

Therefore, the limiting values of p_{ij} are considered

$$1 \leq p_{ij} \leq 5; i, j = 1, \ 2, \ 3, \ 4 \qquad (38)$$

Now, the executable IVGP model of the problem can be obtained by following the expression in (8).

In model execution, four priority factors P_1, P_2, P_3 and P_4 are accounted for incorporation to model goals concerning achievements of target levels. Again, three priority structures with changes in goal priorities are made to execute the model under three Runs. Then, sensitivity analysis is conducted to obtain changes in solutions in the premises of best solution identification.

The executable IVGP model under vth priority $(v = 1, 2, 3)$ structure is presented below.

Determine $\{p_{ij}; \ i, j = 1, 2, 3, 4\}$ so as to
Minimize Z_v
and satisfy goal Eqs. (31)–(36),
subject to system constraints (37) and (38),
with

$$d_{r,kL}^{-} + d_{r,kU}^{+} \leq V_r, r \in \{1, 2, 3, 4\}; k \in \{1, 2, \ldots, 24\} \qquad (39)$$

The achievement functions Z_v, $(v = 1, 2, 3)$, defined for the Runs are presented in Table 2.

Now, for simplicity and without loss of generality, $\lambda_r = 0.5$ is taken into account and equal weight distribution is induced to model goals at each of the four priority factors under the three Runs in course of executing the problem in (39).

Table 2 Descriptions of achievement functions under three Runs

Run	Priority achievement function (Z_v)
1	$Z_1 =$ $\left\{ \begin{array}{l} P_1\left(\lambda_1 \sum\limits_{k=1}^{6} w_{1,k}(d_{1,kL}^- + d_{1,kU}^+) + (1-\lambda_1)V_1 \right), \ P_2\left(\lambda_2 \sum\limits_{k=7}^{12} w_{2,k}(d_{2,kL}^- + d_{2,kU}^+) + (1-\lambda_2)V_2 \right), \\ P_3\left(\lambda_3 \sum\limits_{k=13}^{18} w_{3,k}(d_{3,kL}^- + d_{3,kU}^+) + (1-\lambda_3)V_3 \right), \ P_4\left(\lambda_4 \sum\limits_{k=19}^{24} w_{4,k}(d_{4,kL}^- + d_{4,kU}^+) + (1-\lambda_4)V_4 \right) \end{array} \right\}$
2	$Z_2 =$ $\left\{ \begin{array}{l} P_1\left(\lambda_1 \sum\limits_{k=19}^{24} w_{1,k}(d_{1,kL}^- + d_{1,kU}^+) + (1-\lambda_1)V_1 \right), \ P_2\left(\lambda_2 \sum\limits_{k=1}^{6} w_{2,k}(d_{2,kL}^- + d_{2,kU}^+) + (1-\lambda_2)V_2 \right), \\ P_3\left(\lambda_3 \sum\limits_{k=7}^{12} w_{3,k}(d_{3,kL}^- + d_{3,kU}^+) + (1-\lambda_3)V_3 \right), \ P_4\left(\lambda_4 \sum\limits_{k=13}^{18} w_{4,k}(d_{4,kL}^- + d_{4,kU}^+) + (1-\lambda_4)V_4 \right) \end{array} \right\}$
3	$Z_3 =$ $\left\{ \begin{array}{l} P_1\left(\lambda_1 \sum\limits_{k=7}^{12} w_{1,k}(d_{1,kL}^- + d_{1,kU}^+) + (1-\lambda_1)V_1 \right), \ P_2\left(\lambda_2 \sum\limits_{k=1}^{6} w_{2,k}(d_{2,kL}^- + d_{2,kU}^+) + (1-\lambda_2)V_2 \right), \\ P_3\left(\lambda_3 \sum\limits_{k=13}^{18} w_{3,k}(d_{3,kL}^- + d_{3,kU}^+) + (1-\lambda_3)V_3 \right), \ P_4\left(\lambda_4 \sum\limits_{k=19}^{24} w_{4,k}(d_{4,kL}^- + d_{4,kU}^+) + (1-\lambda_4)V_4 \right) \end{array} \right\}$

The GA program is adopted in Language C++. The model (variable size 190, constraint size 200) is carried out in Pentium IV CPU with 2.66 GHz clock pulse and 2 GB RAM. The generation numbers $= 300$ are initially considered. The genetic parameter values: $p_c = 0.8$, $p_m = 0.08$, and the P size $= 100$ along with the chromosome length $= 150$ are taken into account to conduct the experiments under three Runs.

The decisions obtained under the Runs are displayed in Table 3.

Now, following the procedure and using the results in Table 3, the ideal solution point is obtained as (5, 5, 4, 3, 5, 5, 5, 1, 2, 5, 1, 1, 5, 1, 5, 2).

Here, it is worthy to note that since the executable goal equations and system constraints are linear in nature, the notion *Euclidean distance* might be used to find distances of different solution achievements from the ideal one.

However, *Euclidean distances* of the solutions obtained under the successive Runs from the ideal solution are obtained as $D^{(1)} = 4.2452$, $D^{(2)} = 3.464$, $D^{(3)} = 4.123$.

The results indicate that the minimum distance corresponds to $D^{(2)}$. As such, the priority structure under Run 2 would be the proper one to reach the ideal point dependent solution.

The resulting solution is obtained at generation number $= 200$ in the genetic search process.

The resultant patrolmen deployment decision is presented in Table 4.

The patrolmen allocation is diagrammatically depicted in Fig. 7.

Now, the solution achievement under the proposed IVGP method is discussed as follows:

Table 3 Solution achievements under three Runs

Run	Decision																	
	p_{11}	p_{12}	p_{13}	p_{14}	p_{21}	p_{22}	p_{23}	p_{24}	p_{31}	p_{32}	p_{33}	p_{34}	p_{41}	p_{42}	p_{43}	p_{44}
1	5	3	4	3	3	5	5	1	2	2	1	1	4	1	5	2
2	5	4	4	3	5	4	5	1	1	5	1	1	5	1	2	2
3	3	4	4	3	4	3	4	1	2	4	1	1	3	1	3	2

Table 4 Resultant patrolmen deployment decision under the proposed method

Road segment area	Shiftwise deployment			
	6 am to 10 am	10 am to 2 pm	2 pm to 6 pm	6 pm to 10 pm
BT road	5	4	4	3
AJC road	5	4	5	1
Stand road	1	5	1	1
EM Bypass	5	1	2	2

Fig. 7 Diagrammatic representation of patrolmen allocation under IVGP method

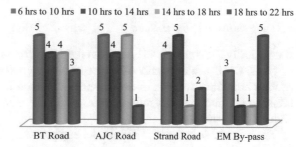

Patrolmen allocation under the proposed method

- The optimal decision for deploying patrol personnel to all road segment areas during all shifts of the time period is 60.
- The success rate of AR during the time period is found to be 94.68%, i.e. under-achievement from the estimated upper bound of target interval is only 5.32%.
- The achievements of goal values associated with the PC and SC functions against the present patrolmen deployment strategy are also found to be achieved nearly the upper bounds of their target intervals.
- It is to noted there that the utilization of the estimated budget during the time period for regular expenditure = Rs. 10,750, i.e. underutilization of the total allocated budget is 25.96%.

The results indicate that the proposed method is pragmatic one for obtaining patrolmen deployment strategy in city road traffic control system.

Now, to illustrate that the method presented here is more effective, the model solution is compared with solution obtained by employing conventional *minsum* IVGP approach and presented in Sect. 5.3.

Table 5 Patrolmen deployment decision under *minsum* IVGP method

Road segment area	Shift-wise deployment			
	6 am to 10 am	10 am to 2 pm	2 pm to 6 pm	6 pm to 10 pm
BT road	5	4	4	3
AJC road	1	5	4	1
Stand road	2	3	1	1
EM Bypass	3	1	5	2

5.3   Performance Comparison

• *Using minsum IVGP method*

If no priority structure model goals are taken into account for achieving target values of goals, i.e. if achievements of target levels of goals are considered on the basis of their numerical weights of importance only, then the model in (8) is reasonably transformed into the conventional *minsum* IVGP model studied by Inuiguchi and Kume (1991) previously.

The executable *minsum* IVGP model appears as

Determine $\{p_{ij}; \ i, j = 1, 2, 3, 4\}$ so as to

$$\text{Minimize} \quad Z' = \{\lambda \sum_{k=1}^{24} w_k(d_{kL}^- + d_{kU}^+) + (1 - \lambda)V\}, 0 < \lambda < 1$$

and satisfy goal Eqs. (31)–(36),
subject to system constraints (37) and (38),
with

$$d_{kL}^- + d_{kU}^+ \leq V, k = 1, 2, \ldots, 24 \tag{40}$$

where $V = [\underset{k=1}{\overset{24}{Max}}(d_{kL}^- + d_{kU}^+)]$, and where $w_k(>0), \ \forall k$, with $\sum_{k=1}^{24} w_k = 1$.

The resulting solution of the model in (40) obtained by taking $\lambda = 0.5$ and introducing equal weight ($w_k = \frac{1}{24}, k = 1, 2, \ldots, 24$) to the model goals in the same premises of making decision is presented in Table 5.

The diagrammatic representations of patrolmen allocation under *minsum* IVGP method are depicted in Fig. 8.

It is to be noted that the achievement of AR is 91.56% (i.e. underachievement of the success rate of AR is 8.44%), which is lower than the solution achievement made for the use of the proposed method.

Again, it may be mentioned that although the goal achievements concerned with PC and SC operations on traffic are found to be within the ranges of their respective

Fig. 8 Graphical representation of patrolmen allocation under *minsum* IVGP method

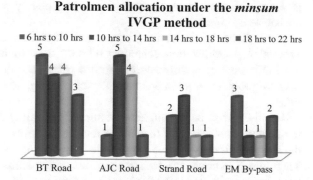

Patrolmen allocation under the *minsum* IVGP method

■ 6 hrs to 10 hrs ■ 10 hrs to 14 hrs ■ 14 hrs to 18 hrs ■ 18 hrs to 22 hrs

Fig. 8 Graphical representation of patrolmen allocation under *minsum* IVGP method

target intervals, the achieved values are lower than that obtained under the proposed method. Further, underutilization of allocated budget is found to be 32.48%, i.e. utilization of budget is lower than that incurred in the case of using *priority-based* IVGP approach.

Remark: The above discussion and solution comparison show that the proposed method is more effective than the conventional one concerning the deployment of patrol manpower properly to control metro-traffic manually in traffic management system.

6 Conclusions and Future Scope

The main merit of employing the method presented here is that interval character-istics regarding attainment of objective values are preserved there in all possible instances of executing the model and interval data is communicated directly to the optimization problem without involving any distributional information unlike SP and FP methods. Further, since the deployment of patrolmen is based upon subjective judgments of accident frequency and traffic density statistics, *priority-based* IVGP method is more advantageous, because DM can effectively generate an alternative deployment decision with a change of priority structure as and when the situation arises in a decision situation.

However, in some practical decision environments, imprecision on the bounds of intervals, i.e. fuzziness of bounded intervals would have to be considered to explore search space for achievement of possible solution in inexact environment. Such an interval representation is called fuzzy interval (Dubois and Prade 2000), and it would have to be viewed as the extension of conceptual frame of conventional interval. It can be conceived that fuzzy interval convey more information than a usual crisp interval, when the source of information is human being and hence subjective one. The constructive solution procedure for solving such problems in the area of MODM is yet to circulate widely in literature. It is probably one of the future studies in decision-making arena.

Again, hybridization of fuzzy interval and GP along with implementation of hybrid evolutionary optimization techniques (Grosan and Abraham 2007) towards promoting traffic control performances and also improving quality of solutions to other MODM problems is an interesting alley of investigation in the avenue of optimization.

Finally, it may be mentioned that micro-level study (Shi et al. 2008) pertaining to qualitative measures for road infrastructural development in contrast to patrolmen deployment strategy is also a sensitive issue in road traffic management system. Therefore, a joint venture for simultaneous optimization of both the aspects, traffic organization and patrolmen deployment, might be considered in future to satisfy the needs of traffic demand in urban society and also to control traffic flow on day-to-day basis for healthy road environment in the current inexactness of MODM world.

Acknowledgements The author is thankful to the Hon'ble Editors and anonymous reviewers for their insightful and helpful comments and suggestions which have led to improve the quality of presentation of the chapter. The author would also like to thank Kolkata traffic police department of West Bengal in India for providing support to work out the chapter.

References

Annual Review Bulletin, Kolkata traffic police, Kolkata, Government of West Bengal, India (2012–2016), http://www.kolkatatrafficpolice.gov.in

A. Biswas, B.B. Pal, Application of fuzzy goal programming technique to land use planning in agricultural system. Omega **33**(5), 391–398 (2005)

G.R. Bitran, Linear multiobjective problems with interval coefficients. Manag. Sci. **26**(7), 694–706 (1980)

C.D. Buchanan, *Traffic in towns—a study of the long term problems of traffic in urban areas* (Penguin Book Publishers, Harmondsworth, London, 1964)

B.C.W. Craenen, A.E. Eiben, E. Marchiori, How to handle constraint with evolutionary algorithms, in *The Practical Handbook of Genetic Algorithms Applications,* ed. by L. Chamber, 2nd edn. (Chapman & Hall/CRC, USA, 2001)

K. Deb, Multi-objective genetic algorithms: problem difficulties and construction of test problems. Evol. Comput. **7**(3), 205–230 (1999)

K. Deb, *Multi-objective optimization using evolutionary algorithms* (Wiley, New York, 2009)

D. Dubois, H. Prade (eds.), *Fundamentals of Fuzzy Sets* (Springer, New York, 2000)

D.E. Goldberg, *Genetic Algorithms in Search, Optimization, and Machine Learning* (Addison-Wesley Longman Publishing Company, Boston, MA, USA, 1989)

C. Grosan, A. Abraham, Hybrid evolutionary algorithms: methodologies, architectures, and reviews. Stud. Comput. Intell. (SCI) **75**, 1–17 (2007)

E.L. Hannan, Note—effects of substituting a linear goal for a fractional goal in the goal programming problem. Manag. Sci. **24**(1), 105–107 (1977)

E.R. Hansen, Global Optimization Using Interval Analysis (Marcel Dekker Inc. and Sun Microsystem Inc. Pub., New York, 2003)

B.H.F. Hasan, M.S.M. Saleh, Evaluating the effectiveness of mutation operators on the behaviour of genetic algorithms applied to non-deterministic polynomial problems. Informatica **35**(4), 513–518 (2011)

M. Ida, Portfolio selection problem with interval coefficients. Appl. Math. Lett. **16**(5), 709–713 (2003)

J.P. Ignizio, *Goal programming and extensions* (Lexington Books, Lexington, MA, 1976)

M. Inuiguchi, Y. Kume, Goal programming problems with interval coefficients and target intervals. Eur. J. Oper. Res. **52**(3), 345–360 (1991)

C. Jiang, X. Han, G.R. Liu, G.P. Liu, A nonlinear interval number programming method for uncertain optimization problems. Eur. J. Oper. Res. **188**(1), 1–13 (2008)

B.S. Kerner, *Introduction to Modern Traffic Flow Theory and Control: The Long Road to Three-Phase Traffic Theory* (Springer, New York, 2009)

F.H. Knight, *Risk, Uncertainty, and Profit* (Houghton Mifflin Co. Ltd., Boston, 1921)

J.S.H. Kornbluth, R.E. Steuer, Goal programming with linear fractional criteria. Eur. J. Oper. Res. **8**(1), 58–65 (1981)

Z. Kulpa, Diagrammatic representation for interval arithmetic. Linear Algebra Appl. **324**(1–3), 55–80 (2001)

R.C. Larson, *Urban Police Patrol Analysis* (MIT Press, Cambridge, 1972)

S.M. Lee, L.S. Franz, A.J. Wyne, Optimizing state patrol manpower allocation. J. Oper. Res. Soc. **30**(10), 885–896 (1979)

B.D. Liu, *Theory and Practice of Uncertain Programming*, 3rd edn. (Springer, Berlin, 2009)

Z. Michalewicz, *Genetic Algorithms + Data Structures = Evolution Programs*, 3rd edn. (Springer, Berlin, 1996)

R.E. Moore, R.B. Kearfort, M.J. Cloud, *Introduction to Interval Analysis* (SIAM, Philadelphia, 2009)

C. Olivera, C.H. Antunes, Multiple objective linear programming models with interval coefficients–an illustrated overview. Eur. J. Oper. Res. **181**(3), 1434–1463 (2007)

B.B. Pal, D. Chakraborti, P. Biswas, A Genetic Algorithm Based Goal Programming Method for Solving Patrol Manpower Deployment Planning Problems with Interval-Valued Resource Goals in Traffic Management System: A Case Study (IEEE Digital Library, *IEEE Press,* 2009), pp. 61–69, https://doi.org/10.1109/icadvc.2009.5378215

B.B. Pal, D. Chakraborti, P. Biswas, A. Mukhopadhyay, An application of genetic algorithm method for solving patrol manpower deployment problems through fuzzy goal programming in traffic management system: a case study. Int. J. Bio-Insp. Comput. **4**(1), 47–60 (2012)

B.B. Pal, B.N. Moitra, A goal programming procedure for solving problems with multiple fuzzy goals using dynamic programming. Eur. J. Oper. Res. **144**(3), 480–491 (2003)

C. Romero, A general structure of achievement function for goal programming model. Eur. J. Oper. Res. **153**(4), 675–686 (2004)

E. Rosenblueth, Two-point estimates in probabilities. Appl. Math. Model. **5**(2), 329–335 (1981)

F. Shi, E.H. Huang, Y.Z. Wang, Study on the functional characteristics of urban transportation micro-circulation system. Urban Stud. **15**(3), 34–36 (2008)

R. Slowinski, A multicriteria fuzzy linear programming method for water supply system development planning. Fuzzy Sets Syst. **19**(3), 217–237 (1986)

W.B. Taylor III, J.L. Moore, E.R. Clayton, K.R. Devis, T.R. Rakes, An integer nonlinear goal programming model for the deployment of state highway patrol units. Manag. Sci. **31**(11), 1335–1347 (1985)

S. Tong, Interval number and fuzzy number linear programming. Fuzzy Sets Syst. **66**(3), 301–306 (1994)

P.L. Yu, A class of solutions for group decision problems. Manag. Sci. **19**(8), 936–946 (1973)

L.A. Zadeh, Fuzzy sets. Inf. Control **8**(3), 338–353 (1965)

H.-J. Zimmermann, Fuzzy programming and linear programming with several objective functions. Fuzzy Sets Syst. **1**(1), 45–55 (1978)

H.-J. Zimmermann, *Fuzzy Set Theory and Its Application* (Kluwer Academic Publishers, Boston, 1991)

Multi-objective Optimization to Improve Robustness in Networks

R. Chulaka Gunasekara, Chilukuri K. Mohan and Kishan Mehrotra

1 Introduction

There exists no unique definition for network robustness. Multiple robustness measures have been introduced to evaluate the capability of a network to withstand failures or attacks. All such measures aim to capture features such as (1) connectivity—robust networks are expected to remain connected even when a set of nodes or edges fail during targeted or natural node/edge failures, (2) distance—distances between the nodes of robust networks should remain minimally affected during node/edge failures, and (3) network properties—the network properties such as degree distribution and distance distribution should change very little when nodes/edges fail. In this section, we discuss three categories of network robustness measures that have been proposed and are widely used in the literature.

1.1 Robustness Measures Based on the Eigenvalues of the Adjacency Matrix

Let A be the adjacency matrix of the network $G = (V, E)$ with n nodes, and let $\lambda_1 \geq \lambda_2 \geq \lambda_3 \geq \cdots \geq \lambda_n$ be the set of eigenvalues of A.

1. *Spectral radius (SR)*: The largest or the principal eigenvalue, λ_1, is called the spectral radius. This has been used as a measure of quantifying network robustness in multiple studies (Chan and Akoglu 2016; Le et al. 2015; Tong et al. 2012, 2010).

R. C. Gunasekara (✉)
IBM T. J. Watson Research Center, Yorktown Heights, New York 10598, USA
e-mail: Chulaka.Gunasekara@ibm.com

C. K. Mohan · K. Mehrotra
Department of EECS, Syracuse University Syracuse, New York 13244, USA
e-mail: mohan@syr.edu

K. Mehrotra
e-mail: mehrotra@syr.edu

© Springer Nature Singapore Pte Ltd. 2018
J. K. Mandal et al. (eds.), *Multi-Objective Optimization*,
https://doi.org/10.1007/978-981-13-1471-1_5

SR is inversely proportional to the epidemic threshold of a network (Chakrabarti et al. 2008).

2. *Spectral gap (SG)*: The difference between the largest and the second largest eigenvalues, $(\lambda_1 - \lambda_2)$, is called the spectral gap. This has also been used to measure the robustness of the graph against attacks (Chan and Akoglu 2016; Malliaros et al. 2012; Yazdani et al. 2011). Spectral gap is related to the expansion properties of the graph; networks with good expansion properties provide excellent communication platforms due to the absence of bottlenecks (Watanabe and Masuda 2010).

3. *Natural connectivity (NC)*: Natural connectivity characterizes the redundancy of alternative paths in the network by quantifying the weighted number of closed walks of all lengths (Chan et al. 2014). This is an important measure because redundancy of routes between the nodes ensures that communication between nodes remains possible during an attack to the network. Denoted by $\bar{\lambda}$, natural connectivity is defined as follows:

$$\bar{\lambda} = \ln \left(\frac{1}{n} \sum_{j=1}^{n} e^{\lambda_j} \right) \tag{1}$$

and is widely used as a measure of robustness in complex networks (Chan and Akoglu 2016; Jun et al. 2010; Chan et al. 2014). A network created by optimizing the natural connectivity is found to exhibit a roughly "eggplant-like" topology, where there is a cluster of high-degree nodes at the head and other low-degree nodes are scattered across the body of the "eggplant" (Peng and Wu 2016).

1.2   Measures Based on the Eigenvalues of the Laplacian Matrix

The topology of a network G with n nodes can also be represented by the $n \times n$ Laplacian matrix $\mathscr{L} = D - A$, where $D = diag(d_u)$ and d_u is the degree of node u. Let the set of eigenvalues of $\mathscr{L}$ be $\mu_1 = 0 \leq \mu_2 \leq \mu_3 \leq \cdots \leq \mu_n$. These eigenvalues are used to define the following measures:

1. *Algebraic connectivity (AC)*: The second smallest eigenvalue of the Laplacian matrix, (μ_2) is also known as algebraic connectivity. The algebraic connectivity is 0 if the network is disconnected and $0 < \mu_2 \leq n$ when the network is connected (Fiedler 1973). The larger the AC, the more difficult it is to cut a graph into disconnected components (Alenazi and Sterbenz 2015); hence, this has been used by many studies to determine the robustness of networks (Jamakovic and Mieghem 2008; Sydney et al. 2013; Chan and Akoglu 2016).

2. *Normalized effective resistance (nER)*: Introduced in (Ellens 2011), nER is defined as

$$nER = \frac{n-1}{\left(n \sum_{i=2}^{n} \frac{1}{\mu_i}\right)}. \tag{2}$$

The usefulness of this measure can be seen when the network is viewed as an electrical circuit with an edge representing a resistor with electrical conductance equal to the edge weight. The effective resistance (R_{vu}) between a pair of nodes u and v is small when there are many paths between nodes u and v with high conductance edges, and R_{uv} is large when there are few paths, with lower conductance, between nodes u and v (Ghosh et al. 2008). Effective resistance is equal to the sum of the (inverse) nonzero Laplacian eigenvalues and has been used in multiple studies to define network robustness (Chan and Akoglu 2016; Ellens 2011). "Network criticality" is a similar robustness metric defined to capture the effect of environmental changes such as traffic variation and topology changes in networks (Bigdeli et al. 2009; Tizghadam and Leon-Garcia 2010).

1.3 Measures Based on Other Properties

In this section, we introduce the other measures that are defined to identify the network robustness.

1. *Harmonic diameter (HD)*: This robustness measure is defined as follows:

$$HD(G) = \frac{n(n-1)}{\left(\sum_{u \neq v \in V} \frac{1}{d(u, v)}\right)}, \tag{3}$$

where n is the number of nodes in the network and $d(u, v)$ is the shortest distance between the nodes u and v (Marchiori and Latora 2000). HD has been used to evaluate network robustness in multiple studies (Boldi and Rosa 2013; Boldi et al. 2011). This measure is analogous to the average distance between all the nodes, but better because this can also be applied to disconnected networks. For ease of comparison with other measures, we use the reciprocal of the harmonic diameter (rHD) in this study, which increases with robustness.

2. *Size of the largest connected component (LCC)*: This measure identifies the size of the largest component during all possible malicious attacks. LCC is defined as follows:

$$R = \frac{1}{n+1} \sum_{Q=0}^{n} s(Q), \tag{4}$$

where n is the number of nodes in the network and $s(Q)$ is the fraction of nodes in the largest connected cluster after attacking Q nodes. LCC was proposed in (Herrmann et al. 2011) and is widely used (Schneider et al. 2011; Tanizawa et al.

2012; Wu and Holme 2011; Zeng and Liu 2012) as a robustness measure in networks. The normalization factor $\frac{1}{n+1}$ ensures that the robustness of networks with different sizes can be compared. The attacks often consist of a certain fraction of node attacks, and after the attack, the measure identifies the number of nodes in the largest connected component. It has been found that the robust networks that optimize this measure form a unique "onion-like" structure consisting of a core of highly connected nodes hierarchically surrounded by rings of nodes with decreasing degree (Herrmann et al. 2011).

3. *Clustering coefficient (CC)*: The abundance of triangles in the network is identified by the clustering coefficient (Watts and Strogatz 1998). The clustering coefficient of a network is calculated based on the local clustering coefficient of each node. The clustering coefficient of the node u is defined as

$$CC_u = \frac{\lambda_G(u)}{\tau_G(u)}, \tag{5}$$

where $\lambda_G(u)$ is the number of triangles connected to node u and $\tau_G(u)$ is the number of triples centered around node u. A triple centered around node u is a set of two edges connected to node u. The overall clustering coefficient of the network is calculated as the average CC_u. A high clustering coefficient indicates high robustness, because the number of alternative paths grows with the number of triangles (Ellens and Kooij 2013).

In addition to the aforementioned methods, more robustness measures have also been proposed in literature, e.g., vertex/edge connectivity, network diameter, average distance between the nodes, vertex/edge betweenness, and number of spanning trees. These measures are excluded in this study due to poor performance in some trivial networks or high computational cost needed for real-world large networks (Ellens and Kooij 2013).

In the following section, we discuss some of the properties of the aforementioned network robustness measures.

2 Properties of Network Robustness Measures

In this section, the aforementioned robustness measures are compared using the following approaches:

1. Robustness values of a few elementary networks are calculated and compared.
2. The similarities and dissimilarities of the robustness measures are compared using the correlation of these measures for a set of generated networks that follow the power law degree distribution.

2.1 Robustness of Elementary Networks

Six elementary networks that we considered are shown in Fig. 1. The networks are ordered by increasing robustness intuitively, i.e., we believe that the network Fig. 1a is the least robust, the network Fig. 1b is more robust than network Fig. 1a, the network Fig. 1c is more robust than Fig. 1b, etc., and the network Fig. 1f is the most robust.

Table 1 shows the robustness values obtained by each robustness measure discussed in Sect. 1 for each of the six elementary networks. The summary of the results is as follows:

1. NC and nER order the networks in the expected order.
2. rHD and AC also order the networks correctly, but fail to distinguish between some of the elementary networks.
3. CC gives a value of 0 to all networks with no triangles and evaluates the empty network to be as robust as the grid network.
4. LCC, SR, and SG order the networks differently than our intuition.
5. LCC gives the same robustness value to both the ring and grid networks, defying intuition. In addition, the star network gets a low LCC value than the path network.
6. SR and SG identify the networks which enable fast communication as robust networks; thus, the star network gets a high robustness value.
7. All the six robustness measures identify the empty network as the least robust network and the fully connected network as the most robust.

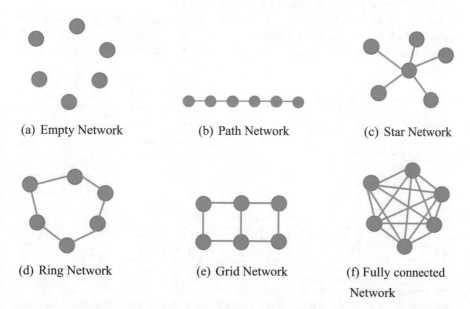

(a) Empty Network   (b) Path Network   (c) Star Network

(d) Ring Network   (e) Grid Network   (f) Fully connected Network

Fig. 1 Six elementary networks considered for robustness calculation; the networks are arranged in the increasing order of robustness assessed intuitively

Table 1 Robustness values of the elementary networks

Network	rHD	LCC	CC	SR	SG	NC	AC	nER
Empty	0	0	0	0	0	0	0	0
Path	0.58	0.33	0	1.80	0.55	0.71	0.27	0.14
Star	0.66	0.17	0	2.24	2.24	0.81	1.0	0.20
Ring	0.66	0.37	0	2.0	1.0	0.83	1.0	0.28
Grid	0.70	0.37	0	2.41	1.41	1.01	1.0	0.35
Full	1.00	0.50	1.0	5.0	6.0	3.22	6.0	1.0

2.2 Correlation of Robustness Measures

In this section, we study similarities in the overall behavior of the robustness measures that were discussed in Sect. 1, using Pearson's correlation coefficient. A high correlation between two measures suggests that a network that shows high robustness in terms of one measure would also show high robustness in terms of the other measures as well.

To evaluate the correlations among the robustness measures, 100 scale-free networks were generated with number of nodes in range (500–5000) and with power law parameters in range (2.0–3.0). For each of the generated networks, the robustness values were calculated. Then, the Pearson product–moment correlation coefficient was calculated between the robustness measures. The correlation coefficients are shown in Table 2 and the associated scatter plots in Fig. 2.[1]

According to the results, some of the robustness measures are highly correlated with each other. Some of these highly correlated pairs include (CC, NC), (SG, NC), and (SG, CC) ($p < 0.001$). Also, the three robustness measures that are calculated

Table 2 Correlations of the robustness measures

	rHD	LCC	CC	SR	SG	NC	AC	nER
rHD	1							
LCC	−0.41	1						
CC	0.90	−0.61	1					
SR	0.88	−0.65	0.88	1				
SG	0.89	−0.58	0.98	0.87	1			
NC	0.90	−0.62	0.99	0.89	0.98	1		
AC	0.86	−0.34	0.75	0.86	0.78	0.76	1	
nER	0.37	0.32	0.02	0.28	0.09	0.04	0.69	1

---

[1]Similar results were obtained when the experiments were carried out for 100 generated scale-free networks by: (1) fixing the number of nodes and changing power law parameter in the aforementioned range, and (2) changing the number of nodes in the aforementioned range and fixing the power law parameter.

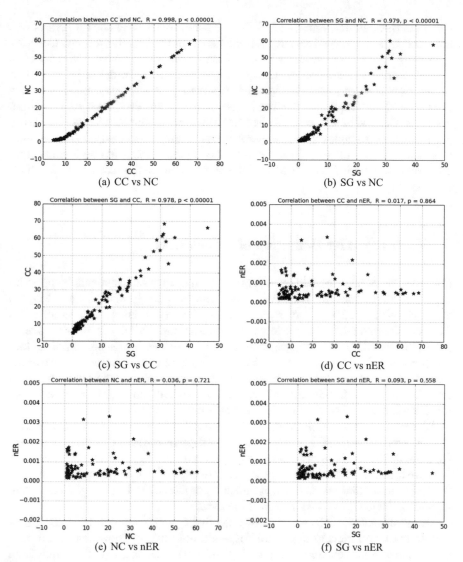

Fig. 2 Scatter plots between the robustness measures

using the eigenvalues of the adjacency matrix (spectral radius, spectral gap, and natural connectivity) are highly correlated.

Some robustness measures are highly uncorrelated. For example, the pairs (CC, nER), (NC, nER), and (SG, nER) show this behavior. The scatter plots of some of the robustness measure pairs are shown in Fig. 2d–f.

Interestingly, LCC negatively correlates with most other robustness measures (except for nER). The negative correlation suggests that when the robustness of the network is increased in terms of LCC, the robustness in terms of other measures will not increase. The observed negative correlation can be explained as follows. Consider improving the LCC measure. As LCC is focused on keeping most of the nodes in a single connected component during node attacks made in the order of degree centrality, new edges (that get added when optimizing LCC) connect nodes with low degree and in different communities in the network. However, such edge addition decreases CC (and other measures), because although the number of edges increases, the number of triangles in the network remains almost unchanged. The number of triangles is less likely to increase, because (1) the number of edges connecting different communities is small, and (2) the nodes to which the edges are added have low degrees.

3   Multi-objective Definition of Robustness

Multiple studies have focused on improving the robustness of a network by optimizing a single robustness measure (Wang and Mieghem 2008; Watanabe and Masuda 2010; Ghosh et al. 2008; Sydney et al. 2013; Chan et al. 2014; Tong et al. 2012; Chan and Akoglu 2016). But the low correlation among some of the robustness measures (shown in Sect. 2.2) suggests that when a single measure is optimized, it does not guarantee the improvement of robustness in terms of other measures. We argue that edges should be added to a network in a manner such that multiple robustness measures improve. In this section, we propose a methodology to improve multiple robustness measures when new edges are added to the network.

We select three uncorrelated robustness measures (largest connected component (LCC), spectral gap (SG), and normalized effective resistance (nER)), one each from the three categories of robustness measures discussed in Sect. 1. Our goal is to improve all three of these measures in the network by edge addition. We formulate the problem as a multi-objective maximization problem: *Find the set of k edges that increase all three robustness measures the most.*

Many methods exist to solve multi-objective optimization problems. The most widely used evolutionary algorithm for multi-objective optimization is the non-dominated sorting genetic algorithm-II (NSGA-II) (Deb et al. 2000), which is an improved version of NSGA (Srinivas and Deb 1994). We use NSGA-II as the multi-objective optimization algorithm in this study, since it has been shown to exhibit superior performance in multiple applications.

We represent the network as a bit string, in which each possible edge that can be added to the network is assigned an index. The number of bits in the bit string is equal to the number of possible edges (m) that can be added to the network. Initially, before any extra edge is added to the network, the bit string consists of all 0s. When a certain edge is selected to be added to the network, the bit value corresponding to the index of the selected edge will be changed to 1.

The key steps of the NSGA-II algorithm to identify the k edges to add to the network are as follows:

1. Initial population—In each individual (bit string) in the initial population, k random bits are assigned the value of 1 to represent that they are selected to be added to the network, and the remaining $(m - k)$ bits are assigned 0, where m is the total number of edges that can be added to the network.
2. Fitness—To calculate the fitness value of each individual, we first add the selected set of k edges to the initial network. Then, the three robustness measures (LCC, SG, and nER) are calculated for the amended network.
3. Crossover—One-point crossover is applied to a fraction P_c of selected individuals to generate offspring.[2]
4. Mutation—Mutation is performed with probability P_m by inverting two bits of different values.
5. Repair—Let the number of bits assigned the value of 1 after crossover and mutation be n_b.

   i If $n_b > k : (n_b - k)$, bits with value 1 are randomly selected and assigned the value 0.
   ii If $n_b < k : (k - n_b)$, bits with value 0 are randomly selected and assigned the value 1.

The number of solutions obtained depends on the network on which the optimization is performed and the objectives selected.

3.1 Fast Calculation of Robustness Measures

The computation of the LCC, SG, and nER is costly for large networks; we use approximation techniques and the results of matrix perturbation theory for fast calculation of these robustness measures.

1. *Size of the largest connected component (LCC)*
   Computing LCC requires calculation of the fraction of nodes in the largest connected cluster after attacking nodes in the order of degree centrality. In many real networks, the degree distribution follows the power law. Hence, the attacks made on the high-degree nodes have the biggest impact on the network. We approximate LCC by attacking only the top $l\%(\ll n)$ nodes with the highest degree centrality. For 100 generated scale-free networks with number of nodes in range [500, 5000] and scale-free parameter in range [2.0, 3.0], the LCC calculated by removing all n nodes in the network has a correlation of 0.87 ($p < 0.0001$) with the LCC approximated by removing only the top 20% degree centrality nodes. This approximation reduces the running time of LCC by 79.9% on average. Hence, in our experiments we approximate the LCC by attacking the top 20% nodes in the network in the order of degree centrality.

---

[2]We have experimented with other crossover operators such as two-point crossover, and the results were similar compared to the one-point crossover. Hence, the results with one-point crossover are reported here.

2. *Spectral gap (SG)*

   For a perturbation ΔA in the adjacency matrix A of the original network G, the new eigenvalues and eigenvectors of the new network G' can be approximated (Stewart and Sun 1990; Chan and Akoglu 2016). The update to the ith eigenvalue can be written as $\Delta \lambda_i \approx x_i{}^T \Delta A x_i$, and the approximated change in the ith eigenvector is $\Delta x_i \approx \displaystyle\sum_{j=1, j \neq i}^{n} \left(\dfrac{x_j^T \Delta A x_i}{\lambda_i - \lambda_j} x_j \right)$. Thus, the change in SG when a new edge (u, v) is added can be approximated by

$$x_1^T \Delta A x_1 - x_2^T \Delta A x_2 = 2(x_{1u} x_{1v} - x_{2u} x_{2v}),$$

   where x_1 and x_2 are the eigenvectors corresponding to two largest eigenvalues λ_1 and λ_2, and x_{ij} denotes the jth element of the ith eigenvector. Since we avoid calculating all the eigenvalues of a large adjacency matrix, the running time is substantially reduced.

3. *Normalized effective resistance (nER)*

   Effective resistance is equal to the sum of reciprocals of the nonzero Laplacian eigenvalues and can be approximated by the first $l - 1$ nonzero eigenvalues instead of all $n - 1$ of them (Ellens 2011). According to matrix perturbation theory, when an edge (u, v) is added to a network, the change in its Laplacian eigenvalue μ_t can be written as $\Delta \mu_t = v_t^T \Delta \mathscr{L} v_t = (v_{tu} - v_{tv})^2$, where μ_t is the tth eigenvalue of the Laplacian matrix $\mathscr{L}$, v_t is the corresponding eigenvector of μ_t and v_{ti} corresponds to the ith element of the tth eigenvector. Using this eigenvalue approximation and matrix perturbation theory, the change in nER when an edge (u, v) is added can be written as

$$\Delta nER \approx \frac{l - 1}{n \left(\frac{1}{\mu_2 + \Delta \mu_2} + \frac{1}{\mu_3 + \Delta \mu_3} + \cdots + \frac{1}{\mu_l + \Delta \mu_l} \right)} - nER$$

$$\Delta nER \approx \frac{l - 1}{n} \left(\sum_{i=2}^{l} \frac{1}{\mu_i + (v_{iu} - v_{iv})} \right)^{-1}.$$

4 Selecting Solutions from Multi-objective Optimization

Multi-objective optimization identifies multiple sets of solutions which fall on the Pareto front. One issue with regard to $\mathcal{O}$-objective optimization is that we obtain a large number of solutions when the value of $\mathcal{O}$ is high and the objectives are uncorrelated (Garza-Fabre et al. 2009; Farina and Amato 2002; Bentley and Wakefield 1998). But the decision-makers who use multi-objective optimization in their applications usually require one or two solutions to be used in their applications. Multiple methods have been proposed in literature to prune the Pareto optimal set of solutions. This section discusses some of the proposed methods.

The methods proposed to select solutions from the Pareto optimal set can be divided into three categories.

4.1 Ranking Methods

In ranking methods, after executing the multi-objective optimization algorithm, the set of Pareto optimal solutions obtained are ranked according to a user-specified criterion. Once the ranking is done, the decision-maker can pick the solutions that are best ranked for the desired applications. Some of the proposed ranking methods include the following:

1. *Weighted sum approach (WS):*
   This is the most widely used approach for pruning solutions from the Pareto optimal set. For an $\mathcal{O}$-objective optimization problem, the weighted sum rank of the Pareto optimal solution X_i is given by

$$WS(X_i) = \sum_{j=1}^{\mathcal{O}} w_j O_j(X_i), \tag{6}$$

   where w_j is the weight assigned to the objective O_j. The weight assignment to the objectives is domain-dependent, and the decision-maker should determine the appropriate weight assignment to the objectives. The result of the ranking depends on the weight assignment. Hence, in applications where the proper weight assignment is unknown, the results of the weighted sum approach are questionable (Garza-Fabre et al. 2009).

2. *Average ranking (AR):*
   This method uses the average of the ranking positions of a solution X_i given by all the objective functions and is calculated as follows:

$$AR(X_i) = \frac{\sum_{j=1}^{\mathcal{O}} R_j(X_i)}{\mathcal{O}}, \tag{7}$$

   where $R_j(X_i)$ is the rank given to the solution X_i by the objective O_j.

3. *Maximum rank (MR):*
   This approach does not assign a rank to each of the solutions in Pareto set. The main steps of the MR are as follows:

   i. Solutions in the Pareto set are ranked separately for each objective.
   ii. The best ranked k points from each objective are extracted.

As this approach selects the best solutions for each objective independently, this tends to extract solutions from extreme points in the Pareto surface (Wismans et al. 2014).

4.2   Pruning Methods

The pruning methods proposed in the literature can be divided into two categories.

1. Clustering:
   The clustering method assumes that the output of the pruning process should be the distinct solutions in the objective space. The number of clusters can either be determined by the decision-maker or can be optimized according to the Pareto set of solutions. From each cluster, one representative solution is chosen, usually the solution nearest to the center of the cluster. The number of clusters is optimized using the average silhouette width (Rousseeuw 1987). For a solution X_i, this approach calculates the average distance $a(X_i)$ to all other points in its cluster and the average distance $b(X_i)$ to all other points in the nearest neighbor cluster.

$$Silhouette(X_i) = \frac{b(X_i) - a(X_i)}{max[a(X_i), b(X_i)]} \tag{8}$$

A silhouette value close to 1 indicates that the solution was assigned to an appropriate cluster. If the silhouette value is close to 0, it means that the solution could be assigned to another cluster; and if it is close to -1, the solution is considered to be misclassified. The overall silhouette width is the average of the silhouette values of all solutions. The largest silhouette width indicates the best clustering, and therefore, the number of clusters associated with this best clustering is taken as the optimal number of clusters. The following are two approaches used to select the representative points from each cluster:

   a. *Cluster centers (CC)*:
      In this method, after the clustering algorithm is executed, the centroids of the clusters are chosen as the representative points from each cluster. In (Chaudhari et al. 2010), k-means (Hartigan and Wong 1979) is used as the clustering algorithm, and the cluster centroids are picked as the representative points.
   b. *Points closest to the ideal point (IP)*:
      The main steps of this approach are as follows (Cheikh et al. 2010):
      i. For each cluster, the ideal point is identified. The ideal point of a subset of points is the virtual point that has a minimal evaluation for each objective.
      ii. Then, for each point in each cluster, the distance to the ideal point of the cluster is calculated.
      iii. From each cluster, the point with the smallest distance to the ideal point is selected.

However, clustering methods do not necessarily guarantee an even spread of solutions, as they are sensitive to the presence of outliers. Also, in cases where the Pareto optimal set does not form any clusters, identifying solutions based on clustering is not ideal.

2. Angle-based pruning:

In this method, the geometric angle between each pair of solutions is calculated, for each objective. A threshold angle is defined for each objective, in order to identify the subset of desirable solutions. The idea is to remove the solutions that only improve some objectives marginally while significantly worsening other objectives (Sudeng and Wattanapongsakorn 2015). For a pair of non-dominating solutions, the geometric angle (denoted by θ_n) is defined as follows, where n is the nth objective, N is the number of objective functions and Δf_m is the difference between the mth objective values of the two non-dominating solutions.

$$\theta_n = tan^{-1}\left[\frac{\sqrt{\sum_{m=1,m\neq n}^{N}(\Delta f_m)^2}}{\Delta f_n}\right] \tag{9}$$

A threshold angle (δ) is defined as the reference angle which is compared to the calculated geometric angle between the pair of solutions. For any pair of solutions (i, j), if θ_n is smaller than the threshold angle, the solution j will be discarded, and the algorithm will keep only solution i. This method may identify the knee points (Bechikh et al. 2011) in the Pareto set.

4.3 Subset Optimality

Each point in the Pareto optimal set is non-dominated by any other point in the same Pareto optimal set with regard to the $\mathcal{O}$ objectives on which the multi-objective optimization algorithm is run. But, when a subset of the $\mathcal{O}$ objectives is considered, some of the points in the Pareto optimal set may dominate other points. Some methods have been proposed to use a subset of $\mathcal{O}$ (subset optimality) to reduce the number of solutions in the Pareto optimal set. Some such methods include:

1. *Favor relation (FR):*

A solution X_i is favored over the solution X_j if and only if X_i is better than X_j on more objectives (Drechsler et al. 2001; Corne and Knowles 2007). Depending on the favor relation between the solutions of the Pareto set, the following steps are followed to create a directed network and prune the Pareto set:

  i. If X_i favored over X_j, an edge from the node X_i to X_j is created.
 ii. The favor relation may not be transitive; thus, the network may have cycles. Collapse all the nodes in each cycle to a single pseudo-node; each node inside a pseudo-node is not better than another in the same cycle.
iii. The nodes with $in\_degree = 0$ are pruned.

As cycle identification is computationally expensive, there are computational limitations in applying this algorithm to Pareto sets that create large directed networks.

2. *K-optimality (KO)*:
   The concept of k-optimality was introduced in (Pierro 2006) and was used to prune solutions from the Pareto optimal solutions. A point X_i in a set of non-dominated $\mathcal{O}$ objective points is efficient with order k, where $1 < k < \mathcal{O}$, if and only if X_i is non-dominated in every k objective subset of the $\mathcal{O}$ objectives. The points that show the highest order k-optimality are selected from the Pareto optimal set.

One issue with the all aforementioned methods for pruning Pareto optimal solutions is that these algorithms need to be run after $\mathcal{O}$ objective optimization is completed. Hence, the decision-makers have to incur more computational cost in addition to the computational cost of $\mathcal{O}$ objective optimization algorithm. In the following subsection, we propose an algorithm which not only reduces the number of solutions in the Pareto set but also reduces the computational cost compared to previously proposed algorithms.

5 *Leave-k-out* Approach for Multi-objective Optimization

The *leave-k-out* approach for an $\mathcal{O}$ objective optimization problem is described below:

i. Select $(\mathcal{O} - k)$ objectives from the set of objectives and run the multi-objective optimization algorithm.
ii. Obtain the Pareto set, and evaluate each solution in the Pareto set on the objectives that were left out.
iii. Select the solutions in the Pareto set which are non-dominated on the evaluation of the objectives which were left out.

Compared to the other approaches proposed, the *Leave-k-out* approach has the following advantages:

1. A high percentage of solutions obtained constitute a subset of the Pareto surface obtained by $\mathcal{O}$ objective optimization.
2. The running time of the optimization reduces, as $(\mathcal{O} - k)$ objective optimization requires less computational effort than the original $\mathcal{O}$ objective optimization problem.
3. This method does not require any additional processing (such as ranking and clustering) after the Pareto set identification, unlike the other approaches.

6 Experimental Results

We use four commonly used real-world network datasets in our experiments. A brief description of the datasets is provided in Table 3.

6.1 *Improving Robustness by Edge Addition*

As discussed before, the Pareto set usually contains a large number of solutions. For example, in identifying 10 new edges to add to the EuroRoad network to maximize all three objectives, the Pareto surface contained 91 solutions, i.e., 91 sets of 10 edges. In order to choose a single solution, we use the *leave-k-out* approach discussed in Sect. 5.

For each case of optimization, we provide four solutions. The first value corresponds to the average of all solutions in the Pareto surface. The other three values correspond to the solutions obtained by the *leave-k-out* approach for $k = 1$ (leave-one-out approach), as described below:

1. $\mathscr{O}_{ave}$—It represents the average value obtained by all the solutions in the Pareto front of the three objective optimizations.
2. $(\mathscr{O} - 1)_{nER}$—In this case, we first perform the optimization on SG and LCC. Then from the non-dominated solutions obtained, the solution that maximizes nER is selected.
3. $(\mathscr{O} - 1)_{LCC}$—The initial optimization is performed on SG and nER. Then from the non-dominated solutions obtained, the solution that maximizes LCC is selected.

Table 3 Statistics and description of the networks used

Network	Nodes	Edges	Description
EuroRoad[a]	1174	1417	International road network in Europe. The nodes represent cities, and an edge indicates cities connected by a road
US airports[b]	1574	28236	The network of flights between the US airports in 2010. Each edge represents a connection from one airport to another
OpenFlights[c]	2939	30501	The network of flights between airports in the world. An edge represents a flight from one airport to another
US power grid[d]	4941	6594	The power grid of the Western States of the US. A node denotes either a generator or a power station, and an edge represents a power line

[a] *Source* http://konect.uni-koblenz.de/networks/subelj_euroroad
[b] *Source* http://konect.uni-koblenz.de/networks/opsahl-usairport
[c] *Source* http://konect.uni-koblenz.de/networks/opsahl-openflightsd
[d] *Source* http://konect.uni-koblenz.de/networks/opsahl-powergrid

4. $(\mathscr{O} - 1)_{SG}$—First, the optimization on LCC and nER is performed. Then from the non-dominated solutions obtained, the solution that maximizes SG is selected.

We use the solutions obtained by multiple other edge addition methods to compare the results. First, we consider single objective optimization to improve network robustness. We obtain sets of edges that optimize SG, LCC, and nER, respectively.

Then, we consider the following heuristic approaches to add edges to the network, which would also improve network robustness:

   i. *Rich—Rich*: The edges are added among the nodes with high degree.
  ii. *Poor—Poor*: The edges are added among the nodes with lowest degree.
 iii. *Rich—Poor*: The edges are added between the nodes with high degree and nodes with low degree.
  iv. *Random*: In this case, we add edges randomly to the network.

We present the results in Figs. 3 and 4. In Fig. 3, the robustness improvement obtained by the proposed multi-objective approach is compared with the robustness improvement obtained by optimizing single robustness measures. In Fig. 4, we compare the robustness improvement by the multi-objective approach with the heuristic edge addition methods. In the figures, the Y axis corresponds to the robustness measure achieved by the network, upon edge addition. Higher values along the Y axis corresponds to more robust networks. The X axis represents the number of new edges added to the network, as a percentage of the quantity of existing edges in the network.

In Fig. 3a, we show how the robustness value of SG changes with the use of different edge addition algorithms. As expected, the solution obtained by optimizing SG gives the best improvement in SG compared to the other algorithms. Poorer results for SG were obtained by the solutions that were optimized for LCC and nER. A similar pattern is seen in Fig. 3b, c as well. In Fig. 3b where we plot the value of LCC with edge addition, the best performance is obtained by optimizing LCC, whereas the solutions that were obtained by optimizing SG and nER perform poorly. In Fig. 3c where we plot the value of nER with edge addition, the best performance is obtained by optimizing nER, the solutions that were obtained by optimizing SG and LCC do not perform well.

The solutions obtained by multi-objective optimization do not perform the best with respect to all criteria, but perform "well" in all the cases. For example, the solution obtained by $(\mathscr{O} - 1)_{nER}$ performs 3rd best in optimizing SG and LCC and performs 4th best in optimizing ER among all the methods that we considered. The network created by adding a set of new edges to optimize a single objective robustness measure (such as SG), will perform well only with respect to the optimized robustness measure. For example, the network created by optimizing SG performs the best when the value of SG is considered (Fig. 3a), but performs really poorly when the other two robustness measures are considered (Fig. 3b, c).

The solutions obtained by *Leave-k-out* approach perform slightly better than the average of solutions of $\mathscr{O}$ objective optimization in each case. This is because in $\mathscr{O}$ objective optimization, when an extra objective is added to the optimization, we

Fig. 3 Robustness
improvement in OpenFlights
network—comparison
between multi-objective
approach and single
objective approaches

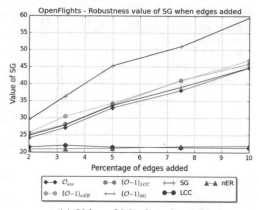

(a) Value of SG after edges added

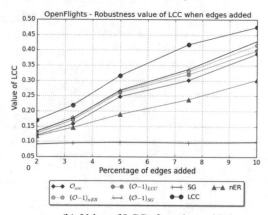

(b) Value of LCC after edges added

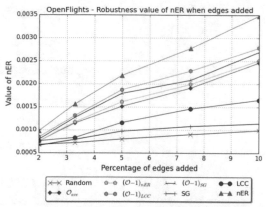

(c) Value of nER after edges added

Fig. 4 Robustness
improvement in OpenFlights
network—comparison
between multi-objective
approach and heuristic
approaches

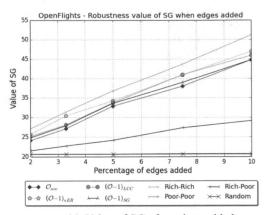

(a) Value of SG after edges added

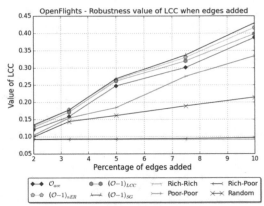

(b) Value of LCC after edges added

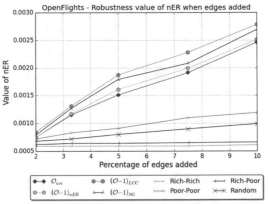

(c) Value of nER after edges added

Table 4 Average robustness ranks of edge addition methods; smaller values represent greater robustness

Method of edge addition	Average rank			
	EuroRoad	OpenFlights	US airports	US power grid
SG	5.3	6.1	6.2	6.1
LCC	5.9	5.1	4.7	5.2
nER	5.6	5.3	5.5	5.2
$(\mathscr{O}-1)_{nER}$	**3.3**	3.4	3.5	3.7
$(\mathscr{O}-1)_{LCC}$	**3.3**	**3.3**	3.5	**3.1**
$(\mathscr{O}-1)_{SG}$	3.4	3.4	**3.3**	3.3
$\mathscr{O}_{ave}$	5.1	5.3	5.5	5.1
Rich–Rich	7.9	7.9	8.0	7.6
Poor–Poor	7.7	8.3	7.5	9.3
Rich–Poor	8.9	8.9	9.3	8.7
Random	9.4	8.9	8.9	8.5

get solutions that perform poorly with regard to the objective of our interest. For example, in 3 objective optimization, some solutions are in the corners of the Pareto surface, perform as well in two objectives, but poorly in the other. These solutions affect the average performance that we considered in the comparison. In the case of *leave-one-out* approach, although all three robustness measures are considered, we consider them in two steps. For example, in the case of $(\mathscr{O}-1)_{nER}$, some solutions perform quite well in LCC, but not as well in SG (these solutions lie on one side of the Pareto front). Those solutions are unlikely to be picked by the second step, when we pick the solutions that perform best in nER, because of the low correlation between LCC and nER. Hence, a solution that performs well in SG is picked by the *leave-one-out* approach. A similar pattern is seen in Fig. 3b, c as well.

As shown in Fig. 4, when heuristic edge addition methods are considered, adding edges among the nodes with high degrees results in much better performance in terms of SG. In fact, Rich–Rich performs 2nd best among all the edge addition methods explored here when SG is considered. Rich–Rich edge addition performs poorly when LCC and nER are considered.

In Table 4, we show the average rank of each edge addition method. The values in Table 4 are the ranks of all 15 cases (three robustness measures of interest and 5 percentages of edge addition) for each network. The edge addition methods corresponding to low ranks perform well on all 3 robustness measures of interest. According to the results, adding edges based on the *leave-one-out* approach yields the best overall robustness in all networks considered in the study.

6.2 Network Robustness After Node Attacks

In this section, we investigate how the networks (with robustness improved by edge addition) performed during a phase of multiple node attacks. In this study, we consider two types of node attacks. In targeted node attacks, the nodes with the high degrees (and their corresponding edges) get removed from the network, and in random node attacks, a set of nodes selected randomly from the network (and their incident edges) get removed from the network.

After edge addition (to improve the robustness of the network) we attack a set of selected nodes in the network. After the attack, we recalculate the network robustness values. In Table 5, we show the robustness values of the OpenFlights network during targeted node attacks and in Table 6, we show the robustness values of the same network during random node failures. In the column corresponding to 0% node attacks, we show the robustness value after 3.3% edges have been added to the network; each subsequent column refers to robustness values calculated after the percentage of nodes are attacked in the network, as indicated by the column heading.

Prior to any node attacks, the highest value of SG is given by the network to which the edges are added by optimizing SG. In the case of targeted attacks (Table 5), as the number of attacked nodes increases, the highest SG values are obtained by the networks to which the edges were added by the $(\mathcal{O} - 1)_{nER}$ approach. For the network to which the edges were added by optimizing SG, the SG value reduces sharply as the number of nodes attacked increases. As expected, this sharp decrease is seen in the network to which the edges were added by Rich–Rich approach. In the cases of LCC and nER, for all node attack levels, the highest robustness values are shown by the networks to which the edges were added to optimize LCC and nER, respectively. But, even in those two cases, the networks to which the edges were added by the *leave-one-out* approach show high robustness values even when subjected to targeted node attacks.

Table 6 shows the results for the case of random node attacks. In this case, the network created by adding edges to improve SG shows the best SG values during random node attacks. However, the robustness of this network is poor when LCC and nER are considered. A similar behavior is shown by the networks created by adding edges to improve LCC and nER. The networks created by adding edges using *leave-one-out* approach show high overall robustness when all three measures of robustness are considered. Hence, we conclude that the networks created by adding edges using *leave-one-out* approach retain the high overall robustness during a phase of multiple random node attacks as well.

7 Conclusion

When edges are added to a network, the properties of the network change. The amount of change depends on the importance of the set of edges added to the network. In

Table 5 Robustness values during targeted node attacks—OpenFlights network

Robustness value	Edge addition method	Percentage nodes attacked						
		0%	2%	5%	7.5%	10%	15%	
SG	SG	36.292	27.686	14.966	12.632	8.575	4.477	
	LCC	21.984	14.811	9.488	7.141	6.314	3.526	
	ER	20.976	14.132	9.018	6.614	6.002	3.349	
	$(\mathscr{O}-1)_{nER}$	30.391	24.289	12.865	11.478	9.466	5.425	
	$(\mathscr{O}-1)_{LCC}$	28.087	17.300	11.035	8.189	7.365	4.223	
	$(\mathscr{O}-1)_{SG}$	27.883	17.200	10.952	8.238	7.317	4.625	
	Rich–Rich	31.367	23.934	12.757	10.518	7.280	3.727	
	Poor–Poor	20.630	13.893	8.905	6.737	5.915	3.280	
	Rich–Poor	22.736	15.339	9.627	7.254	6.403	3.531	
LCC	SG	0.095	0.087	0.075	0.072	0.066	0.051	
	LCC	0.219	0.209	0.200	0.189	0.169	0.148	
	ER	0.146	0.132	0.112	0.107	0.094	0.064	
	$(\mathscr{O}-1)_{nER}$	0.168	0.168	0.159	0.151	0.134	0.115	
	$(\mathscr{O}-1)_{LCC}$	0.161	0.152	0.137	0.127	0.121	0.094	
	$(\mathscr{O}-1)_{SG}$	0.167	0.165	0.153	0.143	0.132	0.099	
	Rich–Rich	0.092	0.085	0.073	0.071	0.063	0.049	
	Poor–Poor	0.154	0.147	0.139	0.136	0.120	0.096	
	Rich–Poor	0.093	0.086	0.076	0.075	0.064	0.048	
ER	SG	0.0008	0.0007	0.0004	0.0004	0.0001	7.68E-05	
	LCC	0.0008	0.0008	0.0005	0.0005	0.0003	0.0002	
	ER	0.0016	0.0014	0.0011	0.0010	0.0006	0.0004	
	$(\mathscr{O}-1)_{nER}$	0.0012	0.0011	0.0008	0.0007	0.0005	0.0004	
	$(\mathscr{O}-1)_{LCC}$	0.0013	0.0013	0.0008	0.0008	0.0004	0.0003	
	$(\mathscr{O}-1)_{SG}$	0.0013	0.0012	0.0007	0.0007	0.0004	0.0003	
	Rich–Rich	0.0006	0.0005	0.0003	0.0003	2.30E-05	5.78E-06	
	Poor–Poor	0.0008	0.0008	0.0004	0.0001	4.01E-06	2.27E-07	
	Rich–Poor	0.0006	0.0006	0.0003	0.0003	3.68E-05	5.56E-06	

Table 6 Robustness values during random node attacks—OpenFlights network

Robustness value	Edge addition method	Percentage nodes attacked					
		0%	2%	5%	7.5%	10%	15%
SG	SG	36.292	31.046	28.240	27.938	27.021	25.591
	LCC	21.984	20.340	20.131	19.541	17.356	16.931
	ER	20.976	19.333	19.077	18.683	16.777	16.206
	$(\mathcal{O}-1)_{nER}$	30.391	27.124	25.985	24.482	24.072	23.276
	$(\mathcal{O}-1)_{LCC}$	28.087	25.407	24.307	23.971	23.619	21.921
	$(\mathcal{O}-1)_{SG}$	27.883	24.931	24.521	23.897	23.176	21.782
	Rich–Rich	31.367	26.218	24.108	23.361	22.591	21.679
	Poor–Poor	20.630	19.232	19.215	17.994	16.418	16.350
	Rich–Poor	22.736	21.795	21.516	20.316	19.577	18.875
LCC	SG	0.095	0.088	0.087	0.079	0.078	0.076
	LCC	0.219	0.155	0.136	0.131	0.129	0.125
	ER	0.146	0.098	0.095	0.095	0.094	0.092
	$(\mathcal{O}-1)_{nER}$	0.168	0.104	0.102	0.102	0.102	0.096
	$(\mathcal{O}-1)_{LCC}$	0.161	0.099	0.096	0.095	0.098	0.093
	$(\mathcal{O}-1)_{SG}$	0.167	0.114	0.104	0.105	0.103	0.100
	Rich–Rich	0.092	0.090	0.087	0.089	0.086	0.081
	Poor–Poor	0.154	0.115	0.111	0.109	0.116	0.109
	Rich–Poor	0.093	0.089	0.089	0.088	0.083	0.081
ER	SG	0.0008	0.0007	0.0006	0.0006	0.0006	0.0006
	LCC	0.0008	0.0006	0.0006	0.0006	0.0006	0.0005
	ER	0.0016	0.0011	0.0011	0.0011	0.0010	0.0010
	$(\mathcal{O}-1)_{nER}$	0.0012	0.0009	0.0009	0.0008	0.0008	0.0007
	$(\mathcal{O}-1)_{LCC}$	0.0013	0.0010	0.0009	0.0009	0.0008	0.0008
	$(\mathcal{O}-1)_{SG}$	0.0013	0.0009	0.0009	0.0009	0.0008	0.0008
	Rich–Rich	0.0006	0.0005	0.0005	0.0005	0.0004	0.0004
	Poor–Poor	0.0008	0.0006	0.0006	0.0005	0.0005	0.0005
	Rich–Poor	0.0006	0.0005	0.0004	0.0004	0.0004	0.0004

this chapter, we addressed the following problem: *Given a network and a budget, how should a set of "key" edges be selected to be added to the network in order to maximally improve the overall robustness of the network?* Toward this goal, first, we discuss the network robustness measures that have been proposed and widely used. We analyze the properties of these robustness measures and identify their similarities and dissimilarities using correlation analysis. Then, we use the *leave-k-out* approach to optimize multiple robustness measures of interest to improve the overall robustness of a network. Experimental evidence shows the improvement in multiple robustness measures when the new edges are added using our approach. The key edge identification and addition approach proposed in this study improve the multiple robustness measures of interest simultaneously, and this can be extremely important in real-world applications. For example, when funds need to be allocated to add new roads to a road network, the objectives of interest would include reducing the distance between cities, keeping the cities connected even if some central cities become inaccessible, etc. For such an application, one can select the objectives accordingly and use the edge addition approach proposed by this study in order to improve the overall robustness of the underlying system.

References

M.J. Alenazi, J.P. Sterbenz, Comprehensive comparison and accuracy of graph metrics in predicting network resilience, in *2015 11th International Conference on the Design of Reliable Communication Networks (DRCN)* (IEEE, 2015), pp. 157–164

S. Bechikh, L.B. Said, K. Ghédira, Searching for knee regions of the pareto front using mobile reference points. Soft Comput. **15**(9), 1807–1823 (2011)

P.J. Bentley, J.P. Wakefield, in *Finding Acceptable Solutions in the Pareto-Optimal Range using Multiobjective Genetic Algorithms* (Springer London, 1998), pp 231–240. https://doi.org/10.1007/978-1-4471-0427-8_25

A. Bigdeli, A. Tizghadam, A. Leon-Garcia, Comparison of network criticality, algebraic connectivity, and other graph metrics, in *Proceedings of the 1st Annual Workshop on Simplifying Complex Network for Practitioners* (ACM, 2009), p. 4

P. Boldi, M. Rosa, Robustness of social and web graphs to node removal. Soc. Netw. Anal. Min. **3**(4), 829–842 (2013)

P. Boldi, M. Rosa, S. Vigna, *Robustness of Social Networks: Comparative Results Based on Distance Distributions* (Springer, 2011)

D. Chakrabarti, Y. Wang, C. Wang, J. Leskovec, C. Faloutsos, Epidemic thresholds in real networks. ACM Trans. Inf. Syst. Secur. (TISSEC) **10**(4), 1 (2008)

H. Chan, L. Akoglu, in Optimizing network robustness by edge rewiring: A general framework. *Data Mining and Knowledge Discovery (DAMI)*, 2016

H. Chan, L. Akoglu, H. Tong, Make it or break it: manipulating robustness in large networks, in *Proceedings of the 2014 SIAM Data Mining Conference* (SIAM, 2004), pp. 325–333

P. Chaudhari, R. Dharaskar, V. Thakare, Computing the most significant solution from pareto front obtained in multi-objective evolutionary. Int. J. Adv. Comput. Sci. Appl. **1**(4), 63–68 (2010)

M. Cheikh, B. Jarboui, T. Loukil, P. Siarry, A method for selecting pareto optimal solutions in multiobjective optimization. J. Inf. Math. Sci. **2**(1), 51–62 (2010)

D.W. Corne, J.D. Knowles, Techniques for highly multiobjective optimisation: some nondominated points are better than others, in *Proceedings of the 9th annual conference on Genetic and
evolutionary computation* (ACM, 2007), pp 773–780

K. Deb, S. Agrawal, A. Pratap, T. Meyarivan, A fast elitist non-dominated sorting genetic algorithm
for multi-objective optimization: Nsga-ii, in *Parallel Problem Solving from Nature PPSN VI*,
(Springer, 2000), pp. 849–858

F. di Pierro, Many-objective evolutionary algorithms and applications to water resources engineering. Ph.D. thesis, University of Exeter, 2006

N. Drechsler, R. Drechsler, B. Becker, Multi-objective optimisation based on relation favour, in
International Conference on Evolutionary Multi-criterion Optimization (Springer, 2001), pp.
154–166

W. Ellens, Effective resistance and other graph measures for network robustness. Ph.D. thesis,
Master thesis, Leiden University, 2011, https://www.math.leidenuniv.nl/scripties/EllensMaster.
pdf

W. Ellens, R.E. Kooij, *Graph Measures and Network Robustness*, 2013. arXiv:13115064

M. Farina, P. Amato, On the optimal solution definition for many-criteria optimization problems,
in *Fuzzy Information Processing Society, 2002. Proceedings. NAFIPS. 2002 Annual Meeting of
the North American*, 2002, pp. 233–238. https://doi.org/10.1109/NAFIPS.2002.1018061

M. Fiedler, Algebraic connectivity of graphs. Czechoslov. Math. J **23**(2), 298–305 (1973)

M. Garza-Fabre, G.T. Pulido, C.A.C. Coello, Ranking methods for many-objective optimization,
in *Mexican International Conference on Artificial Intelligence*, (Springer, 2009), pp. 633–645

A. Ghosh, S. Boyd, A. Saberi, Minimizing effective resistance of a graph. SIAM Rev. **50**(1), 37–66
(2008)

J.A. Hartigan, M.A. Wong, Algorithm as 136: a k-means clustering algorithm. J. R. Stat. Soc. Ser.
C (Applied Statistics) **28**(1), 100–108 (1979)

H.J. Herrmann, C.M. Schneider, AA.. Moreira, J.S. Andrade Jr., S. Havlin, Onion-like network
topology enhances robustness against malicious attacks. J. Stat. Mech. Theory Exp. **2011**(01),
01–027 (2011)

A. Jamakovic, P. van Mieghem, *On the Robustness of Complex Networks by using the Algebraic
Connectivity* (Springer, 2008)

W. Jun, M. Barahona, T. Yue-Jin, D. Hong-Zhong, Natural connectivity of complex networks. Chin.
Phys. Lett. **27**(7), 078–902 (2010)

L.T. Le, T. Eliassi-Rad, H. Tong, Met: a fast algorithm for minimizing propagation in large graphs
with small eigen-gaps, in *Proceedings of the 2015 SIAM International Conference on Data
Mining*, SDM, SIAM, vol. 15, 2015, pp. 694–702

F.D. Malliaros, V. Megalooikonomou, C. Faloutsos, Fast robustness estimation in large social
graphs: communities and anomaly detection. SDM, SIAM **12**, 942–953 (2012)

M. Marchiori, V. Latora, Harmony in the small-world. Phys. A Stat. Mech. Appl. **285**(3), 539–546
(2000)

G.S. Peng, J. Wu, Optimal network topology for structural robustness based on natural connectivity.
Phys. Stati. Mech. Appl. **443**, 212–220 (2016)

P.J. Rousseeuw, Silhouettes: a graphical aid to the interpretation and validation of cluster analysis.
J. Comput. Appl. Math. **20**, 53–65 (1987)

C.M. Schneider, A.A. Moreira, J.S. Andrade, S. Havlin, H.J. Herrmann, Mitigation of malicious
attacks on networks. Proc. Nat. Acad. Sci. **108**(10), 3838–3841 (2011)

N. Srinivas, K. Deb, Muiltiobjective optimization using nondominated sorting in genetic algorithms.
Evol. Comput. **2**(3), 221–248 (1994)

G. Stewart, J.G. Sun, *Matrix Perturbation Theory (computer science and scientific computing)*,
1990

S. Sudeng, N. Wattanapongsakorn, Post pareto-optimal pruning algorithm for multiple objective
optimization using specific extended angle dominance. Eng. Appl. Artif. Intell. **38**, 221–236
(2015)

A. Sydney, C. Scoglio, D. Gruenbacher, Optimizing algebraic connectivity by edge rewiring. Appl. Math. Comput. **219**(10), 5465–5479 (2013)

T. Tanizawa, S. Havlin, H.E. Stanley, Robustness of onionlike correlated networks against targeted attacks. Phys. Rev. E **85**(4), 046–109 (2012)

A. Tizghadam, A. Leon-Garcia, Autonomic traffic engineering for network robustness. IEEE J. Sel. Areas Commun. **28**(1), 39–50 (2010)

H. Tong, B.A. Prakash, C. Tsourakakis, T. Eliassi-Rad, C. Faloutsos, D.H. Chau, On the vulnerability of large graphs, in *2010 IEEE 10th International Conference on Data Mining (ICDM)* (IEEE, 2010), pp. 1091–1096

H. Tong, B.A. Prakash, T. Eliassi-Rad, M. Faloutsos, C. Faloutsos, Gelling, and melting, large graphs by edge manipulation, in *Proceedings of the 21st ACM International Conference on Information and knowledge Management* (ACM, 2012), pp 245–254

H. Wang, P. van Mieghem, Algebraic connectivity optimization via link addition, in *Proceedings of the 3rd International Conference on Bio-Inspired Models of Network, Information and Computing Sytems, ICST (Institute for Computer Sciences, Social-Informatics and Telecommunications Engineering)*, 2008, p. 22

T. Watanabe, N. Masuda, Enhancing the spectral gap of networks by node removal. Phys. Rev. E **82**(4), 046–102 (2010)

D.J. Watts, S.H. Strogatz, Collective dynamics of small-world networks. Nature **393**(6684), 440–442 (1998)

L.J. Wismans, T. Brands, E.C. Van Berkum, M.C. Bliemer, Pruning and ranking the pareto optimal set, application for the dynamic multi-objective network design problem. J. Adv. Transp. **48**(6), 588–607 (2014)

Z.X. Wu, P. Holme, Onion structure and network robustness. Phys. Rev. E **84**(2), 026–106 (2011)

A. Yazdani, R.A. Otoo, P. Jeffrey, Resilience enhancing expansion strategies for water distribution systems: a network theory approach. Environ. Model. Softw. **26**(12), 1574–1582 (2011)

A. Zeng, W. Liu, Enhancing network robustness against malicious attacks. Phys. Rev. E **85**(6), 066–130 (2012)

On Joint Maximization in Energy and Spectral Efficiency in Cooperative Cognitive Radio Networks

Santi P. Maity and Anal Paul

1 Introduction

The exponential growth in the use of wireless devices to support a large number of users with the data-intensive applications put a high demand on radio frequency spectrum availability and its efficient utilization too. The present wireless communication services operate in static mode, and hence, any emerging or emergency service always faces the spectrum allocation problem. While exploring a new band of spectrum (for example, mm wave) is challenging to the telecommunication service provider, often leasing the licensed spectrum for opportunistic use is cost-effective. Furthermore, the statics of different bands of spectrum utilization shows an interesting picture. The substantial studies made by FCC and OFCOM report that spectrum is underutilized as the majority of the spectrum allocated to the licensed users, called as primary users (PU), are observed to be ideal over a longer duration of time (Federal Communications Commission 2002; OFCOM 2007). To address this spectrum scarcity and underutilization issues, the concept of cognitive radio (CR) appears as an opportunistic access of the unused radio spectrum by the unlicensed users, named as secondary users (SU), by means of cognition of radio environment and reconfigurability in its data transmission (Fette 2009). SU must be obliged to protect the interest of the PUs by maintaining the interference limit (Fette 2009; Hassan et al. 2017).

The first and the foremost issue in design of CR system is the fast and reliable spectrum sensing (SS), a method of determining whether PUs are transmitting data over the licensed band at some time instant. Knowledge of SS available to SUs protects PU from the harmful interference and identifies the unused spectrum for the secondary data transmission (Ostovar and Chang 2017). Reliable SS technique offers the scope of using the idle part of the spectrum and enhances the spectrum

S. P. Maity (✉) · A. Paul
Department of Information Technology, Indian Institute of Engineering Science
and Technology Shibpur, Howrah 711103, India
e-mail: santipmaity@it.iiests.ac.in

A. Paul
e-mail: ap.rs2015@it.iiests.ac.in

© Springer Nature Singapore Pte Ltd. 2018
J. K. Mandal et al. (eds.), *Multi-Objective Optimization*,
https://doi.org/10.1007/978-981-13-1471-1_6

efficiency (SE), a measure of data transmission rate over a given spectrum. Several SS techniques, for example, energy detection (Yang et al. 2016), cyclostationary feature detection (Yang et al. 2015), eigenvalue-based detection (Yang et al. 2017), generalized likelihood ratio test (Sedighi et al. 2013), matched filter based (Zhang et al. 2015), etc. are reported in the literature to improve the SE performance (Cicho et al. 2016; Akyildiz et al. 2011; Yucek and Arslan 2009). The relative pros and cons of those techniques are also investigated and reported in (Akyildiz et al. 2011; Yucek and Arslan 2009). Among the different SS techniques, energy detection scheme is widely used due to its low computational complexity, low power consumption, very less sensing time, and generic implementation (Paul and Maity 2016; Cicho et al. 2016). Performance of SS is typically measured by the probability of PU detection (P_d) and the probability of false alarm (P_{fa}). The measure P_d indicates the probability of correct detection of PU transmission when it actually transmits over a frequency band. The other measure P_{fa} represents the probability of false detection of PU transmission when it is not actually active over the channel (Paul and Maity 2016).

Recently, the cooperative spectrum sharing technique in CR network (CRN) is studied and reported a lot (Chatterjee et al. 2015, 2016a, b). In the cooperative CRN (CCRN), a PU can involve SUs as the cooperative relays for its own transmission. It is known as PU cooperation (Chatterjee et al. 2016a). In return, the SUs achieve an opportunistic access of the wireless channel for their own transmissions. Thus, throughput improvement is possible by power saving operation of SU for its own as well as in PU cooperation data transmission leading to an increase in energy efficiency (EE). However, SUs are mostly operated by the limited battery sources, and hence, maximization in EE is highly demanding for enhancing the network lifetime. Different parameters, for example, the optimal values of transmission and cooperation power, the number of relays involved, and sensing and transmission slots on frame-based system design, have varied impact on EE. It is observed that majority of the power (stored energy) is consumed at SUs due to the PU cooperation and the secondary transmission rather than its use on data processing, for example, SS (Chatterjee et al. 2016a). It indicates that the enhancement in SE is obtained at the cost of high energy consumption due to participation of several SU nodes in the system and this significantly reduces the EE. Therefore, improvement in EE along with SE becomes a major concern in CCRN (Chatterjee et al. 2015, 2016a; Ren et al. 2016; Awin et al. 2016).

The authors in (Chatterjee et al. 2015, 2016a) study the PU cooperation which is based on SS decision. The proposed approach ensures that SU acts as amplify-and-forward (AF) relay in PU transmissions and as a reward, SUs are allowed to access the spectrum when PU is found to be idle (Chatterjee et al. 2015, 2016a). The authors in (Hu et al. 2016) separately optimized the SE and EE via joint optimization of sensing duration and final decision threshold in SS. Based on the solutions of the two cases, the general problem of SE–EE trade-off is solved. Bicen et al. (2015) proposed an analytical framework to study the communication and distributed sensing efficiency impacts of common control interface (CCI) utilization for spectrum hand-off in CRN. The developed framework evaluates the achievable delay, spectrum, and energy efficiency in CRN with and without CCI.

1.1 Machine Learning in CR

The discussion on the previous section highlights that CR works on the basis of two principles: cognition and reconfigurability. This makes CR brain empowered intelligent communication which learns its radio environment before transmission to start and then adaptively adjust its resources, for example, bandwidth, transmission power, signaling, data rate, etc. Much in success on CR system design lies on efficient use of cognitive logic tools such as machine learning (ML). Many different ML algorithms that involve tools like neural networks (Jiang et al. 2017), hidden Markov models (Choi and Hossain 2013), and genetical algorithms (Jiao and Joe 2016; Lang et al. 2016; Celik and Kamal 2016) are studied in CR to meet the high quality of services (QoS) (Clancy et al. 2007). The recent literature on CRs report the scope and contributions on both supervised and unsupervised learning techniques to resolve various optimization problems. The authors in (Jiang et al. 2017; Zhang et al. 2017) proposed supervised learning techniques based on neural networks and support vector machines (SVMs) to find the optimal spectrum handover solutions in CRNs. A modification in particle swarm optimization, called as fast convergence particle swarm optimization (FC-PSO), is explored to address the sensing-throughput trade-off in various signal-to-noise ratio (SNR) conditions under the constraint of PU protection (Rashid et al. 2015). The formulated optimization problem reduces P_{fa} and consequently enhances the spectrum usability of SUs. The genetic algorithms (GAs), a kind of stochastic search algorithm, work on the mechanics of natural gene selection procedure. It works on the concept of "survival of the fittest" with a random exchange of information, but in a more structured way (Lang et al. 2016). GA is also explored to resolve the spectrum allocation problem in CRNs (Huang et al. 2016; Morabit et al. 2015). The authors in (Hojjati et al. 2016) used GA to develop an energy-efficient cooperative SS technique. The authors in (Wen et al. 2012) discussed a spectrum allocation problem based on GA for CR and proposed a Max-Overall-Performance algorithm, which offers the best result in spectrum allocation with a target of maximizing systems' overall performance. The main advantage of GA over other soft computing techniques is its capability of handling multi-objective functions (Morabit et al. 2015).

Differential evolution (DE), another powerful stochastic searching optimization algorithm, finds extensive use in recent works on CR research (Ye et al. 2013; Maity et al. 2016; Paul and Maity 2016; Zhang and Zhao 2016; Ng 2015). The key difference in working principles of DE from GA or PSO is its mechanism for generating the new solutions (Das and Suganthan 2011). The solutions in DE evolve through the iterative cycles of three essential DE operators: mutation, crossover, and selection. The authors in (Ye et al. 2013) applied DE algorithm for power allocation to a subcarrier for maximizing the secondary transmission rate under the constraint of PU interference limit. The unsupervised learning algorithms such as clustering techniques are also investigated in CR to improve the SE and EE. In (Maity et al. 2016), the authors applied fuzzy C-means (FCM) clustering technique along with DE algorithm to improve the probability of PU detection and significant reduction on energy consumption at fusion center (FC) in CSS. However, performance of FCM deteriorates when the data structure of input patterns is nonspherical. This problem is solved by the kernel-fuzzy c-means (KFCM) technique (Paul and Maity 2016). KFCM maps nonlinear input space into the high-dimensional space. This projection helps to apply a linear classifier for partitioning the data in respective clusters. The DE algorithm is also applied with the KFCM clustering strategy to improve the system performance (Paul and Maity 2016) over the FCM-based DE technique (Maity et al. 2016).

1.2 Scope and Contributions

The literature review on CR research is rich with optimization problems solved using classical techniques (Hu et al. 2016; Bicen et al. 2015; Cicho et al. 2016; Chatterjee et al. 2015, 2016a) as well as using various ML techniques (Ye et al. 2013; Zhang and Zhao 2016; Ng 2015; Huang et al. 2016; Morabit et al. 2015). It is often observed that the traditional optimization techniques require high computational cost and implementation of those algorithms is much complex. In general, all classical approaches are found to be efficient for solving linear problems and provide regular solutions. The main difficulties occur for the nonlinear problems where the solutions are not always regular (Gao et al. 2014). Nonlinear optimization problems with several constraints often suffer from the local optima issue. Though some techniques try to obtain the feasible solutions using the penalty functions (Duan et al. 2012; Kaur et al. 2017), the performance is not adequate due to the complex selection of the penalty parameters. In all such cases, it is found that ML techniques often offer low-cost, tractable yet efficient solutions. Among the diverse ML techniques, DE algorithm is found to be very efficient to obtain a global solution in several optimization problems (Das and Suganthan 2011) where the objective function and constraints are nonlinear in nature. DE is also extensively explored in recent CR research (Maity et al. 2016; Paul and Maity 2016; Ye et al. 2013). The high convergence characteristics of DE along with a powerful searching technique and the presence of only few control parameters make it easy to implement (Das and Suganthan 2011).

Cooperative CRN system model is considered here in a sensing—cooperation—transmission framework. Based on the sensing result, SU either cooperates in PU data transmission or transmits its own data. The work jointly maximizes the SE and EE in an integrated platform using DE algorithm under the constraints of sensing duration, probability of PU detection, PU cooperation rate, and limited SU power budget. The objective is to find the optimal SS time along with the maximum power allocation for PU cooperation and SU data transmission which maximize the SE and EE. It is worth mentioning that DE algorithm is applied for obtaining a global solution as the constraints and objective function in the present problem are nonlinear. The combined issues of SS, PU cooperation, and opportunistic secondary data transmissions in a single time frame form a nontrivial optimization problem. For this kind of conflicting optimization problem, there needs a set of Pareto-optimal solutions for overall performance improvement of the system. Efficient heuristic approach of DE finds the globally optimized solution from the set of solutions. A large set of simulation results validates the effectiveness of the proposed technique.

One may also suggest to use other evolutionary algorithms (EAs) like genetic algorithms (GAs), ant colony optimization (ACO), evolutionary programming (EP), cultural algorithm (CA), genetic programming (GP), multi-objective particle swarm optimization (PSO), etc. for such random search. However, in the proposed work, the calculated energy values are in the form of real numbers. DE offers the better system stability and reliability over GAs when initial population set contains real numbers. As a result, the ideas of mutation and crossover are substantially different in DE over the other existing algorithms. In various real-time applications, EAs encounter various flaws arising out from optimization in noisy fitness functions, deviation in the environmental parameters due to the dynamical adoption, approximation errors in fitness function, aberration of the global optimal point after termination of the searching process, etc. It is worth mentioning that DE effectively overcomes the above problems due to its inherent nature of iterative and adaptive global searching technique. Hence, DE provides an enhanced performance in respect of inevitability, convergence speed, accuracy, and robustness irrespective of nature of the problem to be solved.

The rest of the chapter is structured as follows: Section 2 presents the proposed system model and Sect. 3 describes the problem formulation and its solution using the DE algorithm. Numerical results for the proposed solution are then demonstrated in Sect. 4 to highlight the spectrum and energy-efficient system design and finally Sect. 5 concludes the paper.

2   System Model

The typical system model is depicted in Fig. 1 and consists of SU and PU networks. Primary network contains a PU transmitter (PU_T) and a PU receiver (PU_R), while the secondary network consists of a SU transmitter (SU_T) and a SU receiver (SU_R). It is assumed that PU_R is located far apart from PU_T and there is no direct reliable path for data transmission over PU_T-PU_R link. Hence, SU_T cooperates with PU_T for primary data transmission and as a reward SU obtains a data transmission opportunity of its own. Figure 2 represents the typical time frame structure which is divided in two different time slots. During the first time slot (τ_s), SU_T node senses the particular channel of PU data transmission to detect the spectrum occupancy or free availability of PU. In the second time slot ($T - \tau_s$), depending on the sensing decision, SU_T either participates in the relaying process of PU data transmission in cooperation or it transmits own data to SU_R.

Fig. 1 Typical system model

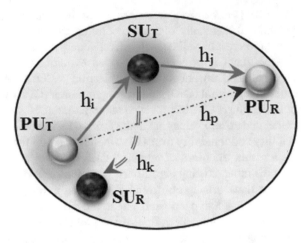

Fig. 2 Typical time frame
model

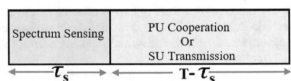

2.1 Signal Model

It is considered that PU_T-PU_R link, SU_T-SU_R transmission links, SU_T-PU_R cooperation link, and the link between PU_T-SU_T follow quasi-static Rayleigh fading. Let the spatial distance between PU_T-PU_R is represented by d_p, and h_p indicates the fading coefficient of the corresponding channel, then $h_p \sim \mathcal{CN}(0, d_p^{-\alpha})$, where α is the path loss exponent. The distances and fading coefficients of the respective nodes are given in Table 1.

The received signal at SU_T from PU_T during τ_s is expressed as

$$Y_{ps}(n) = \Psi h_i x_p(n) + \upsilon_i(n), \ \forall n = 1, 2, \ldots, N \tag{1}$$

Table 1 Rayleigh channel fading coefficients and distances

Wireless link	Fading coefficients	Distance
PU_T-PU_R	h_p	d_p
SU_T-SU_R	h_k	d_k
SU_T-PU_T	h_i	d_i
SU_T-PU_R	h_j	d_j

In Eq. 1, $x_p(n)$ denotes the PU transmitted signal which is modeled as circularly symmetric complex Gaussian (CSCG) random sequence with zero mean and variance $E[|x_p(n)|^2] = P_p$. Here, P_p represents PU transmission power. The v_i represents the noise at SU_T, where v_i follows CSCG along with zero mean and the variance is P_{ns}. Let the binary PU indicator is symbolized by Ψ, where $\Psi = 1$, or $\Psi = 0$, which indicates that PU is either active or ideal over a particular frequency band, respectively. The total number of PU samples is denoted by N and $N = \tau_s * f_s$, where f_s is the sampling frequency. It is considered that an energy detector is embedded in SU_T and SU_T performs local SS during the sensing process. The test statistics $(T_{Y_{ps}})$ of PU samples is denoted by

$$T_{Y_{ps}} = \frac{1}{N} \sum_{n=1}^{N} |Y_{ps}(n)|^2$$

The detection probability and the false alarm probability of the PU at SU_T are mathematically formulated as (Chatterjee et al. 2016a)

$$P_d = Q\left\{ \left(\frac{\varepsilon}{P_{ns}} - \gamma_P - 1 \right) \frac{\sqrt{\tau_s f_s}}{\gamma_p + 1} \right\} \tag{2}$$

$$P_{fa} = Q\left\{ \left(\frac{\varepsilon}{P_{ns}} - 1 \right) \sqrt{f_s \tau_s} \right\} \tag{3}$$

The Q-function of Eqs. 2 and 3 is expressed as

$$Q(x) = \frac{1}{\sqrt{2\pi}} \int_{x}^{\infty} \left(e^{-\frac{t^2}{2}} \right) dt$$

The symbol ε represents the detection threshold for energy detection and γ indicates average SNR at SU_T.

$$\gamma = \frac{d_i^{-\alpha} P_p}{P_{ns}} \tag{4}$$

If SS decision indicates that PU is active, then SU_T cooperates in PU transmission. The PU cooperation time duration is divided into two equal sub-slots. At the first sub-slot, PU_T signal is received at SU_T (as no effective data transmission occurs at PU_T to PU_R over the direct link). During the second sub-interval, SU_T amplifies the received signal with an amplification power gain ω_c and forward the same to PU_R. The received signal at PU_R from SU_T is represented as

$$Y_{ipr} = \sqrt{\omega_c} Y_{ps} h_j + \upsilon_{p1} \tag{5}$$

The noise at PU_R is υ_{p1} during cooperation and follows CSCG random variable along with zero mean and the variance is P_{np}. The symbol "R_{1p}" represents the PU cooperation SE if PU is correctly detected and R_{2p} denotes the SE when missed detection is occurred. In case of R_{2p}, SU_T starts data transmission which causes a harmful interference at PU_R. Since $R_{1p} >> R_{2p}, \therefore R_P \approx R_{1p}$. So, R_P is expressed as follows (Chatterjee et al. 2016a):

$$R_P = P(H_1)P_d\left(\frac{T - \tau_s}{2T}\right)\log_2\left(1 + \frac{P_c|h_j|^2|h_i|^2 P_p}{P_{np}}\right) \tag{6}$$

The parameters $P(H_1) = P(\Psi = 1)$ and $P(H_0) = P(\Psi = 0)$ denote the stationary probabilities of PUs presence and absence, respectively. If PU is found to be idle, SU_T starts transmission of its data to the SU_R. The signal received at SU_R from SU_T is represented as

$$Y_{isr} = h_k x_k + \upsilon_{p2} \tag{7}$$

The signal transmitted by SU_T is denoted by the symbol x_k and it is assumed to be CSCG random process with zero mean and the variance $E[|x_k(n)|^2] = P_T$. The P_T is the transmission power of SU_T. The symbol υ_{p2} represents the additive noise at SU_R and is considered as CSCG random variable ($\upsilon_{p2} \sim \mathcal{CN}(0, P_{ns})$). The SU_T transmits in two cases, in the first case when PU is found to be idle and is correctly detected (i.e., no false alarm) and in the second case when PU is active the missed detection is occurred. However, missed detection creates the unwanted interference at SU_R due to PU data transmission. In such situation, SU's transmitted data packet is assumed to be lost. So, SE between SU_T and SU_R is obtained as

$$\eta_{SE} = \left\{P(H_0)(1 - P_f)\left(\frac{T - \tau_s}{2T}\right)\right\}\log_2\left(1 + \frac{|h_k|^2 P_T}{P_{ns}}\right) \tag{8}$$

The average power consumption of SU_T during PU cooperation and SU transmission is denoted by P_{av}. Now P_{av} can be expressed as

$$P_{av} = \left\{P(H_0)(1 - P_f) + P(H_1)(1 - P_d)\right\}\left(\frac{T - \tau_s}{T}\right)P_T + \left(\frac{T - \tau_s}{2T}\right)P_c \tag{9}$$

P_c denote the power consumption at SU_T for PU cooperation, and is derived as

$$P_c = (P(H_0)P_{fa}P_{ns} + P(H_1)P_d(d_i^{-\alpha}P_p + P_{ns}))\omega_c \tag{10}$$

The EE of the secondary network is expressed as (Chatterjee et al. 2016b)

$$\eta_{EE} = \frac{\eta_{SE}}{P_{av}} \tag{11}$$

3 Problem Formulation and Proposed Solution

The EE–SE trade-off in CRN is illustrated in this section. The objective of the present work is to obtain the minimal sensing time (τ_s), required cooperation power (P_c), and the transmission power (P_T) at SU_T to study the trade-off. The objective function is formed by maintaining the constraints of SS duration, PU cooperation rate (R_{th}), certain threshold of P_d, and a predefined power budget P_{max} for SU_T. Hence, the optimization problem is formulated as

$$\max_{\tau_s, P_c, P_T} \left(\eta_{SE} + \eta_{EE} \right)$$

while satisfying the constraints,

C1: Sensing time duration, $0 \leq \tau_s \leq T$

C2: PU detection probability threshold, $P_d = \overline{P_d}$ $\qquad$ (12)

C3: PU cooperation rate, $P(H_1) P_d \left(\frac{T-\tau_s}{2T} \right) \log_2 \left(1 + \frac{P_c |h_j|^2 |h_i|^2 P_p}{P_{np}} \right) \geq R_{th}$

C4: Power constraint, $P_c + P_T \leq P_{max}$

Considering Eqs. 8, 9, and 11, the objective function (12) is rewritten as

$$\Rightarrow \max_{\tau_s, P_c, P_T} \left\{ \left\{ P(H_0)(1 - P_f) \left(\frac{T-\tau_s}{2T} \right) \log_2 \left(1 + \frac{|h_k|^2 P_T}{P_{ns}} \right) \right\} + \frac{\eta_{SE}}{P_{av}} \right\}$$

$$\Rightarrow \max_{\tau_s, P_c, P_T} \left[P(H_0)(1 - P_f) \left(\frac{T-\tau_s}{2T} \right) \log_2 \left(1 + \frac{|h_k|^2 P_T}{P_{ns}} \right) \right.$$

$$+ \frac{P(H_0)(1 - P_f) \left(\frac{T-\tau_s}{2T} \right) \log_2 \left(1 + \frac{|h_k|^2 P_T}{P_{ns}} \right)}{\left\{ P(H_0)(1 - P_f) + P(H_1)(1 - P_d) \right\} \left(\frac{T-\tau_s}{T} \right) P_T + \left(\frac{T-\tau_s}{2T} \right) P_c} \right] \qquad (13)$$

Using Eq. 2 in the constraint $C2$, the required τ_s can be obtained as

$$\overline{P_d} = P_d \equiv Q \left\{ \left(\frac{\varepsilon}{P_{ns}} - \gamma_P - 1 \right) \frac{\sqrt{\tau_s f_s}}{\gamma_P + 1} \right\}$$

$$\Rightarrow \left\{ \left(\frac{\varepsilon}{P_{ns}} - \gamma_P - 1 \right) \frac{\sqrt{\tau_s f_s}}{\gamma_P + 1} \right\} = Q^{-1}(\overline{P_d})$$

$$\Rightarrow \frac{\sqrt{\tau_s f_s}}{\gamma_P + 1} = \left\{ Q^{-1}(\overline{P_d}) \left(\frac{P_{ns}}{\varepsilon - P_{ns}(\gamma_p + 1)} \right) \right\}$$

$$\Rightarrow \tau_s = \frac{(\gamma_P + 1)^2}{f_s} \left\{ Q^{-1}(\overline{P_d}) \left(\frac{P_{ns}}{\varepsilon - P_{ns}(\gamma_p + 1)} \right) \right\}^2 \qquad (14)$$

Simplifying the constraint $C3$, P_c value is expressed as

$$P(H_1)P_d\left(T - \tau_s\right)\log_2\left(1 + \frac{P_c|h_j|^2|h_i|^2 P_p}{P_{np}}\right) \geq 2T R_{th}$$

$$\Rightarrow \log_2\left(1 + \frac{P_c|h_j|^2|h_i|^2 P_p}{P_{np}}\right) \geq \frac{2T R_{th}}{P(H_1)P_d(T - \tau_s)}$$

$$\Rightarrow \left(\frac{P_c|h_j|^2|h_i|^2 P_p}{P_{np}}\right) \geq \left\{e^{\frac{2T R_{th}}{P(H_1)P_d(T-\tau_s)}} - 1\right\}$$

$$\Rightarrow P_c \geq \frac{P_{np}}{|h_j|^2|h_i|^2 P_p}\left\{e^{\frac{2T R_{th}}{P(H_1)P_d(T-\tau_s)}} - 1\right\} \qquad (15)$$

Due to the several constraints and nonlinear nature, the optimization problem provides an infinite number of nontrivial Pareto-optimal solutions. Hence, the conflicting inequality constraints must be evaluated simultaneously to find the global optimal solution. This problem is solved using the DE algorithm, as DE is widely accepted to solve the several nonlinear and complex optimization problems (Das and Suganthan 2011). A brief introduction about the working principle of DE algorithm is given below.

A set $\Re$ of optimization parameters (τ_s, P_c, P_T) is called an individual in DE, and it is represented by $\Re$-dimensional parameter vector. The population set consists of NP parameter vectors $(z_i^G, i = 1, 2, \ldots, NP)$, G indicates the number of generation. NP represents the maximum number of members in the population set, and it cannot be changed throughout the evolution process. The initial population (IP) is randomly selected from the search space with a uniform distribution. Generally, each decision parameter in every vector of IP is assigned with a random value from the selected boundary constraints:

$$z_{i,j}^0 = p_j + rand_j \cdot (q_j - p_j)$$

where $rand_j \sim U(0, 1)$ denotes the uniform distribution between $(0, 1)$ and it creates a new value for each of the decision parameters. The symbols p_j and q_j indicate the lower and the upper boundary values for the jth decision parameter, respectively. The large number of population explores all aspects of the search space for low-dimensional problems. However, due to the time constraints, the population size must be restricted. After several experimental analyses, it is found that for $Np^* = 50$ the optimization function obtains the optimal solution for the present problem. Typically, DE has three control operators: mutation, crossover, and selection. The main idea behind DE is to generate the trial vectors. Mutation and crossover operators are simultaneously applied to obtain the trial vectors, and finally selection operator determines the appropriate vectors for the next generation.

A. Mutation

For each target vector z_i^G, a mutant vector v_i^{G+1} is created as follows:

$$v_i^{G+1} = z_{r1}^G + F * |z_{r2}^G - z_{r3}^G|, r1 \neq r2 \neq r3 \neq i \tag{16}$$

with randomly chosen indices $r_1, r_2, r_3 \in 1, 2, \ldots, NP$. F is a real number to control the amplification of the difference vector $|z_{r2}^G - z_{r3}^G|$, where $F \in [0, 2]$. The premature convergence takes place due to the small values of F, while the very high values of F delay the overall searching process. The F-value is widely varied along with the patterns of Np^* in different application problems (Gao et al. 2014). According to the (Gao et al. 2014), the F-value is fixed at 0.5 in the proposed work while $Np^* > 10$.

B. Crossover

The target vector is now mixed with the mutated vector and the following process is used to produce the trial vector u_i^{G+1}:

$$u_{ij}^{G+1} = \begin{cases} v_{ij}^G, rand(j) \leq C_{ross}, & or, \quad j = rand\{n(i)\} \\ z_{ij}^G, rand(j) > C_{ross}, & or, \quad j \neq rand\{n(i)\} \end{cases} \tag{17}$$

where $j = 1, 2, \ldots, \Re, rand(j) \in [0, 1]$ is the jth generation of an uniform random number generator. Here, $C_{ross} \in [0, 1]$ denotes the crossover probability constant and it is selected by the user. The parameter $rand\{n(i)\} \in 1, 2, \ldots, \Re$ selects the random index, and it ensures that u_i^{G+1} will get at least one element from v_i^{G+1}; otherwise, no new parent vector would be produced for the next. The selection of an efficient C_{ross} value is totally based on the nature of the problem. It is observed that $C_{ross} \in [0.9, 1]$ is appropriate for the non-separable objective functions, while $C_{ross} \in [0, 0.2]$ is effective for the separable functions (Gao et al. 2014). In this case, C_{ross} is fixed at 0.9.

C. Selection

DE adapts a greedy selection strategy. If the trial vector u_i^{G+1} produces a better fitness function value than z_i^G, then u_i^{G+1} is set to z_i^{G+1}; otherwise, the old vector z_i^G is continued. The selection technique works as follows:

$$z_i^{G+1} = \begin{cases} u_i^{G+1}, f\left(u_i^{G+1}\right) > f(z_i^G) \\ z_i^j, f\left(u_i^{G+1}\right) \leq f(z_i^G) \end{cases} \tag{18}$$

The sequential steps of DE algorithm are given in Algorithm 1.

Input : A population of NP^* individuals $z_i^G, i = 1, 2, ..., NP$, are determined to satisfy the constraints in (12)

Output: Optimal z_i^{*G} is obtained at the final stage.

The sequential steps of DE algorithm are as follows:

---

begin

    repeat

        Step I: Mutation: Initiate a mutated vector $\vec{v}_i^G$ in respect to the target vector $\vec{z}_i^G$ for every i.

        for $i = 1$ *to* NP^* **do**

$$v_i^{G+1} = z_{r1}^G + F * |z_{r2}^G - z_{r3}^G|, r1 \neq r2 \neq r3 \neq i$$

        where r_1, r_2, and r_3 are randomly selected vectors. Scaling factor: $F = 0.5$, $F \in [0, 1]$

        end

        Step II: Crossover: Every target vector $\vec{z}_i^G$, generates a trial vector u_i^{G+1}, *iff* randomly generated number between [0, 1] $< C_{ross}$.

        for $i = 1$ *to* NP^* **do**

$$u_{ij}^{G+1} = \begin{cases} v_{ij}^G, rand(j) \leq C_{ross}, & or, & j = rand\{n(i)\} \\ z_{ij}^G, rand(j) > C_{ross}, & or, & j \neq rand\{n(i)\} \end{cases} \quad (19)$$

        end

        Step III: Selection: Estimate the trial vector.

        for $i = 1$ *to* NP^* **do**

$$z_i^{G+1} = \begin{cases} u_i^{G+1}, f\left(u_i^{G+1}\right) > f(z_i^G) \\ z_i^j, f\left(u_i^{G+1}\right) \leq f(z_i^G) \end{cases} \quad (20)$$

        end

        Increment the count C=C+1

    until *maximum iteration count is reached*;

end

Algorithm 1: DE algorithm

4 Numerical Results

This section demonstrates simulation results for the performance analysis of the proposed work. Monte Carlo simulations for 10,000 times are performed to consider the variability of the wireless channel. The numerical values of required parameters for simulation are considered as follows: $T = 100$ ms, $f_s = 10$ KHz, $N = 500$, $P(H_0) = 0.7$, $P(H_1) = 0.3$, $P_p = -10$, $P_{np} = -20$, $P_{max} = 0$ dBW, $\varepsilon = 1$ J, $d_i = 1$, $d_k = 1.25$, $d_j = 2$ m, $\alpha = 3$, $\overline{P_d} = 0.9$, $P_f = 0.15$, and $R_{th} = 0.5$ bps/Hz. The obtained EE and SE outcomes are normalized with respect to its maximum EE and SE values, respectively. Maximum EE occurs at $\tau_s = 44.54$ ms but SE value is significantly low at that time. On the other hand, the maximum SE value is obtained at $\tau_s = 1$ ms and as same time EE obtains a minimum value.

Figure 3 illustrates the EE–SE trade-off for the sensing duration τ_s. It is noted from the figure that the maximization of EE and SE cannot be obtained at the same time. It occurs due to the conflicting nature of EE and SE functions under the constraints. It is observed that EE is increased and SE is decreased with the increment in τ_s values. Equation 8 validates such outcomes of SE function. The incremental values of τ_s

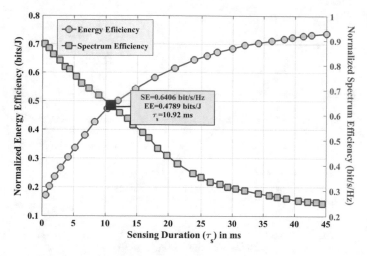

Fig. 3 Normalized EE and SE versus sensing duration

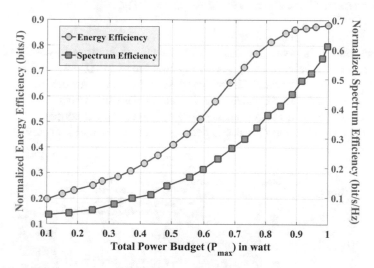

Fig. 4 Normalized EE and SE versus total power budget P_{max}

significantly reduce the value of P_{av}. Hence, EE value is increased. From Fig. 3, it is observed that the sum maximization of both EE and SE is found in $\tau_s = 10.92$ ms when EE and SE values are 0.4789 bits/J and 0.6406 bit/s/Hz, respectively.

Now the graphical results in Fig. 4 represent the variation in the maximum power budget P_{max} on its consequent effect in EE and SE performance. From Fig. 4, it is clearly observed that increase in P_{max} values increases the EE and SE values. The high value of P_{max} signifies that system can allocate more power in SU transmission and PU cooperation. This scenario significantly improves the EE and SE values at

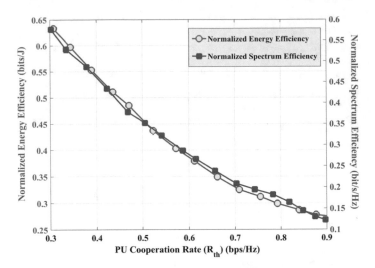

Fig. 5 Normalized EE and SE versus PU cooperation rate (R_{th})

the same time. It is noted that EE and SE values are improved by $\approx$39.14% and $\approx$46.42%, respectively, when P_{max} value is increased from 0.7 w to 0.9 w.

Figure 5 shows the optimal EE and SE of SU network against the PU cooperation rate (R_{th}). The EE and SE values are decreased with the increase in R_{th} values. It takes place due to the requirement of more power for PU cooperation. With an increment in P_c value, the P_T value is decreased. From Eq. 8, it is observed that η_{SE}

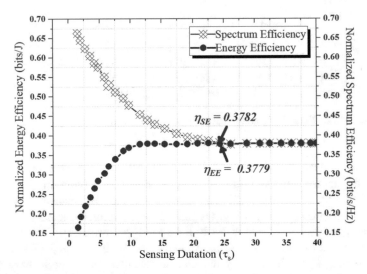

Fig. 6 Maximization of normalized η_{EE} and η_{SE} (while η_{EE} and η_{SE} values are under a predefined constraint) versus sensing duration

depends on P_T value. Therefore, reduction in SE values is justifiable in this case. Similarly, higher values of P_c increase the P_{av} values (Eq. 9), which decrease the η_{EE} values. It is noted that EE and SE both are reduced by $\approx$34.17% and $\approx$63.14%, respectively, when R_{th} value is increased from 0.4 bps/Hz to 0.6 bps/Hz.

The Pareto relationship between η_{EE} and η_{SE} is shown in Fig. 6, where η_{EE} and η_{SE} are set as $\eta_{EE} \geq 0.15$ and $\eta_{SE} \geq 0.375$. It is noted that with the initial increment in τ_s, η_{SE} decreases while η_{EE} increases. However, with the further increment in τ_s, it is observed that no further increment in η_{EE} and decrement in η_{SE} are possible. Moreover, the characteristics of both EE and SE remain unchanged along with τ_s after a certain values of $\eta_{SE} = 0.3782$ and $\eta_{EE} = 0.3779$. The constraint $\eta_{SE} \geq 0.375$ does not allow any further decrement in η_{SE}, and this also restricts the further increment in η_{EE} value. The value of η_{EE} does not change since both η_{SE} and P_{av} in Eq. 11 are constant.

5 Conclusions

The proposed work makes a study on the EE and SE trade-off in cooperative cognitive radio network. The framework is developed to maximize the EE and SE under the constraints of sensing duration, PU detection threshold, PU cooperation rate, and SU power budget. The proposed nonlinear objective function along with the constraints is evaluated through the DE algorithm to obtain the global optimal value. EE–SE trade-off is obtained where EE and SE values are 0.4789 bits/J and 0.6406 bit/s/Hz, respectively. It is observed that EE and SE both are increased by $\approx$39.14% and $\approx$46.42%, respectively, when P_{max} value is increased from 0.7 w to 0.9 w. It is also found that increase in R_{th} value reduces the EE and SE by $\approx$34.17% and $\approx$ 63.14%, respectively, and it is justified. The following point may be a limitation for the present work. Since the present system model involves single user sensing, the high-value target detection (for example, $\overline{P_d} \approx 0.95$ and $\overline{P_f} \approx 0.05$) may not be met at deep fading channel or sensing duration to be large enough failing to meet high cooperation SE (for PU network) and high value of SE and EE maximization for the secondary network.

However, the proposed system model may be extended as follows:

- The objective function EE–SE maximization can be modeled as (optimal) weighted combination or a parametric form where the optimal weight or parameter can also be calculated.
- The present system model can be extended in multiple SU systems to find its optimal number to meet the detection reliability in the presence of malicious operation that is SS.

References

I.F. Akyildiz, B.F. Lo, R. Balakrishnan, Cooperative spectrum sensing in cognitive radio networks: a survey. Phys. Commun. **4**(1), 40–62 (2011)

F.A. Awin, E. Abdel-Raheem, M. Ahmadi, Designing an optimal energy efficient cluster-based spectrum sensing for cognitive radio networks. IEEE Commun. Lett. **20**(9), 1884–1887 (2016)

A.O. Bicen, E.B. Pehlivanoglu, S. Galmes, O.B. Akan, Dedicated radio utilization for spectrum handoff and efficiency in cognitive radio networks. IEEE Trans. Wirel. Commun. **14**(9), 5251–5259 (2015)

A. Celik, A.E. Kamal, Multi-objective clustering optimization for multi-channel cooperative spectrum sensing in heterogeneous green crns. IEEE Trans. Cogn. Commun. Netw. **2**(2), 150–161 (2016)

S. Chatterjee, S.P. Maity, T. Acharya, On optimal sensing time and power allocation for energy efficient cooperative cognitive radio networks, in *Proceedings of the IEEE International Conference on Advanced Networks and Telecommuncations Systems (ANTS)*, Dec 2015, pp. 1–6

S. Chatterjee, S.P. Maity, T. Acharya, Energy efficiency in cooperative cognitive radio network in the presence of malicious users. IEEE Syst. J. **PP**(99), 1–10 (2016a)

S. Chatterjee, S.P. Maity, T. Acharya, Trade-off on spectrum-energy efficiency in cooperative cognitive radio networks, in *Proceedings of the International Conference on Signal Processing and Communications (SPCOM)*, June 2016b, 1–5

K.W. Choi, E. Hossain, Estimation of primary user parameters in cognitive radio systems via hidden markov model. IEEE Trans. Signal Process. **61**(3), 782–795 (2013)

Cicho, K., Kliks, A., Bogucka, H.: Energy-efficient cooperative spectrum sensing: a survey. IEEE Commun. Surv. Tutor. **18**(3), 1861–1886 (thirdquarter 2016)

C. Clancy, J. Hecker, E. Stuntebeck, T. O'Shea, Applications of machine learning to cognitive radio networks. IEEE Wirel. Commun. **14**(4), 47–52 (2007)

S. Das, P.N. Suganthan, Differential evolution: a survey of the state-of-the-art. IEEE Trans. Evol. Comput. **15**(1), 4–31 (2011)

L. Duan, A.W. Min, J. Huang, K.G. Shin, Attack prevention for collaborative spectrum sensing in cognitive radio networks. IEEE J. Sel. Areas Commun. **30**(9), 1658–1665 (2012)

Federal Communications Commission, Spectrum policy task force, Report ET Docket no. 02–135, 2002

B.A. Fette, *Cognitive radio technology* (Academic Press, 2009)

W. Gao, G.G. Yen, S. Liu, A cluster-based differential evolution with self-adaptive strategy for multimodal optimization. IEEE Trans. Cybern. **44**(8), 1314–1327 (2014)

M.R. Hassan, G. Karmakar, J. Kamruzzaman, B. Srinivasan, Exclusive use spectrum access trading models in cognitive radio networks: A survey, in *IEEE Communications Surveys Tutorials*, vol. PP(99), 2017, pp. 1–1

S.H. Hojjati, A. Ebrahimzadeh, M. Najimi, A. Reihanian, Sensor selection for cooperative spectrum sensing in multiantenna sensor networks based on convex optimization and genetic algorithm. IEEE Sens. J. **16**(10), 3486–3487 (2016)

H. Hu, H. Zhang, Y.C. Liang, On the spectrum- and energy-efficiency tradeoff in cognitive radio networks. IEEE Trans. Commun. **64**(2), 490–501 (2016)

R. Huang, J. Chang, Y. Ren, F. He, C. Guan, Spectrum allocation of cognitive radio network based on optimized genetic algorithm in underlay network, in *Proceedings of the IEEE International Conference on Communication Software and Networks (ICCSN)*, June 2016, pp. 418–422

C. Jiang, H. Zhang, Y. Ren, Z. Han, K.C. Chen, L. Hanzo, Machine learning paradigms for next-generation wireless networks. IEEE Wirel. Commun. **24**(2), 98–105 (2017)

Y. Jiao, I. Joe, Energy-efficient resource allocation for heterogeneous cognitive radio network based on two-tier crossover genetic algorithm. J. Commun. Netw. **18**(1), 112–122 (2016)

A. Kaur, S. Sharma, A. Mishra, Sensing period adaptation for multiobjective optimisation in cognitive radio using jaya algorithm. Electron. Lett. **53**(19), 1335–1336 (2017)

H.S. Lang, S.C. Lin, W.H. Fang, Subcarrier pairing and power allocation with interference management in cognitive relay networks based on genetic algorithms. IEEE Trans. Veh. Technol. **65**(9), 7051–7063 (2016)

S.P. Maity, S. Chatterjee, T. Acharya, On optimal fuzzy c-means clustering for energy efficient cooperative spectrum sensing in cognitive radio networks. Digit. Signal Process. **49**(C), 104–115 (Feb 2016)

Y.E. Morabit, F. Mrabti, E.H. Abarkan, Spectrum allocation using genetic algorithm in cognitive radio networks, in *Proceedings of the Third International Workshop on RFID and Adaptive Wireless Sensor Networks (RAWSN)*, May 2015, pp. 90–93

P.C. Ng, Optimization of spectrum sensing for cognitive sensor network using differential evolution approach in smart environment, in *Proceedigns of the IEEE International Conference on Networking, Sensing and Control*, Apr 2015, pp. 592–596

OFCOM, Digital Dividend Review, A statement on our approach towards awarding the digital dividend, 2007

A. Ostovar, Z. Chang, Optimisation of cooperative spectrum sensing via optimal power allocation in cognitive radio networks. IET Commun. **11**(13), 2116–2124 (2017)

A. Paul, S.P. Maity, Kernel fuzzy c-means clustering on energy detection based cooperative spectrum sensing. Digit. Commun. Netw. **2**(4), 196–205 (2016)

R.A. Rashid, A.H.F.B.A. Hamid, N. Fisal, S.K.Syed-Yusof, H. Hosseini, A. Lo, A. Farzamnia, Efficient in-band spectrum sensing using swarm intelligence for cognitive radio network. Can. J. Electr. Comput. Eng. **38**(2), 106–115 (Spring 2015)

J. Ren, Y. Zhang, Q. Ye, K. Yang, K. Zhang, X.S. Shen, Exploiting secure and energy-efficient collaborative spectrum sensing for cognitive radio sensor networks. IEEE Trans. Wirel. Commun. **15**(10), 6813–6827 (2016)

S. Sedighi, A. Taherpour, S.S.M. Monfared, Bayesian generalised likelihood ratio test-based multiple antenna spectrum sensing for cognitive radios. IET Commun. **7**(18), 2151–2165 (2013)

K. Wen, L. Fu, X. Li, *Genetic Algorithm Based Spectrum Allocation for Cognitive Radio Networks* (Springer, Berlin, Heidelberg, 2012), pp. 693–700

M. Yang, Y. Li, X. Liu, W. Tang, Cyclostationary feature detection based spectrum sensing algorithm under complicated electromagnetic environment in cognitive radio networks. China Commun. **12**(9), 35–44 (2015)

G. Yang, J. Wang, J. Luo, O.Y. Wen, H. Li, Q. Li, S. Li, Cooperative spectrum sensing in heterogeneous cognitive radio networks based on normalized energy detection. IEEE Trans. Veh. Technol. **65**(3), 1452–1463 (2016)

X. Yang, K. Lei, L. Hu, X. Cao, X. Huang, Eigenvalue ratio based blind spectrum sensing algorithm for multiband cognitive radios with relatively small samples. Electron. Lett. **53**(16), 1150–1152 (2017)

F. Ye, J. Liu, W.Y. Lv, H. Gu, Resource allocation based differential evolution algorithm for cognitive radio system, in *Proceedings of the IEEE International Conference on Measurement, Information and Control*, vol. 02, Aug 2013, pp. 1460–1463

T. Yucek, H. Arslan, A survey of spectrum sensing algorithms for cognitive radio applications. IEEE Commun. Surv. Tutor. **11**(1), 116–130 (First 2009)

S. Zhang, X. Zhao, Power allocation for sensing-based spectrum sharing cognitive radio system with primary quantized side information. China Commun. **13**(9), 33–43 (2016)

X. Zhang, F. Gao, R. Chai, T. Jiang, Matched filter based spectrum sensing when primary user has multiple power levels. China Commun. **12**(2), 21–31 (2015)

M. Zhang, M. Diao, L. Guo, Convolutional neural networks for automatic cognitive radio waveform recognition. IEEE Access **5**, 11074–11082 (2017)

Multi-Objective Optimization Approaches in Biological Learning System on Microarray Data

Saurav Mallik, Tapas Bhadra, Soumita Seth, Sanghamitra Bandyopadhyay
and Jianjiao Chen

1 Introduction

In the data mining field, microarray data analysis is one of the most challenging tasks due to the curse of dimensionality problem. Microarray (Bandyopadhyay et al. 2014) is a technique in which a supporting material (made of either a glass slide or bead) is utilized onto which thousands of molecules or their fragments like fragments of DNA or proteins are conjoined in a regular manner for the genetic study or biochemical experiment. The main goal of microarray data analysis consists of discovering genes that share the same type of expression patterns across a subset of conditions. Notably, the extracted information is sub-matrices of the microarray data that satisfy an integrity constraint. Cancer investigation in microarray data plays a major role in cancer analysis and its treatment (Bandyopadhyay et al. 2014, 2016; Mallik et al. 2016; Maulik et al. 2015). The complex gene expression patterns of cancer form cancer microarray data. DNA microarray technologies help to observe the functionalities of the thousands of genes together which can be applied as one of the most robust tools to realize the regulatory schemes of gene expression in cells (Spieth et al. 2005). These techniques also assist researchers to study cells under various conditions, for example, different environmental influences, medical treatment, etc. Microarray experiments generally produce time series of measured

S. Mallik (✉) · S. Bandyopadhyay
Indian Statistical Institute (ISI), Kolkata 700108, West Bengal, India
e-mail: sauravmtech2@gmail.com; saurav.mallik@med.miami.edu

S. Bandyopadhyay
e-mail: sanghami@isical.ac.in

S. Mallik · J. Chen
University of Miami, Miami 33136, FL, USA
e-mail: joycjj83@gmail.com

T. Bhadra · S. Seth
Aliah University, Kolkata 700156, West Bengal, India
e-mail: tapas.bhadra@gmail.com

S. Seth
e-mail: soumita.seth@gmail.com

© Springer Nature Singapore Pte Ltd. 2018
J. K. Mandal et al. (eds.), *Multi-Objective Optimization*,
https://doi.org/10.1007/978-981-13-1471-1_7

values to specify the activation level of every experienced gene in a genome. To analyze DNA microarray data, first it is required to produce new biological assumptions by a preprocessing phase including checking data distribution, normalization, and gene filtering, discretization. More than one objective function is set to obtain optimal solutions in multi-objective optimization. In such optimization technique, basically single-objective function with a bargain among different objectives is optimized. In mathematics or computer science, an optimization problem is nothing but finding the best solution from all feasible solutions. Thus, to get this optimal solution, we need to do maximization or minimization of each objective function. However, in those clustering algorithms, a single-objective function is not much efficient for the categorical data. To overcome this limitation, we utilize multiple objective functions (Mukhopadhyay et al. 2007) that should be optimized simultaneously. The concept of Pareto-optimal (Mukhopadhyay et al. 2007) has a good effect on multi-objective optimization. In the viewpoint of the minimization problem, a decision vector might be defined as Pareto-optimal, if and only if there exists no such vector which dominates it or no such feasible solution that might create reduction on the basis of any benchmark without a concurrent increase of at least one. In addition, multi-objective optimization is carried out by particle swarm optimization (Mandal and Mukhopadhyay 2014). In this chapter, we discuss several multi-objective optimization methods that are applied in general on microarray data. Some of them are based on integrated learning which may use in cancer classification, gene regularity network, a differential evolution that is basically multi-objective evolution based. Some of them are based on a multi-objective genetic algorithm which has been used for HIV detection (Aqil et al. 2014, 2015), gene selection (Mallik and Maulik 2015; Mandal and Mukhopadhyay 2014), etc. A popular genetic algorithm for multi-objective optimization is NSGA-II (non-dominated sorting genetic Algorithm-II) (Deb et al. 2002). Furthermore, some approaches are based on multi-objective particle swarm optimization (MPSO) which is a heuristic, multi-agent, optimization, and evolutionary technique. The prediction of miRNA transcription start sites is an interesting topic of researches for knowing about primary miRNAs (Mallik and Maulik 2015). Recently, Bhattacharyya et al. (2012) developed a miRNA TSS prediction model by integrating a support vector machine with an archived multi-objective simulated annealing-based feature reduction technique (Bhattacharyya et al. 2012; Sen et al. 2017). Moreover, fundamental comparative study and further discussion are included for enlightenment of advantages and limitations of the described methods.

2 Fundamental Terms and Preliminaries

In biological perspective, a gene (Pearson 2006) is stated as the primary physical as well as functional unit of heredity in an organism. Deoxyribonucleic acid (DNA) exists in the chromosomes of all the organisms in each cell. It encodes genetic information in terms of functioning and development of organisms including human. Each gene is associated with a particular function or protein through a specific set of instructions (coding). According to molecular biology (Claverie 2005), in a biological system, the flow of transfer of genetic sequence information (like DNA replication, transcription, and translation) takes place between sequential information-carrying biopolymers (viz., DNA, RNA, and protein). DNA replication is a process where two similar copies of DNA are created by an existing DNA. DNA transcription transfers DNA information to RNA (ribonucleic acid), whereas translation creates protein from RNA. MicroRNAs (miRNAs) are noncoding RNAs tiny in size, i.e., length approximately 22 nucleotides, but it plays a significant role in the transcriptional and post-transcriptional regulation of the gene expression. By surveying existing literature, we have found several strong evidence where miRNAs have a great impact on occurring diseases such as cancers, diabetes, AIDS, etc. Irregular expression of multiple genes/miRNAs causes disease condition in the human body. Such deregulation is generally affected by many genetic and epigenetic factors.

2.1 Microarray

Microarray (Bandyopadhyay et al. 2014) is a technique in which a supporting material (made of either a glass slide or bead) is used onto which numerous molecules or their fragments such as DNA fragments or proteins are attached in a fundamental pattern for biochemical experiment or genetic analysis. In order to measure the activity levels of the thousands of biochips (i.e., genes/miRNAs) simultaneously over different experimental conditions, DNA microarray is a useful technology. Microarray data are highly useful in the experiments through clustering (Maulik et al. 2011), biclustering (Maulik et al. 2015), multi-class clustering (Maulik et al. 2010), classification (Wu 2006), differential gene selection (Joseph et al. 2012), single-nucleotide polymorphism (SNP) detection (Hacia et al. 1999), cancer subtypes selection (Maulik et al. 2010), etc. Epigenetics (Mallik et al. 2017) is the study of rapid changes or phenotypes without alteration in DNA sequence. DNA methylation (Mallik et al. 2017, 2013a, b) is an important epigenetic factor which generally reduces gene expression. DNA methylation is nothing but an inclusion of a methyl group (i.e., "–CH3") to the position of number five of the cytosine pyrimidine ring in genomic DNA.

2.2 Statistical Tests

Interestingly, statistical analysis plays a significant role to identify differentially expressed or differentially methylated genes (Mallik et al. 2013b) in the diseased samples as compared to the normal samples. There are various types of statistical tests (Bandyopadhyay et al. 2014) which can be grouped into two categories, i.e., parametric test, and nonparametric test. The parametric test is used broadly which might have two subcategories: equal variance assumption and unequal variance assumption. Some well-known statistical tests under equal variance assumption are t-test, one-way ANOVA (Anova1) test, and Pearson's correlation test. Under the unequal variance assumption, the popular statistical test is Welch's t-test. Besides, some well-known statistical tests under the nonparametric test are significant analysis of microarrays (SAM), linear models for microarray and RNA-seq data (LIMMA), Wilcoxon rank sum test, etc. In the parametric test, data are assumed to be existed in the form of the normal distribution, whereas the nonparametric test basically performs well for non-normal distributed data.

T-test

T-test (Bandyopadhyay et al. 2014) a fundamental statistical test. The "two sample t-test" compares the difference between the means between the two groups in relation to the variation in the data. In the t-test statistics, p-value is computed from t-table or cumulative distribution function (CDF) (Bandyopadhyay et al. 2014). This p-value indicates the probability of observing a t-value which is either equal to or greater than the actually observed t-value in which the given null hypothesis is true.

Anova 1 Test

Anova 1 (Analysis of Variance 1) (Bandyopadhyay et al. 2014) is a statistical test where the mean in between two or more populations or groups can be tested. By estimating comparisons of variance estimates, we can make comparison among the group means. Variance is partitioned into two components, one is caused by a random error (i.e., square sum within groups) and another is caused by the differences between the means of groups. After that, these variance components are tested for the statistical significance. If significance is true, we reject the null hypothesis of no difference between the means in between the populations and subsequently we accept the alternative hypothesis concerning about the significant difference in between the two populations.

Pearson's Correlation Test

Pearson's Correlation Test (Bandyopadhyay et al. 2014) applied on the data are from bivariate normal population since correlation is a degree to which two variables are covaried in either positive or negative. In order to measure the actual strength of a relationship between two variables, correlation coefficient is used. In linear regression, we use Pearson's correlation coefficient to estimate the intensity of linear association between variables. However, it may be possible to have the nonlinear association. Hence, we need to test the data closely to determine whether there is any linear association or not.

Wilcoxon Rank Sum Test (RST)

In case of small sample size where the groups are not normally distributed, t-test might not provide the valid result. In such case, we use an alternative test which is Wilcoxon rank sum test (RST) (Bandyopadhyay et al. 2014). In this test first, the ranks of the combined samples are computed. Thereafter, the summation of ranks for each different group is computed. RST is almost equivalent to Mann–Whitney test. Of note, RST is much slower than the Student's t-test.

Significance Analysis of Microarray (SAM)

Significance analysis of microarrays (SAM) (Bandyopadhyay et al. 2014) is somewhat same as the t-test. However, it can overcome the limitation of t-test. The only difference is that SAM uses a fudge factor which is added to the standard error in the denominator, whereas t-test does not use such fudge factor. This fudge factor is used to solve the low variance problem. In spite of that, the main problem of using SAM is that its performance is not always consistent.

Linear Models for Microarray Data (LIMMA)

Microarray data normally consists of few numbers of arrays/samples. So when a large number of genes are present in the data, a dimensionality problem may arise. Linear Models for Microarray Data (LIMMA) method (Bandyopadhyay et al. 2014) may overcome this problem. To handle such limitation, an empirical Bayes approach is considered in LIMMA test. This method is one of the best efficient and consistent tests for microarray data as well as RNA-Seq data where a prior knowledge regarding the unknown gene (or, miRNA) specified variances to be the inverse-gamma distribution is assumed.

2.3   Epigenetic Biomarker

In medical science, some medical conditions that are observed in a patient externally and which differ from medical symptoms felt by patients themselves only, are termed as biomarker (Mallik et al. 2017). It is a special subtype of a medical sign. Epigenetic biomarkers discovery is also a popular ongoing research domain.

2.4   Multi-Objective Optimization

However, we generally face some challenges with different benchmark clustering algorithms like k-means, k-medoids, partition around medoids (PAM) to cope up with different types of categorical data. In general, such clustering algorithms are based on single-objective function. In mathematics or computer science, an optimization problem is nothing but finding the best solution from all feasible solutions. Thus, to get this optimal solution, we need to do maximization or minimization of each objective function. However, in those clustering algorithms, the single-objective function is not so efficient for the categorical data. To overcome this limitation, we utilize multiple objective functions (Mukhopadhyay et al. 2007) that should be optimized simultaneously. In a multi-objective scenario, conflicting objective functions are often considered as the underlying objective functions. The single-objective function produces us the best solution whereas multi-objective optimization (MOO) produces a set of Pareto-optimal solutions as a final solution. In formal, multi-objective optimization can be defined as a set of decision parameters which will satisfy some inequality constraints and some equality constraints, and finally optimize the objective functions.

2.5 Pareto-Optimal

The concept of Pareto-optimal (Mukhopadhyay et al. 2007) has a good effect on multi-objective optimization. From the viewpoint of the minimization problem, a decision vector might be defined as Pareto-optimal if and only if there exists no such vector which dominates it or there exists no such feasible solution which might produce a reduction based on any criterion without the concurrent increase in at least one.

3 Method Hierarchy

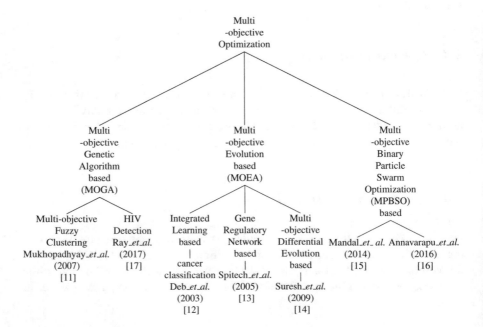

4   Description of Methods

Multi-objective optimization and machine learning techniques are currently utilized on integrated manner for solving several research objectives in the domain of bioinformatics. In this chapter, we elaborate several such algorithms that employ with the aforementioned problem.

4.1   Integrated Learning Approach to Classify Multi-class Cancer Data

Deb and Reddy (2003) proposed a multi-objective optimization-based learning approach for classification of binary and multi-class (Bhadra et al. 2017) microarray data (Deb and Reddy 2003). In the bioinformatic field, finding gene subsets play a responsible role to classify available samples into two or more classes (malignant and/or, benign) through utilizing this model. The availability of a very few disease samples compared to the number of genes and a massive search space of the solutions stand researchers in front of a big challenge. Different combinations of genes may obtain almost identical classification accuracy. The main challenge of researchers is to find a small-sized reliable gene classifier with a high classification accuracy rate. In this multi-objective minimization problem, the authors minimize the classifier size as well as the number of misclassified samples in both training and test datasets. To tackle binary and multi-class classification problems, they used the standard weighted voting method for building a unified model. Their proposed multi-objective evolutionary algorithm can uniquely identify several robust classifiers. It can also classify the smaller sized testing dataset with 100% accuracy rate. The experimental results demonstrate that its accuracy rate is comparatively better than the earlier reported values. Furthermore, the algorithm is found to be flexible and highly effective in nature.

4.2   Multi-Objective Optimization Method on Gene Regularity Networks

Spieth et al. (2005) have combinedly applied a multi-objective optimization model and microarray data technology (Liu et al. 2011) on gene regulatory networks. In systems biology, the study on gene regularity networks is one of the very interesting topics of research. Here, the authors focused on the task of obtaining gene regulatory networks from empirical DNA microarray data. In this regard, authors used reverse engineering sets of the time-series data profile acquired through the artificial expression analysis to obtain the parameters of a nonlinear biological systems from the viewpoint of a multi-objective optimization problem. The first objective, also known

as relative standard error (RSE), is the discrepancy between the simulated and experimental data. On the other hand, the second objective measures the affinity of the
system. Here, the goal is to minimize both objectives to develop a system, fitting the
data as best as possible, and making it connected sparsely, thereby biological believ-
able. This approach can analyze the impact of the affinity of the regulatory network
on the overall interpretation process. Here, authors have developed a mathematical
model which simulates the gene regularity system in the interpretation process with
a multi-objective evolutionary algorithm (MOEA). From the abstract point of view,
here authors have obtained the nature of a cell through a gene regulatory network
that contains N genes where every gene g_i crops a convinced extend of RNA r_i
in expression and then alters the absorption of this RNA level with the changes of
time : $r(t + 1) = h(r(t)); r(t) = (r_1, \ldots, r_n)$. Here, authors use fine documented and
tested S-Sytems to model and to replicate regulatory networks. S-Systems follow the
power-law form which can be depicted through a set of the nonlinear differential
equations:

$$\frac{dr_i(t)}{dt} = \alpha_i \prod_{k=1}^{N} r_k(t)^{G_{i,k}} - \beta_i \prod_{k=1}^{N} r_k(t)_{i,k}^{H} \tag{1}$$

where $G_{i,k}$ and $H_{i,k}$ denote kinetic exponents, α_i and β_i define positive rate constants,
and N is the total number of equations in this system. They split the equations in (1)
into two integrals: an excitatory and an inhibitory component. The kinetic exponents
$G_{i,k}$ and $H_{i,k}$ regulate the framework of the regulatory network. If $G_{i,k} > 0$, gene
g_k starts the fusion of gene g_i. If $G_{i,k} < 0$ gene g_k stops the fusion of gene g_i.
Similarly, a positive value of $H_{i,k}$ reveals that gene g_k generates the deterioration
of the mRNA level of gene g_i whereas a negative value of $H_{i,k}$ proves that gene g_k
conceals the deterioration of the mRNA level of gene g_i. The authors simultaneously
test the affinity and the RSE by using a multi-objective evolutionary algorithm, that
optimizes the parameters denoted by G, H, α_i and β_i. The first objective function to
estimate the RSE fitness of the individuals is equated by

$$f_{obj1} = \sum_{i=1}^{N} \sum_{p=1}^{S} \left\{ \left(\frac{\hat{r}_i(t_p) - r_i(t_p)}{r_i(t_p)} \right)^2 \right\} \tag{2}$$

where N be the number of genes in a system, S be the number of samples observed
in the empirical time-series data, and $\hat{x}$ and x differentiate the estimated data of the
replicated model and sampled data in the experiment. The aim is to minimize the
fitness value f_{obj1}. However, the second optimization objective refers to the minimization of the closeness of the system, since in biology, the gene regulatory network is
treated as a sparse network. The affinity can be presented in two distinct manners:
The first one computes the maximal affinity of the genes which is the total number
of synergies that the system contains. It is formulated as follows:

$$f_{obj2}^1 = \sum_{i=1}^{N}(|sign(\alpha_i)| + |sign(\beta_i)|) + \sum_{i=1}^{N}\sum_{k=1}^{N}(|sign(G_{i,k})| + |sign(H_{i,k})|) \quad (3)$$

And the second one is defined as follows, the median moderate affinity of all genes:

$$f_{obj2}^2 = median\left(\frac{f_{obj2}^1}{N}\right) \quad (4)$$

After that, author has compared their proposed algorithm with various multi-objective techniques to infer gene regulatory networks from the time-series microarray data (Bandyopadhyay et al. 2014). The result shows that the number of interaction between different integrals has a great impact. Here, the increasing number of communications makes the algorithm more difficult since the correct affinity is not previously known by any researcher. However, the authors claim that their result is in a set of the Pareto-optimal solutions from which the researcher can choose the most suitable one.

4.3 Multi-Objective Genetic Algorithm in Fuzzy Clustering of Categorical Attributes

Mukhopadhyay et al. (2010) proposed multi-objective genetic algorithm-based fuzzy clustering method (Mukhopadhyay et al. 2007) designed solely for categorical attributes (Mukhopadhyay et al. 2007). In general, most of the clustering algorithms such as partitioning around medoids (PAM), K-medoids, K-modes, and fuzzy K-modes intend to optimize a single measure or objective function. Genetic algorithms are also often used to design clustering algorithms for many data clustering problems. Most of them generally optimize a single-objective function. However, optimizing a single measure may not be appropriate for different types of categorical datasets. To overcome this limitation, the authors used multiple, often conflicting, objective functions in their proposed multi-objective-based fuzzy clustering approach. Here, authors apply the concept of NSGA-II (non-dominated sorting genetic algorithm-II), the popular elitist multi-objective genetic algorithm (MOGA), for an evolving set of near-Pareto-optimal nondegenerate fuzzy partition matrices. Besides this, the distance measure in between the two feature vectors is selected in this article. The fitness objective vector is made of global compactness of the clusters as well as fuzzy separation which they optimized simultaneously. In this article, authors have adopted their method by following steps:
Step 1: Developing a distance measure.
Step 2: Chromosome encoding and population initialization.
Step 3: Computation of fitness functions.
Step 4: Applying genetic operators conventional uniform crossover and mutation.
Step 5: Obtaining final solution.

Here, authors calculate the discrepancy between two categorical objects which have more than one categorical attributes as a distance measure. Let, $a_k = [a_{k1}, a_{k2}, \ldots, a_{kn}]$ and $a_l = [a_{l1}, a_{l2}, \ldots, a_{ln}]$ be the two categorical objects denoted by n categorical attributes. The distance measure between a_k and a_l is defined by $Dist(a_k, a_l)$ which measures the total dissimilarity of the respective attribute categories of the two categorical objects. Notably,

$$Dist(a_k, a_l) = \sum_{i=1}^{n} \delta(a_{ki}, a_{li}) \tag{5}$$

where

$$\delta(a_{ki}, a_{li}) = \begin{cases} 0 \text{ if } a_{ki} = a_{li} \\ 1 \text{ if } a_{ki} \neq a_{li} \end{cases} \tag{6}$$

Here, authors define chromosome as a chain of characteristic values depicting the K cluster modes. Since the number of characteristics of each categorical object is m $X_1, X_2, \ldots, X_m$, the size of a chromosome is measured by $m \times K$ through the assumption that the first m genes express the m-dimensions of the first cluster mode, and the next m positions depict the same for the second cluster mode, and so on. The initial population of chromosomes is formed by random data points. After defining the initial population, next step is the computation of fitness function. The fitness vector is made of the two fitness functions: i.e., the global compactness π of the clusters, and the fuzzy separation Sp. Here, authors optimize two objective functions simultaneously. To compute the measure first need to extract mode from an encoded chromosome. Let the modes are denoted by $b_1, b_2, \ldots, b_K$. The membership values u_{pr}, p = 1, 2, ..., K and r = 1, 2, ..., m are computed as follows:

$$u_{pr} = \frac{1}{\sum_{q=1}^{K} \left(\frac{Dist(b_p, a_r)}{Dist(b_q, a_r)}\right)^{\frac{1}{w-1}}}, \quad \text{for } 1 \leq p \leq K; 1 \leq r \leq m \tag{7}$$

$Dist(b_p, a_r)$ and $Dist(b_q, a_r)$ are priory defined. w is weighting coefficient. If $Dist(b_q, a_r)$ is equal to zero for some q, $u_{pr} = 0$ for all p = 1, 2, ..., K, $p \neq q$, Otherwise $u_{pr} = 1$. Afterwards, each mode encoded in a chromosome is modified by $b_p = [b_{p1}, b_{p2}, \ldots, b_{pn}]$. The variance σ_p and fuzzy cardinality n_p for the pth cluster are defined by

$$\sigma_p = \sum_{r=1}^{m} u_{pr}^w Dist(b_p, a_r), \quad 1 \leq p \leq K \tag{8}$$

and

$$n_p = \sum_{r=1}^{m} u_{pr} \quad 1 \leq p \leq K \tag{9}$$

Now the global compactness π of the solution is depicted by the chromosome is formulated as

$$\pi = \sum_{p=1}^{K} \frac{\sigma_p}{n_p} = \sum_{p=1}^{K} \frac{\sum_{r=1}^{m} u_{pr}^{w} Dist(b_p, a_r)}{\sum_{r=1}^{m} u_{pr}} \qquad (10)$$

Another fitness function fuzzy separation Sp, the mode b_p is assumed as the center of a fuzzy set of the pth cluster, $\{b_q | 1 \le q \le K, q \ne p\}$. The membership score of each b_q to b_p, $q \ne p$, is formulated as

$$\mu_{pq} = \frac{1}{\sum_{t=1, t\ne q}^{K} \left(\frac{Dist(b_q, b_p)}{Dist(b_q, b_t)}\right)^{\frac{1}{w-1}}}, \qquad p \ne q \qquad (11)$$

The fuzzy separation is referred as

$$Sp = \sum_{p=1}^{K} \sum_{q=1, q\ne p}^{K} \mu_{pq}^{w} Dist(b_p, b_q) \qquad (12)$$

Here, the objective functions are chosen for their contradictory nature because minimizing π increase the compactness of clusters as well as maximizing Sp increase the intercluster separation. Need to balance these two objective function critically to get good solutions. After that, authors obtain the new offspring solutions from the matching set of chromosomes by using the first genetic operator, i.e., conventional uniform crossover with a random mask. Mutation is another genetic operator is applied on a chromosome for selecting it to be mutated. However, the gene position of that chromosome is selected randomly. After that, another random value is selected from respective categorical set of attributes, and the categorical value of a particular gene position is substituted by it. Here, the non-dominated solutions of parent and child populations are reproduced in next generation. The different solutions are provided to the clustering problem by the near-Pareto-optimal strings of the last generation. In case of each non-dominated solution, the clustering label vector is identified from the solution first through assigning every point to the cluster by according to their memberships from highest to lowest. After that, need to reorder the label vectors in a fashion of one after another. Subsequently, the clustering points reveal at least 50% accuracy of solution. Authors use those points as the training set and the remaining points are taken as a test class. Different methods like K-nearest neighbor (K-NN) are applied for classification. Using this process, authors obtain final solution. In this article, authors conclude that their proposed method is significantly superior to other well-known techniques on basis of their developed statistical significance test. This method is also efficiently applicable on various synthetic and real-life categorical datasets.

4.4 Multi-Objective Differential Evolution for Automatic Clustering of Microarray Datasets

Suresh et al. (2009) adopted differential evolution (DE) (Bhadra et al. 2012) to design a multi-objective optimization-based automatic clustering method (Suresh et al. 2009). For this purpose, the authors used two conflicting objective functions : the Xie-Beni index (XiB_r) and a penalized version of fuzzy C means (FCM) function (J_r). The main steps of the multi-objective clustering scheme are mentioned below.

Step 1: Search for the variable representation and plan to obtain the accurate number of clusters: The authors considered n data points where each of them is d-dimensional. A maximum number of clusters, denoted by K_{max}, is defined as $K_{max} + K_{max} * d$. Here, the authors specified a threshold value $Th_{i,j}$ where i represents the vector in DE population and j represents a specific cluster. When $Th_{i,j} > 0.5$ the j-th cluster corresponding to the i-th vector is active; otherwise, it is inactive.

Step 2: (a) Choosing the objective functions: In general, the conflicting type objective functions are chosen while designing the multi-objective clustering algorithm as it directs the path of obtaining globally optimal solutions. The Xie-Beni index XiB_r and a penalized version of the FCM function J_r have chosen as the two objective functions. The FCM measure is referred as

$$J_r = (1+k) \sum_{j=1}^{n} \sum_{i=1}^{k} u_{ij}^r . dist^2(\vec{V_j}, \vec{c_i}), \qquad 1 \leq q \leq \infty \qquad (13)$$

$$\vec{V_j} = \{V_1, V_2, \ldots, V_d\}$$

which is a real-coded search vector of d optimization parameters

$$\vec{c_i} = \{c_1, c_2, \ldots, c_k\}$$

which represents the centers encoded in DE vectors.

Here, r denotes the fuzzy exponent, $dist$ specifies a distance measure between i-th cluster centroid and the j-th pattern vector, k denotes the number of active cluster centroids, and u_{ij} stands for the membership value of j-th pattern in the i-th cluster. The ratio of the total variation σ to the minimum separation Sp of the clusters is revealed by a function which is called XiB index, where σ and Sp was formulated as

$$\sigma = \sum_{t=1}^{n} \sum_{i=1}^{k} u_{it}^2 . dist(\vec{c_i}, \vec{V_t}) \qquad (14)$$

and

$$Sp(V) = \min_{i \neq j} \{dist^2(\vec{c_i}, \vec{c_j})\} \qquad (15)$$

The Xie-Beni Index was then formulated as

$$XiB_r = \frac{\sigma}{n \times Sp(V)} = \frac{\sum_{t=1}^{n} \sum_{i=1}^{k} u_{it}^2 . dist(\vec{c_i}, \vec{V_t})}{n \times \min_{i \neq j}\{dist^2(\vec{V_i}, \vec{V_j})\}} \tag{16}$$

It is noted that when the compactness of partitions is high and all the clusters are appropriately separated, the value of the σ becomes low whereas the value of Sp turns out to be high. This indicates it provides lower values of XiB_r index. So, the objective is minimizing the Xie-Beni index. u_{ij} stands for the membership score of the $j - th$ pattern in $i - th$ cluster, where $i = 1, 2, \ldots, k$ and $j = 1, 2, \ldots, n$. The membership score is depicted as

$$u_{ij} = \frac{1}{\sum_{t=1}^{k} \left(\frac{dist(\vec{c_i}, \vec{V_j})}{dist(\vec{c_t}, \vec{V_j})} \right)^{\frac{2}{r-1}}} \tag{17}$$

To obtain good solutions, it is therefore necessary to simultaneously minimize both J_r and XiB_r indices.

(b) The conventional DE operator is used to create a trial vector of regulatory genes and cluster centroids to simultaneously optimize the said two indices. Hence, the regulatory genes and cluster centroids are matured in nature.

Step 2 is repeated until it meets the stopping condition.

Step 3: Selection of the best solution from Pareto-front: Finally, it obtains the final set of Pareto-optimal solution from which the best one needs to be chosen by using by gap statistic. It is important to say that the actual Pareto-optimal solution here implies the correct number of clusters.

Here, authors have used six artificial and four real-life datasets of a variable range of complexities and obtained promising result during the comparison with some well-known schemes of multi-objective evolution algorithms.

4.5 Multi-Objective Particle Swarm Optimization to Identify Gene Marker

Mandal and Mukhopadhyay (2014) introduced a new approach using particle swarm optimization (PSO) with the properties of the multi-objective function (Liu et al. 2008) to identify nonredundant and relevant gene markers from the microarray data (Mandal and Mukhopadhyay 2014; Sabzevari and Abdullahi 2011). In fact, this article concerns about a feature selection problem that utilizes a graph-theoretic method in which a feature-dissimilarity graph is formed from the given data matrix. A node represents a feature whereas an edge stands for a dissimilarity. Both the nodes and edges are utilized in terms of some weight according to the relevance and dissimilarity, respectively, among the features. The authors map the identification of the nonredundant and relevant gene marker problem into the densest subgraph finding problem. The proposed multi-objective PSO method optimizes two objective functions, viz., average node weights and average edge weights, simultaneously. The algorithm proposes to detect nonredundant and relevant disease-related genes from microarray gene expression data.

4.6 Multi-Objective Binary Particle Swarm Optimization Algorithm for Cancer Data Feature Selection

Rao Annavarapu et al. (2016) proposed an approach of multi-objective binary particle swarm optimization (MOBPSO) for generating the cancer gene markers (Sarkar and Maulik 2014) from the microarray gene expression profile (Rao Annavarapu et al. 2016). Using some preprocessing techniques, the size of the high-dimensional cancer gene expression data is reduced to select proper domain features from the initials. The aim of preprocessing step is to eliminate the redundant genes. During feature subset generation, an appropriate smallest set of deferentially expressed genes are selected across the classes for the efficient and accurate classification. Preliminarily, we convert the values ranging in between zero to one using min-max normalization technique (Bandyopadhyay et al. 2014). The gene-wise min-max normalization is described as follows:

$$ab^1(x_i) = \frac{(ab_j(x_i) - min_j)}{(max_j - min_j)} \quad \forall i \tag{18}$$

where ab_j^1 and ab_j refer to the normalized score and unnormalized score, respectively. Of note, max_j and min_j signify the maximum and the minimum values, successively in each gene-wise vector ab_j. After that, they have taken two thresholds $Th_{initial}$ and Th_{final}, based on the central concept of quartiles in the literature by Banerjee et al. (2007). Then, the gene-wise value table is converted to the Boolean (viz., either 0 or 1) form in the following way:
if $ab^1(x) \leq Th_{initial}$, then fix "0"; else if $ab^1(x) \geq Th_{final}$, then fix "1"; else keep "*" denoting "don't care" condition. The mean of the occurrences of "*" is computed on the threshold Th_{ab}. The genes (attributes) whose number of "*"s are $\geq Th_{ab}$ are discarded from the table. It is the updated (i.e., reduced) attribute (gene) table value. Thereafter, distinction table (i.e., a matrix having the Boolean values) is prepared of which an entry $d((k,j),i)$ denotes the pair of objects (x_k, x_j) along with gene (attribute) ab_i. This discretization is performed by the following step:

$$d((k,j),i) = \begin{cases} 1 & \text{if } ab_i(x_k) \neq ab_i(x_j) \\ 0 & \text{if } ab_i(x_k) = ab_i(x_j) \end{cases} \tag{19}$$

Now, the presence of a "1" refers to the gene (attribute) ab_i's capability to distinguish in between the pair of the objects (x_k, x_j). The authors define a multi-objective function by using two fitness (objective) functions, which are made of the cardinality of feature subsets, and that contain distinctive capability of those selected subsets. First one (depicted as Ft_1) is defined by the number of features (viz., a number of 1's) and the second one (depicted as Ft_2) is defined on the extent to which the feature can recognize among the pair of the objects. The aforementioned objective functions are defined as follows:

$$Ft_1(v_a) = \frac{(N_{v_a} - Ob_{v_a})}{N_{v_a}} \tag{20}$$

where v_a is the subset of attributes or genes, Ob_{v_a} is the number of 1's in the feature vector v_a, and N_{v_a} is the number of conditional features in vector v_a or the length of v_a

$$Ft_2(v_a) = \frac{R_{v_a}}{(cl_1 \times cl_2)} \tag{21}$$

where R_{v_a} stands for the number of the object pairs v_a that can be distinguishable. The two objective functions are conjoined into one through the weighted sum as

$$Ft = Ft_1 \times \alpha + Ft_2 \times (1 - \alpha) \tag{22}$$

where $0 < \alpha < 1$. Let us assume an example where there are seven conditional features, viz., $(f_1, f_2, f_3, f_4, f_5, f_6, f_7)$. Thus, the length of the vector is $N_{v_a} = 7$. The Boolean digit becomes "1" if the respective feature can be distinguished between the two objects, whereas the Boolean digit becomes "0" if the corresponding feature cannot be differentiated between two objects. Suppose, here, there are two classes: cl_1 that contains two objects cl_{11} and cl_{12}, and cl_2 that has three objects cl_{21}, cl_{22}, and cl_{23}. The objective is here to select the minimal number of columns (features or genes) from Table 1 that covers all the rows (i.e., object pairs in Table 1) denoted by the Boolean digit "1". Let us assume a sample input vector depicted as $v_a = (1, 1, 1, 0, 1, 0, 1)$ where $cl_1 = 2$, $cl_2 = 3$, and the number of 1's in v_a denoted by Ob_{v_a} is 5. R_{v_a} is here computed as compare to the input vector v_a matching number of 1's from each row in the distinction table, which signifies that R_{v_a} is equal to 5. Thus,

Table 1 A simple example of a distinction table

Object pairs	f_1	f_2	f_3	f_4	f_5	f_6	f_7
(cl_{11}, cl_{21})	1	1	1	0	1	0	1
(cl_{11}, cl_{22})	0	1	0	1	0	1	0
(cl_{11}, cl_{23})	0	1	1	0	1	0	0
(cl_{12}, cl_{21})	1	0	1	0	1	0	1
(cl_{12}, cl_{22})	0	1	0	0	1	0	0
(cl_{12}, cl_{23})	1	0	1	0	1	0	0

$$Ft_1(v_a) = \frac{(N_{v_a} - Ob_{v_a})}{N_{v_a}} = \frac{(7-5)}{7} = 0.29$$

and

$$Ft_2(v_a) = \frac{R_{v_a}}{(cl_1 \times cl_2)} = \frac{5}{2 \times 3} = 0.84$$

At each formation, the best non-influenced solutions for united populace of the swarm at two successive genesis (current and next) are conserved. The best 50% of solutions are permitted to derive the next genesis. This step has been repeated for a finite number of times (generations). One random solution among top-ranked non-dominated solutions is chosen as best. There might be more than one top-ranked solutions available, but all solutions might have the same priority (Mallik et al. 2015). Authors claim that the algorithm provides more accurate and validate results with compare to the other benchmark methods.

4.7 Multi-Objective Approach for Identifying Coexpressed Module During HIV Disease Progression

Ray and Maulik (2017) have applied a multi-objective methodology in human immunodeficiency virus (HIV) infection progression to find differentially coexpressed modules (Ray and Maulik 2017). Here, authors analyze the gene expression data to generate the co-regulated pattern of a cluster of a specified phenotype. The proposed approach prepares a novel multi-objective framework to identify differentially coexpressed modules in terms of altering coexpression in the gene modules across the various stages of HIV-1 progression. The proposed method is completely depending on the genetic algorithm-based multi-objective approach. Here, the objective function is formulated through the diversity of coexpression pattern of gene pairs over the acute and the chronic stages in the HIV infection. The authors have promoted the two correlation matrices which depict the association of expression among each association over the two epidemic stages. The first objective function is made by the distance between two correlation matrices. Suppose, $Mat_i(s_k)$ stands for the correlation matrix of the module Mat_i in the stage denoted as s_k. Authors assume that the inner product between two matrices $Mat_i(s_1)$ and $Mat_j(s_2)$ of the infection stages s_1 and s_2 satisfy the following expression:

$$[Mat_i(s_1), Mat_j(s_2)] = tr\{Mat_i(s_1)Mat_j(s_2)\} \leq ||Mat_i(s_1)||_F ||Mat_j(s_2)||_F \quad (23)$$

For the two infection stages referred to as s_1 and s_2, the distance metric is formulated as follows:

$$d_1 = Dist_{Mat_i, Mat_j} = 1 - \frac{tr\{Mat_i(s_1)\}tr\{Mat_j(s_2)\}}{||Mat_i(s_1)||_F ||Mat_j(s_2)||_F} \quad (24)$$

where *tr* denotes the trace operator, whereas $||.||_F$ depicts Frobenius norm. The metric $Dist_{Mat_i, Mat_j}$ denotes the gap (i.e., distance) in between the two correlation matrices Mat_i and Mat_j. If the two correlation matrices are same it will be zero, whereas it will be equal to one whenever the matrices vary maximum amount. However, the distance is inspired by Herdin et al. (2005), in which the metric utilized in order to notify the variation of the spatial structure of multiple-input multiple-output (MIMO) channels. Of note, the second objective function promotes the module eigengene oriented measure. Module eigengene is basically treated as a representative of the entire gene module. The first left singular vector of the gene expression matrix with respect to the module is kept as the second objective function. Module eigengene is generally measured by the highest amount of alternations in the module expression data. The Pearson correlation between the expression data of a gene and a module eigengene is measured through the membership score of that gene in a specific module. For a module, the module membership ($M\mu$) values of all genes are obtained by two infection stages. Let, $M\mu_{g_k}^{m_k^s}$ depicts the module membership score of a gene g_k of the module m_k in the infection stage s. For every module, they have calculated a metric for the two infection stages s_1 and s_2. Specially, for individual gene of a module, they have calculated the following metric:

$$diff\_M\mu_{g_k}^{s_1, s_2} = |M\mu_{g_k}^{m_k^{s1}} - M\mu_{g_k}^{m_k^{s2}}| \tag{25}$$

This metric stands for the absolute change between the module membership score of a gene in two individual infection stages. Here, authors compute the mean value of all these values within a module. The second objective function is described by

$$d_2 = \frac{\sum_{Mo_k \in Mo} \sum_{g_k \in Mo_k} diff\_M\mu_{g_k}^{s_1, s_2}}{K} \tag{26}$$

where *Mo* be the module set, and *K* denotes the number of modules. Here, authors also compare their results (i.e., a set of Pareto-optimal solution) with the state-of-the-art methods using simulated data.

4.8   Other Methods

Seridi et al. (2012) propose a novel hybrid multi-objective meta-heuristic depending upon NSGA-II (non-dominated sorting genetic algorithm-II), CC (Cheng and Church) heuristic, and a multi-objective local search PLS-1 (Pareto Local Search 1). The objective of any biclustering algorithm in microarray data is to identify a subset of genes which are expressed similarly in a subset of conditions. Maulik et al. (2008) have used the concept of fuzzy for discovering overlapping biclusters. Here, they have developed a multi-objective genetic algorithm-based approach for probabilistic fuzzy biclustering (Liu et al. 2011) which minimizes the residual but maximizes cluster size as well as the expression profile variance.

5   Discussion

In 2003 (Deb and Reddy 2003), the multi-objective evolutionary algorithm is applied on various disease-related datasets such as leukemia dataset, colon cancer dataset, and multi-class several tissue-specific tumor datasets to find optimal gene subset. Comparative result depicts that the updated version of NSGA-II generates smaller sized classifiers, but provides more accurate classifications than the previous. The generic procedure involves two or more classes. Reliability in classification ensures a prediction strength in the evaluation of a classifier. It is an approach which is capable to figure out multiple different classifiers, each having the same size and classification accuracy. Since there are very few samples as compared to the large number of genes in the gene expression data, it may produce nearly 100% classification accuracy.

In Spieth et al. (2005), the results of the test examples show that the number of the interactions has a significant impact on the ability of an Evolutionary Algorithm (EA) to fit a mathematical model to a given time series of microarray data. The large number of connections between the system components provides a better result as compared to the data fitness. Since there is a large number of model parameters as well as the small number of datasets available, the system of equations is to be under-determined highly. The multi-objective technique shows promising performance as compared to the standard single-objective algorithms. Multi-objective evolutionary algorithm (MOEA) preserves better diversity of the solutions in the respective population since the standard algorithms are able to cope better with the issue of ambiguity than the single-objective optimization algorithms. Furthermore, with the situation of increasing the number of interactions, it would be more difficult to identify the real system with respect to the correct parameter values. MOEA generates the outcome in a set of Pareto-optimal solutions from which the researchers can select the most preferable and appropriate model complying with different biological constraints.

A multi-objective genetic algorithm, NSGA-II utilizes the fuzzy clustering of categorical attributes to obtain an evolving set of near-Pareto-optimal nondegenerate fuzzy partition matrices (Mukhopadhyay et al. 2007). Here, authors have applied their proposed algorithm on two synthetic (artificial) datasets and two real-life datasets (Congressional Votes and Zoo) and compared the performance with other different algorithms. In this article, the performance of the algorithm measured by %CP score which is defined as the percentage of pairs of points which are clustered together correctly. In addition, one way ANOVA test is applied to compare the mean %CP in different algorithms. Authors produce extremely strong evidence against the null hypothesis which states that the multi-objective fuzzy clustering generates better %CP as compared to the other benchmark algorithms only for the goodness of algorithm, but not by chance. For all other datasets, the proposed algorithm proves same.

In Suresh et al. (2009), authors have proposed two multi-objective algorithms with variants of DE and compared their proposed algorithm with two other well-known multi-objective clustering algorithms. Here they test their method on six artificial datasets and four real-life datasets. They have compared their results with the result of NSGA-II method for the same dataset. DE-variant multi-objective optimization algorithms produce better final clustering solutions as compared to NSGA-II or MOCK using the same objective function. In fact, they not only find out the correct partitions in the data but also, in all cases, they are able to determine an optimal number of classes with minimum standard deviations. Here, the experimental outcome indicates that DE holds immense promise as a candidate optimization technique for the multi-objective clustering. Apart from this, in this book chapter, we have discussed some other multi-objective optimization methods that are based on particle swarm optimization (PSO).

In Mandal and Mukhopadhyay (2014), authors have extended particle swarm optimization (PSO) with the properties of the multi-objective function to identify nonredundant and relevant gene markers from microarray data (Sarkar and Maulik 2014). The performance of the method is evaluated using several evaluation metrics such as sensitivity, specificity, accuracy, fscore, AUC, and average correlation. Here, the entire dataset is divided into two different sets, viz., training set and test set. First, they apply their proposed approach on the training data. Thus, a set of non-dominated candidate solutions are obtained. After that, for finding final genetic markers, they employ with the BMI score which considers the discriminative power of each gene by incorporating the true positive rate from the logistic regression. They basically utilize graph-based multi-objective particle swarm optimization (MPSO) where they have identified subgraph having nonredundant and relevant feature nodes. They basically model their multi-objective framework by non-dominated sorting and crowding distance sorting. Three real-life datasets are used here in performance analysis. Their proposed method provides better performance as compared to the corresponding single-objective versions. They further apply various statistical tests, viz., t-test, Wilcoxon's rank sum test to obtain the significance and verification of proposed method. In Rao Annavarapu et al. (2016), authors applied multi-objective binary particle swarm optimization approach for the cancer feature selection of the microarray data. They use different types of cancer microarray datasets like colon cancer, lymphoma, and leukemia and obtain minimal subsets of features. For lymphoma dataset, 100% correct classification score is achieved by using Bayes logistic regression, whereas Bayes net classifier yields 95.84% and naive Bayes classifiers produce 97.42% on 22 feature subsets. Similarly, for the leukemia data, the percentages of classification accuracy are 92.1% with Bayes Logistic Regression, 89.48% with the other two classifiers on the 14 feature subsets. The result of the three benchmark cancer datasets demonstrates the feasibility and effectiveness of the proposed method. The performance of the proposed method along with the existing methods is compared with the standard classifiers, and it reports that the proposed method is better in competitive performance than the other methods.

In Ray and Maulik (2017), authors develop a multi-objective framework to identify differential coexpression modules from two microarray dataset corresponding to two different phenotypes. They optimize their two objective functions and compare their algorithm with some state-of-the-art algorithms for measuring the differential coexpression. It is evident that the proposed algorithm performs better than the others. They measure the performance of their algorithm on simulated dataset. The simulated study validates the correctness. Through the simulated data, it represents that the mean value is increasing with Standard Deviation (SD) values ranging from 0.1 to 0.5 as expected. Hence, it shows a strong increasing pattern of mean with increasing SD values. This proposed algorithm produces essential symptoms of HIV in human body which ultimately causes AIDS. However, a tiny part of HIV infected gene remains clinically stable for a long time span, and they refer to as long-term nonprogressors. Since in nonprogressor stage, HIV infected individual remains clinically stable for a long time span, we need to know how the regulation pattern of genes changes across the three stages of progression, viz., acute to chronic stages along with acute to nonprogressor stage. This is the future scope of this article.

Throughout the analysis of all existing multi-objective optimization techniques on microarray data, it is clear that most of them are based on the approach of either Genetic Algorithm or PSO or differential evolution. From the various dimensions, those methods can win over other or can be lost by other one.

6 Conclusion

In this chapter, we have reviewed a number of multi-objective optimization techniques toward solving several biological problems dealing with the microarray data. In this context, three popular multi-objective optimization techniques, viz., MOGA, MOEA, and MPSO, have been reviewed to show their effectiveness in tackling the various aspects of biological problems. It could conclusively be argued that multi-objective optimization techniques are one of the most useful techniques and thus, are highly recommended to design sophisticated learning systems related to various biological problems such as gene marker detection, gene-module identification, development of rule-based classifier, etc., in microarray data as well as multi-omics data containing several profiles from different directions.

Hence, in this chapter, to analyze the correct applicable method under some constraints in terms of fulfilling specified biological functional objectives, we have covered all such works along with their scope, advantages, limitations, and comparative studies.

References

M. Aqil, S. Mallik, S. Bandyopadhyay, U. Maulik, S. Jameel, Transcriptomic analysis of mrnas in human monocytic cells expressing the HIV-1 nef protein and their exosomes. BioMed Res. Int. **2015**, 1–10 (2015). Article id: 492395

M. Aqil, A.R. Naqvi, S. Mallik, S. Bandyopadhyay, U. Maulik, S. Jameel, The HIV Nef protein modulates cellular and exosomal miRNA profiles in human monocytic cells. J. Extracell. Vesicles **3**, 23129 (2014)

S. Bandyopadhyay, S. Mallik, Integrating multiple data sources for combinatorial marker discovery : a study in tumorigenesis. IEEE/ACM Trans. Comput. Biol. Bioinform. (2016) (accepted). https://doi.org/10.1109/TCBB.2016.2636207

S. Bandyopadhyay, S. Mallik, A. Mukhopadhyay, A survey and comparative study of statistical tests for identifying differential expression from microarray data. IEEE/ACM Trans. Comput. Biol. Bioinf. **11**(1), 95–115 (2014)

M. Banerjee, S. Mitra, H. Banka, Evolutionary rough feature selection in gene expression data. IEEE Trans. Syst. Man Cybern. Part C (Appl. Rev.) **37**, 622–32 (2007)

T. Bhadra, S. Bandyopadhyay, U. Maulik, Differential evolution based optimization of SVM parameters for meta classifier design. J. Proced. Technol. **4**, 50–57 (2012)

T. Bhadra, S. Mallik, S. Bandyopadhyay, Identification of multi-view gene modules using mutual information based hypograph mining. IEEE Trans. Syst. Man Cybern. Syst (2017) (accepted). https://doi.org/10.1109/TSMC.2017.2726553

M. Bhattacharyya, L. Feuerbach, T. Bhadra, T. Lengauer, S. Bandyopadhyay, MicroRNA transcription start site prediction with multi-objective feature selection. J. Stat. Appl. In Genet. Mol. Biol. **11**(1), 1–25 (2012)

J.M. Claverie, Fewer genes, more noncoding RNA. Science **309**(5740), 1529–1530 (2005)

K. Deb, A. Pratap, S. Agarwal, T. Meyarivan, A fast and elitist multi-objective genetic algorithm: NSGA-II. IEEE Trans. Evol. Comput. **6**(2), 182–197 (2002). https://doi.org/10.1109/4235.996017

K. Deb, A.R. Reddy, Classification of Two and Multi Class Cancer Data Reliably Using Multi-Objective Evolutionary Algorithms. KanGAL Report Number 2003006 (2003)

J. Hacia, J. Fan, O. Ryder, L. Jin, K. Edgemon, G. Ghandour, R. Mayer, B. Sun, L. Hsie, C. Robbins, L. Brody, D. Wang, E. Lander, R. Lipshutz, S. Fodor, F. Collins, Determination of ancestral alleles for human single-nucleotide polymorphisms using high-density oligonucleotide arrays. Nat. Genet. **22**, 164–167 (1999)

M. Herdin, N. Czink, H. Ozcelik, E. Bonek, Correlation matrix distance, a meaningful measure for evaluation of non-stationary mimo channels, in *IEEE 61st Vehicular Technology Conference, 136140. 2005 IEEE 61st Vehicular Technology Conference* 30 May–1 June 2005 (IEEE)

Z. Joseph, A. Gitter, I. Simon, Studying and modelling dynamic biological processes using time-series gene expression data. Nat. Rev. Genet. **13**, 552–564 (2012)

J. Liu, Z. Li, X. Hu, Y. Chen, E. Park, Dynamic biclustering of microarray data by multi-objective immune optimization. BMC Genomics **12**(suppl 2)(11) (2011). https://doi.org/10.1186/1471-2164-12-S2-S11

J. Liu, Z. Li, F. Liu, Multi-objective particle swarm optimization biclustering of microarray data, in *IEEE International Conference on Bioinformatics and Biomedicine(BIBM 2008)*, Hyatt Regency Philadelphia, PA, USA, (IEEE Computer Society, 2008), pp. 363–366. https://doi.org/10.1109/BIBM.2008.17

S. Mallik, T. Bhadra, U. Maulik, Identifying epigenetic biomarkers using maximal relevance and minimal redundancy based feature selection for multi-omics data. IEEE Trans. Nanobiosci. 1536–1241 (2017)

S. Mallik, U. Maulik, MiRNA-TF-gene network analysis through ranking of biomolecules for multi-informative uterine leiomyoma dataset. J. Biomed. Inform. **57**, 308–319 (2015)

S. Mallik, A. Mukhopadhyay, U. Maulik, S. Bandyopadhyay, Integrated analysis of gene expression and genome-wide DNA methylation for tumor prediction: an association rule mining-based approach, in proceedings of the ieee symposium on computational intelligence in bioinformatics and computational biology (CIBCB), in *IEEE Symposium Series on Computational Intelligence—SSCI 2013*, Singapore, 16 Apr 2013 (2013a), pp. 120–127

S. Mallik, S. Sen, U. Maulik, IDPT: Insights into potential intrinsically disordered proteins through transcriptomic analysis of genes for prostate carcinoma epigenetic data, in *Gene*, vol. 586 (Elsevier, 2016), pp. 87–96

S. Mallik, A. Mukhopadhyay, U. Maulik, Integrated statistical and rule- mining techniques for DNA methylation and gene expression data analysis. J. Artif. Intell. Soft Comput. Res. **3**(2), 101–115 (2013b)

S. Mallik, A. Mukhopadhyay, U. Maulik, RANWAR: rank-based weighted association rule mining from gene expression and methylation data. IEEE Trans. Nanobiosci. **14**(1), 59–66 (2015)

M. Mandal, A. Mukhopadhyay, A graph-theoretic approach for identifying non-redundant and relevant gene markers from microarray data using multi-objective binary PSO, PLOS One, **9**(3) (2014)

U. Maulik, S. Bandyopadhyay, A. Mukhopadhyay, *Multi-objective Genetic Algorithms for Clustering: Applications in Data Mining and Bioinformatics* (Springer, New York Inc., 2011), p. 281. https://doi.org/10.1007/978-3-642-16615-0

U. Maulik, S. Mallik, A. Mukhopadhyay, S. Bandyopadhyay, Analyzing gene expression and methylation data profiles using StatBicRM: statistical biclustering-based rule mining. PLoS One **10**(4), e0119448 (2015). https://doi.org/10.1371/journal.pone.0119448

U. Maulik, A. Mukhopadhyay, S. Bandyopadhyay, M.Q. Zhang, X. Zhang, Multi-objective fuzzy biclustering in microarray data: Method and a new performance measure, in *IEEE World Congress on Computational Intelligence, Evolutionary Computation*, CEC, Hong Kong, China (2008). Print ISSN: 1089-778X Electronic ISSN: 1941-0026

U. Maulik, S. Bandyopadhyay, A. Mukhopadhyay, Multi-class clustering of cancer subtypes through SVM based ensemble of pareto- optimal solutions for gene marker identification. PLoS One **5**(11), e13803 (2010)

A. Mukhopadhyay, U. Maulik, S. Bandyopadhyay, Multi-objective evolutionary approach to fuzzy clustering of microarray data. book chapter in analysis of biological data, in *A soft computing approach, science, engineering, and biology informatics*, vol. 3, pp. 303–328 (2007)

A. Mukhopadhyay, U. Maulik, S. Bandyopadhyay, Multi-objective genetic fuzzy clustering of categorical attributes, in *10th International Conference on Information Technology* (IEEE Computer Society, 2007)

A. Mukhopadhyay, U. Maulik, S. Bandyopdhyay, On biclustering of gene expression data. Curr. Bioinf. **5**(3), 204–216 (2010)

H. Pearson, Genetics: what is a gene? Nature **441**(7092), 398–401 (2006)

C.S. Rao Annavarapu, S. Dara, H. Banka, Cancer microarray data feature selection using multi-objective binary particle swarm optimization algorithm. EXCLI J. **1611–2156**(15), 460–473 (2016)

S. Ray, U. Maulik, Identifying differentially coexpressed module during HIV disease progression: a multi-objective approach. Technical Report, Nature (2017)

S. Sabzevari, S. Abdullahi, Gene selection in microarray data from multi-objective perspective, in *IEEE 3rd International Conference on Data Mining and Optimization (DMO)*, Putrajaya, Malaysia, Electronic, pp. 199–207 (2011). ISSN: 2155-6946, Print ISSN: 2155-6938

A. Sarkar, U. Maulik, Cancer biomarker assessment using evolutionary rough multi-objective optimization algorithm, in *Artificial Intelligent Algorithms and Techniques for Handling Uncertainties: Theory and Practice, ACIR series*, ed. by P. Vasant (IGI Global, 2014), pp. 1–23 . https://doi.org/10.4018/978-1-4666-7258-1.ch016.

S. Sen, U. Maulik, S. Mallik, S. Bandyopadhyay, Detecting TF-MiRNA-gene network based modules for 5hmC and 5mC brain samples: a intra- and inter-species case-study between human and rhesus. BMC Genet (2017). https://doi.org/10.1186/s12863-017-0574-7

K. Seridi, L. Jourdan and E. Talbi, Hybrid metaheuristic for multi-objective biclustering in microarray data, in *IEEE Symposium on Computational Intelligence in Bioinformatics and Computational Biology (CIBCB)*, San Diego, CA, USA (2012). Electronic ISBN: 978-1-4673-1191-5, Print ISBN: 978-1-4673-1190-8

C. Spieth, F. Streichert, N. Speer, A. Zell, Multi-objective model optimization for inferring gene regulatory networks, in *International Conference on Evolutionary Multi-criterion Optimization, EMO 2005*, LNCS 3410, (Springer, Heidelberg, 2005), pp. 607–620

K. Suresh, D. Kundu, S. Ghosh, S. Das, A. Abraham, S.Y. Han, Multi-Objective Differential Evolution for Automatic Clustering with Application to Micro-Array Data Analysis, in *Sensors*, vol. 9 (2009), pp. 3981–4004. ISSN 1424-8220

B. Wu, Differential gene expression detection and sample classification using penalized linear regression models. Bioinformatics **22**, 472–476 (2006)

Application of Multiobjective Optimization Techniques in Biomedical Image Segmentation—A Study

Shouvik Chakraborty and Kalyani Mali

1 Introduction

Among many problems, image segmentation is considered as one of the major challenges and a necessary stage of image analysis process. Segmentation is one of the important stages of medical image analysis, and it has a high impact on the quality and accuracy of the final results obtained. In biomedical domain, the accuracy of the final result is very important (Demirci 2006). The process of segmentation is nothing but the clustering of similar pixels. The similarity is determined depending on one or more than one properties. The major goal of the segmentation process is to find similar regions that indicate different objects in an image. To achieve this goal, multiple objectives may need to be considered.

Multiobjective optimization in the decision-making process of image segmentation can be considered as a new area of research (Saha et al. 2009; Nakib et al. 2009a; Shirakawa and Nagao 2009; Maulik and Saha 2009). In general, different objectives for a problem are conflicting in nature. Therefore, it is not possible to optimize all objectives simultaneously. In general, most of the real-world problems have different objectives to be optimized. For example, maximize intercluster distance, minimize intra-cluster distance, minimize feature count, maximize feature similarity index, etc. Optimization of these metrics as a whole is a difficult issue because there are different types of images of different modalities. Multiobjective optimization techniques are useful and effective to solve these kinds of problems (Guliashki et al. 2009; Jones et al. 2002; Coello 1999).

S. Chakraborty (✉) · K. Mali
University of Kalyani, Kalyani, Nadia, West Bengal, India
e-mail: shouvikchakraborty51@gmail.com

K. Mali
e-mail: kalyanimali1992@gmail.com

© Springer Nature Singapore Pte Ltd. 2018
J. K. Mandal et al. (eds.), *Multi-Objective Optimization*,
https://doi.org/10.1007/978-981-13-1471-1_8

181

The objective of this chapter is to study the literature related to the multiobjective segmentation methods. In this work, the problem of image segmentation is discussed in the light of multiobjective optimization. Moreover, study about the application of multiobjective optimization methods in classification approaches is also provided. Some clustering methods with different objectives are also discussed.

In general, image segmentation methods consist of various stages. The process generally begins with the representation of the pattern that can be done by means of cluster count, estimating the available patterns, and different properties found in the clusters. Some features may be controllable. But some features may not be controllable by the user. After selecting the appropriate representation of the pattern, the next step is to select feature. Feature selection is one of the important tasks and it needs some careful investigation of the image. In this stage, identification of the most suitable set of feature that can be taken from the actual feature set is performed. These features are useful in the clustering process. After feature selection, feature extraction is performed. Feature extraction is one of the necessary steps where some new features are derived by employing some algorithms on the selected features (from the previous stage) (Chakraborty et al. 2017a).

Any one or both of the discussed methods can be employed to get useful collection of features that can be used in clustering. The main goal of feature selection and extraction process is to select appropriate features and discard irrelevant features. In biomedical image analysis, images of different modalities are considered, and some set features that can be useful in one modality may not be applicable for other modalities. Moreover, some features are very useful for a particular modality. For example, different features based on intensity values, shape, and spatial association are considered for the segmentation process of the images that are obtained using CT scan method.

Similarity among patterns is computed with the help of selected and extracted features. In general, the similarity is computed using a distance function that computes the distance between two patterns. Various well-known distances are used to compute the similarity between two patterns. For example, common distances like Euclidean, City-Block, etc., are frequently used to represent similarity or dissimilarity among different patterns. Now, distance measures are not sufficient to reflect conceptual similarity among different patterns. So, different other similarity measures are used to find the conceptual similarity among patterns. The selection of similarity measure generally depends on the type of the image under test. For example, in the segmentation process of biomedical images, features like spatial coherence and homogeneity are widely used. Some other commonly used criteria are connectedness of different regions versus compactness within a region (Chin-Wei and Rajeswari 2010).

The whole process discussed above can be greatly affected by the grouping approach. Different approaches can be considered for grouping. The obtained results can be represented using two methods, namely hard/crisp and fuzzy. In case of crisp representation, the boundary of a particular region is determined rigidly, i.e., one pixel can be a part of only one region. But in case of fuzzy representation, one region is not rigidly restricted within a certain boundary, i.e., one pixel can be a member of different region with a degree of membership. Figure 1 illustrates the difference

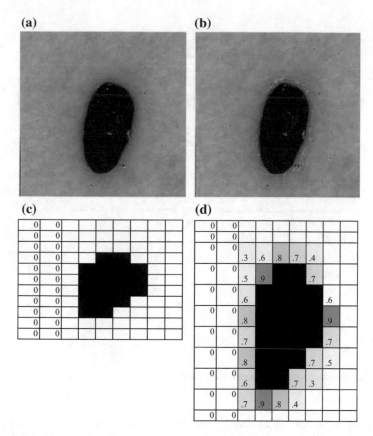

Fig. 1 **a** Original image of angioma skin disease. **b** Naïve segmentation provided in ISIC dataset. **c** Crisp representation. **d** Fuzzy representation

between crisp and fuzzy representation. Here, the boundary of a skin lesion image (International Skin Imaging Collaboration Website n.d.) is considered for illustration. In this figure, both crisp and fuzzy clustering has been demonstrated. In Fig. 1c, it can be observed that each pixel belongs to a particular class only, i.e., either 0 or 1. This is the hard or crisp representation. On the other hand, Fig. 1d shows the fuzzy representation of the same. Here, one pixel has a variable degree of participation in one class. Hence, each pattern has a variable membership.

To measure the quality of the output obtained by a clustering method, the validity analysis must be performed. It is an important step that helps us to assess the result and sometimes the algorithm. Sometimes this analysis is performed subjectively, however, several indexes are available for the quantitative assessment of the results obtained from a clustering method. These methods basically employ some objectives (can be one or more than one) that can be optimized. Quality measures are generally objective in nature and are used to judge the quality of the output. Different objectives

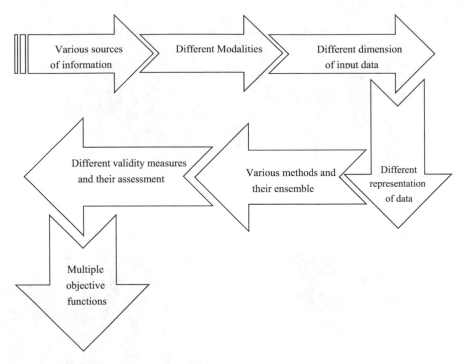

Fig. 2 Multiobjective nature of the problems from input to output

are used in this process which is needed to be optimized to obtain an accurate number of clusters which leads to the better segmentation results.

Multiple objectives (i.e., criteria) may be useful in the segmentation process as well as in the evaluation scheme. In the analysis of the biomedical images, there may be various sources of data that is used for a particular problem. There can be multidimensional data that is needed to be processed. Sometimes, multiple techniques are combined to solve a particular instance of a problem (i.e., ensemble). These issues lead us towards the multiobjective methods of problem-solving. Figure 2 illustrates the multiobjective nature of the problems from input to output.

One of the popular and widely used multiobjective optimization techniques is to convert the different objectives in such a way so that they can be treated and processed as a single objective. This is generally performed by imposing some numerical weights on different evaluation parameters (i.e., criteria). After assigning suitable weights, these values are multiplied with the objective function and aggregated to form a single objective. Apart from this, another well-known technique is concurrent optimization of different objectives. These techniques are also known as Pareto optimization approaches.

In the Pareto optimization methods, a simple transformation of multiple criteria into a single one using some weights is not sufficient. Here, the computation is performed on the basis of the dominance of attributes (Miettinen 2008). In the method, the relationship among different objectives is considered to compute intermediate results prior to achieve a Pareto-optimal result (Bong and Wang 2006). In this context, the Pareto-optimal output is a collection of results that are non-dominated with respect to each other. If we deviate from Pareto-optimal solution, then we may achieve some gain for a certain objective but we have to compromise some other objectives. This kind of solutions is preferred over single solutions because they are more realistic and can be applied on many real-life problems. Different methods have been developed to analyze biomedical images considering multiple objectives to be optimized. In this chapter, a comprehensive review is provided on different multiobjective methods for image segmentation.

2 Multiobjective Optimization

In this section, some basic ideas of multiobjective optimization method are illustrated.

In various real-life problems, different objectives have to be optimized simultaneously. It is in contrast to the situations handled by the conventional techniques where a single objective needs to be optimized. The major problem in multiobjective optimization is lack of well-defined optimum. Therefore, it is difficult to compare obtained solution with each other. Generally, these kinds of problems can have multiple solutions. Each solution can be accepted where the relative significance of different objectives is not known. In these cases, the optimum solution is somewhat subjective and dependent upon the user (Deb 2001).

Multiobjective optimization problems can be defined as follows (Deb 2001):

$$optimize \quad \vec{F}(\vec{x}) = [f_1(\vec{x}), f_2(\vec{x}), \ldots, f_k(\vec{x})]$$

Satisfying the constraints,

$$g_i(\vec{x}) \leq 0 \quad i = 1, 2, 3, 4, \ldots, m$$
$$h_i(\vec{x}) = 0 \quad i = 1, 2, 3, 4, \ldots, p$$

Here,

$$\vec{x} = [x_1, x_2, \ldots, x_n]$$

f_i (i = 1, 2, 3, ..., k) are the objective functions, g_i (i = 1, 2, 3, ..., m) and hi (i = 1, 2, 3, ..., p) are the constraint functions.

To define optimality, some definitions must be discussed.

Definition 1 (*Dominating Solutions*) A set of solutions $\vec{x} = [x_1, x_2, x_3, \ldots, x_k]$ is said to be dominating over another vector $\vec{y} = [y_1, y_2, y_3, \ldots, y_k]$ if and only if $\vec{x}$ is partially less than $\vec{y}$. It is denoted by $\vec{x} \prec \vec{y}$.

Definition 2 (*Pareto-optimal solutions*) A solution is said to be Pareto-optimal solution if and only if the solution is not dominated by any other solution.

Definition 3 (*Pareto-optimal front*) The graphical representation of the fitness functions whose non-dominated vectors belongs to the set of Pareto-optimal solutions is known as Pareto-optimal front.

One point can be noted that all Pareto-optimal solutions may not be useful or sometimes, not feasible.

3   Application of Multiobjective Optimization in Biomedical Images

One of the oldest approaches that use multiobjective optimization for image analysis can be found in Bhanu et al. (1993). In this work, television image analysis was performed. Later, many works has been done using multiobjective optimization for image analysis. Various image analysis tasks like segmentation, classification, etc., have been performed using the multiobjective optimization technique. A summary of multiobjective optimization techniques used in biomedical image analysis is summarized in Table 1.

From Table 1, it can be observed that multiobjective-based methods are widely used for medical image analysis. In general, medical images fall under the multispectral image analysis category. Therefore, problem representation using multiobjective methods is suitable for medical images (Niu and Shen 2006). In some works, a fusion of various features obtained from different images has been proposed (Zhang and Rockett 2005). Wavelet (Seal n.d.; Mali et al. 2015)-based fusion has been used in (Niu and Shen 2006). Multiobjective techniques for medical images analysis can be tested on various datasets. One of the popular ones is the UCI repository (Newman et al. 1998). In this repository, one can get some images on skin, Iris, Wine, Breast cancer, etc., and so on. Another popular one is the BrainWeb (Cocosco et al. 1997). In (Saha and Bandyopadhyay 2010), various types of datasets have been used to test the effectiveness of the multiobjective techniques in digital image segmentation.

Segmentation of digital images can be considered as the classification of pixels. The classification of the pixels is done on basis of different features (Hore et al. 2016a; Chakraborty et al. 2016). In case of classification, multiobjective approach is used to extract features. In case of medical image analysis and segmentation, the classifiers are trained in such a way so that edges of different regions can be efficiently tacked, and it is one of the important jobs for different applications (Hore et al. 2015, 2016b; Chakraborty et al. 2017a, b). Multiobjective approaches are also useful in constructing an ensemble of various classifiers.

In general, two types of approaches are considered for biomedical image classification. The first one is called supervised, where the classifier is trained earlier. And, the second one is called unsupervised classification, where the process begins

Table 1 Summary of multiobjective optimization methods used in biomedical image analysis

Method	Description	Dataset
Nakib et al. (2009b)	In this work, a multiobjective optimization-based technique was proposed for thresholding. Pareto-based method was employed to perform the job. This method is capable of solving multiple criteria simultaneously. This feature helps to enhance the quality of the segmented image. NSGA-II method was used to generate the Pareto front and desired optimal solution	Berkley dataset (Berkeley University 2007)
Omran et al. (2005a)	Differential Evolution (DE)-based method was proposed in this work. It is basically a clustering technique that computes the user given number of clusters	Imaginary dataset contains three types: synthetic, MRI, and LANDSAT 5 MSS (79 m GSD)
Mukhopadhyay et al. (2009b)	A clustering method based on multiobjective genetic fuzzy clustering technique was used in this work. It takes help from the NSGA-II for searching purpose. This method was applied on different medical images like T1-weighted, T2-weighted, MRI brain images, etc.	BrainWeb database (n.d.)
Faceli et al. (2008)	In this work, a selection technique was developed. This method was based on corrected Rand index. This method was applied on the collection of solutions collected from the MOCLE method	Synthetic and datasets like iris, glass obtained from UCI Machine learning repository (Newman et al. 2006)
Pulkkine and Koivisto (2008)	In this work, a hybrid technique was developed to identify Pareto-optimal fuzzy classifiers. This method uses an initialization technique that consists of two stages. So, no early knowledge is needed. Moreover, no random initialization is performed. In the first step, a decision tree is constructed. To construct the decision tree, C4.5 method was employed. Depending on the decision tree, input space is divided. The population is then created by substituting parameters of fuzzy classifiers. One thing should be kept in mind that the initial population should be well distributed. NSGA-II was employed to optimize initial population. One of the advantages of this technique is, it does not need the count of the fuzzy sets along with distribution	Wisconsin breast cancer, Pima Indians diabetes, Glass Cleveland heart disease, Sonar Wine (Newman et al. 2006)
Ahmadian et al. (2007)	This work focuses on the ensemble of classifiers. Multiobjective optimization method was employed to support the classifier. In this method, a "bagging-and-boosting-like" technique was used. It is also advantageous in smaller dimensional feature space. In lower dimension of the feature space, ambiguity can be generated which can be removed by this method. Multiobjective GA has been used to combine and create an ensemble	Wisconsin Breast Cancer, Wisconsin Breast Cancer, Iris, Pima (Newman et al. 2006)

(continued)

Table 1 (continued)

Method	Description	Dataset
Cococcioni et al. (2007)	In this work, a method has been proposed that can choose most appropriate binary classifier based on fuzzy rule for a particular problem. An algorithm consisting of three objectives has been developed. Sensitivity and specificity were used to compute the accuracy of the system. The ROC convex hull technique was used to find the most appropriate classifier	20 clinical cases, containing a total of about 7,000 slices. The CT scans were collected during a mass screening performed in an Italian Hospital
Erin (2001)	In this work, a feature selection method was proposed. Multiobjective GA is applied to choose some features that can give the optimal segmentation output. This method was tested on medical images. Here, self-organizing map is employed to perform classification. Quantization error and topology preservation are the two attributes of the SOM that is used as the objective function of the GA. This method was tested on MRI image data and 3D simulation model	MRI dataset from the whole brain atlas (Johnson and Becker 1999)
Nakib et al. (2008)	The main motivation behind this method is—single objective optimization methods may not always be useful for the segmentation job. In this method, two objectives have been considered. These are within-class attribute and the overall probability of error attribute. It is used to compute optimal threshold value. Moreover, a modified simulated annealing method was proposed for the "histogram Gaussian curve fitting" problem	Berkley dataset (Berkeley University 2007)
Wang and Wang (2006)	This work focuses on building an ensemble of classifiers. Here, each classifier is trained depending on specific weights. Weighting is associated with the input set. A genetic algorithm was used for the searching purpose. It can find suitable chromosomes. One of the major differences from the traditional GA is that it does not only consider the best solution. It exploits all solutions generated during the process. This method was tested on the UCI benchmark datasets	This breast cancer database is obtained from the University of Wisconsin Hospitals, Madison from Dr. William H. Wolberg and some other datasets
Krishna and Kumar (2016)	In this work, a hybrid approach has been proposed for color image segmentation. Genetic algorithm and differential evolution has been used as a hybrid method for color image segmentation	Berkley dataset (Berkeley University 2007)

with the partitioning process the input image into different clusters, i.e., set of related objects.

Multiobjective methods may use some different criteria than the conventional unsupervised methods. In (Optiz 1999), the fitness criteria are defined as the accuracy and diversity of the classifier used instead of conventional fitness criterion of the unsupervised classifiers. In (Cococcioni et al. 2007), accuracy and complexity of the used classifier have been considered as the objective. In this method, approximate Pareto-optimal solutions have been computed. In some of the cases, multiobjective optimization methods have been used along with semi-supervised classification methods (Ghoggali and Melgani 2007; Ghoggali et al. 2009).

Sometimes, fuzzy segmentation method gives better performance. In case of images that contain ambiguity, fuzzy methods perform better than crisp techniques. Imprecise data can be easily handled using fuzzy methods (Karmakar and Dooleya 2002). Fuzzy rule-based methods are useful in various occasions because they can interpret both linguistic and numeric variables. So, multiobjective techniques coupled with fuzzy rule-based systems can also be used for classification purpose. Fuzzy rules can be used to find a collection of non-dominated rules with the help of some meta-heuristic (Chakraborty et al. 2017a; Chakraborty and Bhowmik 2013; Chakraborty and Bhowmik 2015a, b; Chakraborty et al. 2015, 2017; Roy et al. 2017) methods. A multiobjective method based on the fuzzy rule has been tested in (Ishibuchi and Nojima 2005). In (Cococcioni et al. 2007), binary classifiers based on fuzzy rule were investigated. In this approach, an optimization method was tested consisting of three objective functions. These methods show that fuzzy-based methods along with multiobjective techniques give faster convergence.

In some of the cases, multiobjective fuzzy clustering techniques are used with artificial neural networks. In (Mukhopadhyay et al. 2008), a probabilistic classifier has been developed based on this technique that gives better performance. This concept was first introduced in (Kottathra and Attikiouzel 1996). In (Abbass 2003), a learning method based on multiobjective ANN was proposed where the main objectives were network complexity and the error involved in training. A summary of some multiobjective approaches that have been applied in different domains along with biomedical image analysis is given in Table 2.

4 Conclusion

In general, image segmentation problems consist of several objectives that have to be optimized simultaneously. So, multiobjective solutions are very helpful in determining the segments by optimizing different objective functions. Evolutionary algorithms are very useful in this context. Multiobjecive evolutionary segmentation methods perform well on different modalities of biomedical images. This chapter investigates some of the multiobjective methods that have been used for biomedical image segmentation as well as some other techniques that can be applied on biomedical images. Different multiobjecive methods along with their features have been

Table 2 Summary of some multiobjective approaches

Method	Multiobjective technique	Number of objectives	Brief description
Mukhopadhyay and Maulik (2009a)	NSGA-II	2	In this method, a fuzzy voting method was developed along with SVM classifier
Mukhopadhyay et al. (2007)	NSGA-II	2	A multiobjective GA-based method was developed for image segmentation
Faceli et al. (2008)	NSGA-II	2	In this work, a method has been proposed for selecting
Mukhopadhyay et al. (2009b)	NSGA-II	3	A clustering method based on multiobjective genetic fuzzy clustering technique was used in this work
Saha and Bandyopadhyay (2010)	Simulated annealing	3	In this work, multiobjective clustering method was developed. The result of this method was the solution that has the minimum Minkowski score
Collins and Kopp (2008)	Simulated annealing	2	In this work, a synthetic data was prepared along with a new performance metric to test the method
Saha and Bandyopadhyay (2008)	Simulated annealing	2	Multiobjective SA-based method was proposed. A parameter to compute the validity of a cluster output was employed for selection
Handl and Knowles (2007)	PESA-II	2	In this work, Silhouette Width along with the attainment score was used to determine the optimum solution and
Shirakawa and Nagao (2009)	PESA-II	2	This method computes different segmentation result using an evolutionary approach. A basic heuristic technique was used for selecting a single solution from the set of Pareto solutions
Matake et al. (2007)	SPEA-2	2	A multiobjective method has been developed to determine the "K" value
Bhanu et al. (1993)	Genetic algorithm and hill climbing	5	Image segmentation method with hybrid search mechanism
Omran et al. (2005b)	Differential evolution	2	DE-based method for unsupervised classification
Paoli et al. (2009)	PSO	2	Multiobjective PSO-based image clustering method was developed. The nearest result to the source of the performance space has been selected

studied in this paper that can help in taking the decision about the algorithm that can solve a particular problem of segmentation more efficiently. Choice of the algorithm is highly dependent on the nature of the problem and the complexity involved. Moreover, problems, as well as algorithms, may have to be customized for an appropriate formation that can handle different constraints appropriately. The determination of all Pareto fronts is not necessary for most of the cases. In general, a portion of the complete solution space is sufficient if it can be determined efficiently. In future, different evolutionary approaches (e.g., firefly, bat, etc.) along with multiobjective solutions can be investigated and compared with others to test the efficiency. Improvement in biomedical image analysis is highly required for accurate and efficient analysis of the data to diagnose various diseases in a timely manner that can save numerous lives and enhance the quality of living.

References

H.A. Abbass, Speeding up backpropagation using multiobjective evolutionary algorithms. Neurocomputing **15**, 2705–2726 (2003)

K. Ahmadian, A. Golestani, M. Analoui, M.R. Jahed, Evolving ensemble of classifiers in low-dimensional spaces using multi-objective evolutionary approach, in *6th IEEE/ACIS International Conference on Computer and Information Science, ICIS* (2007)

Berkeley University, The Berkeley segmentation dataset and benchmark. Grouping and Ecological Statistics. June 2007, http://www.eecs.berkeley.edu/Research/Projects/CS/vision/grouping/

B. Bhanu, S. Lee, S. Das, Adaptive image segmentation using multiobjective evaluation and hybrid search methods. *Machine Learning in Computer Vision* (1993)

Brainweb: simulated brain database, http://www.bic.mni.mcgill.ca/brainweb

C.W. Bong, Y.C. Wang, A multi-objective hybrid metaheuristic for zone definition procedure. Int. J. Serv. Oper. Inf. **1**(1–2), 146–164 (2006)

S. Chakraborty, S. Bhowmik, An efficient approach to job shop scheduling problem using simulated annealing. Int. J. Hybrid Inf. Technol. **8**(11), 273–284 (2015a)

S. Chakraborty, S. Bhowmik, Blending Roulette wheel selection with simulated annealing for job shop scheduling problem, in *Proceedings of Michael Farady IET International Summit MFIIS 2015*, Kolkata, India, vol. 2, pp. 579–585 (2015b)

S. Chakraborty, A. Seal, M. Roy, An elitist model for obtaining alignment of multiple sequences using genetic algorithm, in *Proceedings of 2nd National Conference NCETAS 2015*, MIET, Bandel, India, pp. 61–67 (2015)

S. Chakraborty, S. Chatterjee, N. Dey, A. Ashour, A.S. Ashour, F. Shi, K. Mali, Modified cuckoo search algorithm in microscopic image segmentation of hippocampus. Microsc. Res. Tech. **00**, 1–22 (2017a). Wiley

S. Chakraborty, K. Mali, S. Chatterjee, S. Banerjee, K.G. Mazumdar, M. Debnath, K. Roy, Detection of skin disease using metaheuristic supported artificial neural networks, in *2017 8th Annual Industrial Automation and Electromechanical Engineering Conference (IEMECON)*, pp. 224–229. IEEE, Aug 2017 (2017b)

S. Chakraborty, M. Roy, S. Hore, A study on different edge detection techniques in digital image processing, in *Proceedings of Feature Detectors and Motion Detection in Video Processing*, IGI Global, pp. 100–122 (2016)

S. Chakraborty, S. Bhowmik, Job shop scheduling using simulated annealing, in *Proceedings of 1st International Conference IC3A 2013*, JIS College of Engineering, Kalyani, India (2013)

S. Chakraborty, S. Chatterjee, A.S. Ashour, K. Mali, N. Dey, Intelligent Computing in Medical Imaging: A Study. *Advancements in Applied Metaheuristic Computing*, vol. 143 (2017a)

S. Chakraborty, S. Chatterjee, N. Dey, A.S. Ashour, F. Shi, Gradient approximation in retinal blood vessel segmentation, in *4th IEEE Uttar Pradesh Section International Conference on Electrical, Computer and Electronics (UPCON)*, Mathura, India, 2017, pp. 618–623 (2017b)

B. Chin-Wei, M. Rajeswari, Multiobjective optimization approaches in image segmentation–the directions and challenges. Int. J. Adv. Soft Comput. Appl. **2**(1) (2010)

C.A. Cocosco, V. Kollokian, R.K.S. Kwan, A.C. Evans, BrainWeb: online interface to a 3D MRI simulated brain database. NeuroImage **5** (1997)

C.A.C. Coello, A comprehensive survey of evolutionary-based multiobjective optimization techniques. Knowl. Inf. Syst. **1**(3), 129–156 (1999)

M.J. Collins, E.B. Kopp, On the design and evaluation of multiobjective single-channel SAR image segmentation. IEEE Trans. Geosci. Remote Sens. (2008)

M. Cococcioni, P. Ducange, B. Lazzerini, F. Marcelloni, Evolutionary multi-objective optimization of fuzzy rule-based classifiers in the ROC space, in *FUZZ-IEEE 2007*, pp. 1–6 (2007)

K. Deb, *Multi-objective Optimization Using Evolutionary Algorithms* (Wiley, England, 2001)

R. Demirci, Rule-based automatic segmentation of color images. AEU Int. J. Electron. Commun. **60**(6), 435–442 (2006)

R.H. Erin, Feature selection for self-organizing feature map neural networks with applications in medical image segmentation, MSc thesis, University of Louisville, 2001

K. Faceli, M.C.P. De-Souto, A.C. De-Carvalho (2008) A strategy for the selection of solutions of the Pareto front approximation in multi-objective clustering approaches, in *10th Brazilian Symposium on Neural Networks, SBRN 2008*

N. Ghoggali, F. Melgani, Semi-supervised multitemporal classification with support vector machines and genetic algorithms, in *Proceedings of the IEEE-International Geoscience and Remote Sensing Symposium IGARSS-2007*, Barcelona, Spain, pp. 2577–2580 (2007)

N. Ghoggali, F. Melgani, Y. Bazi, A multiobjective genetic SVM approach for classification problems with limited training samples. IEEE Trans. Geosci. Remote Sens. **47**, 1707–1718 (2009)

V. Guliashki, H. Toshev, C. Korsemov, Survey of evolutionary algorithms used in multiobjective optimization. Probl. Eng. Cybern. Robot. **60**(1), 42–54 (2009)

J. Handl, J. Knowles, An evolutionary approach to multiobjective clustering. IEEE Trans. Evol. Comput. **11**, 56–76 (2007)

S. Hore, S. Chakraborty, A.S. Ashour, N. Dey, A.S. Ashour, D.S. Pistolla, T. Bhattacharya, S.R. Chaudhuri, Finding contours of hippocampus brain cell using microscopic image analysis. J. Adv. Microsc. Res. **10**(2), 93–103 (2015)

S. Hore, S. Chakraborty, S. Chatterjee, N. Dey , A.S. Ashour, L.V. Chung, D.N. Le, An integrated interactive technique for image segmentation using stack based seeded region growing and thresholding. Int. J. Electr. Comput. Eng. **6**(6) (2016a)

S. Hore, S. Chatterjee, S. Chakraborty, R.K. Shaw, Analysis of different feature description algorithm in object recognition, in *Feature Detectors and Motion Detection in Video Processing*, IGI Global, pp. 66–99 (2016b)

International Skin Imaging Collaboration Website (n.d.). http://www.isdis.net/index.php/isic-project. Accessed 12 July 2017. (19:52 GMT+05:30)

H. Ishibuchi, Y. Nojima, Performance evaluation of evolutionary multiobjective approaches to the design of fuzzy rule-based ensemble classifiers, in *Fifth International Conference on Hybrid Intelligent Systems (HIS'05)*. IEEE (2005)

D.F. Jones, S.K. Mirrazavi, M. Tamiz, Multi-objective meta-heuristics: an overview of the current state-of-the-art. Eur. J. Oper. Res. **137**(1), 1–9 (2002)

K.A. Johnson, J.A. Becker, The whole brain atlas (1999), http://www.med.harvard.edu/AANLIB/home.html

G.C. Karmakar, L.S. Dooleya, A generic fuzzy rule based image segmentation algorithm. Pattern Recogn. Lett. **23**

K. Kottathra, Y. Attikiouzel, A novel multicriteria optimization algorithm for the structure determination of multilayer feedforward neural networks. J. Netw. Comput. Appl. **19**, 135–147 (1996)

R.V.V. Krishna, S.S. Kumar, Hybridizing differential evolution with a genetic algorithm for color image segmentation. Eng. Technol. Appl. Sci. Res. **6**(5), 1182–1186 (2016)

N. Matake, T. Hiroyasu, M. Miki, T. Senda, Multiobjective clustering with automatic k-determination for large-scale data, in *Genetic and Evolutionary Computation Conference*, London, England, pp. 861–868 (2007)

K. Mali, S. Chakraborty, A. Seal, M. Roy, An efficient image cryptographic algorithm based on frequency domain using haar wavelet transform. Int. J. Secur. Appl. **9**(12), 265–274 (2015)

U. Maulik, I. Saha, Modified differential evolution based fuzzy clustering for pixel classification in remote sensing imagery. Pattern Recogn. **42**(9), 2135–2149 (2009)

K. Miettinen, Introduction to multiobjective optimization: noninteractive approaches. Multiobjective Optim. **5252**, 1–26 (2008)

A. Mukhopadhyay, S. Bandyopadhyay, U. Maulik, Clustering using multi-objective genetic algorithm and its application to image segmentation, in *IEEE International Conference on Systems, Man and Cybernetics*, vol. 3 (2007)

A. Mukhopadhyay, S. Bandyopadhyay, U. Maulik, Combining multiobjective fuzzy clustering and probabilistic ANN classifier for unsupervised pattern classification: application to satellite image segmentation, in *Congress on Evolutionary Computation*. IEEE, pp. 877–883 (2008)

A. Mukhopadhyay, U. Maulik, S. Bandyopadhyay, Multiobjective genetic clustering with ensemble among Pareto front solutions: application to MRI brain image segmentation, in *7th International Conference on Advances in Pattern Recognition* (2009b)

A. Mukhopadhyay, U. Maulik, Unsupervised pixel classification in satellite imagery using multi-objective fuzzy clustering combined with SVM classifier. IEEE Trans. Geosci. Remote Sens. **47** (2009a)

A. Nakib, H. Oulhadj, P. Siarry, Non-supervised image segmentation based on multiobjective optimization. Pattern Recogn. Lett. **29** (2008)

A. Nakib, H. Oulhadj, P. Siarry, Fractional differentiation and non-Pareto multiobjective optimization for image thresholding. Eng. Appl. Artif. Intell. **22**(2), 236–249 (2009a)

A. Nakib, H. Oulhadj, P. Siarry, Image thresholding based on Pareto multiobjective optimization. Eng. Appl. Artif. Intell. (2009b)

Y. Niu, L. Shen, A novel approach to image fusion based on multi objective optimization, in *6th World Congress on Intelligent Control and Automation* (2006)

D. Newman, S. Hettich, C. Blake, C. Merz, UCI repository of machine learning databases, University of California Irvine, Dept. of Information and Computer Sciences (1998), http://www.ics.uci.edu/~mlearn/MLRepository.html. Acessado em 06/07/2006 1998

M.G.H. Omran, A.P. Engelbrecht, A. Salman, Differential evolution methods for unsupervised image classification, in *Congress on Evolutionary Computation* (2005a)

D.W. Optiz, Feature Selection for ensembles, in *16th National Conference on Artificial Intelligence (AAAI)* (1999)

M.G.H. Omran, A.P. Engelbrecht, A. Salman, Differential evolution methods for unsupervised image classification, in *Proceedings of Congress on Evolutionary Computation* (2005b)

A. Paoli, F. Melgani, E. Pasolli, Clustering of hyperspectral images based on multiobjective particle swarm optimization. IEEE Trans. Geosci. Remote Sens. Accepted for publication (2009)

P. Pulkkine, H. Koivisto, Fuzzy classifier identification using decision tree and multiobjective evolutionary algorithms. Int. J. Approx. Reason. **48**, 526–543 (2008)

M. Roy, S. Chakraborty, K. Mali, S. Chatterjee, S. Banerjee, A. Chakraborty, K. Roy, Biomedical image enhancement based on modified cuckoo search and morphology, in *2017 8th Annual Industrial Automation and Electromechanical Engineering Conference (IEMECON)*, pp. 230–235. IEEE, Aug 2017

I. Saha, U. Maulik, S. Bandyopadhyay, An improved multi-objective technique for fuzzy clustering with application to IRS image segmentation. *Applications of Evolutionary Computing*, pp. 426–431 (2009)

S. Saha, S. Bandyopadhyay, A multiobjective simulated annealing based fuzzy-clustering technique with symmetry for pixel classification in remote sensing imagery, in *19th International Conference on Pattern Recognition* (2008)

S. Saha, S. Bandyopadhyay, A new symmetry based multiobjective clustering technique for automatic evolution of clusters. Pattern Recogn. (2010)

A. Seal, S. Chakraborty, K. Mali, A new and resilient image encryption technique based on pixel manipulation, value transformation and visual transformation utilizing single–level haar wavelet transform, in *Proceedings of Advances in Intelligent Systems and Computing* (Springer), pp. 603–611

S. Shirakawa, T. Nagao, Evolutionary image segmentation based on multiobjective clustering, in *IEEE Congress on Evolutionary Computation, 2009. CEC'09*, pp. 2466–2473, May 2009

X. Wang, H. Wang, Classification by evolutionary ensembles. Pattern Recogn. **39**, 595–607 (2006)

Y. Zhang, P.I. Rockett (2005) Evolving optimal feature extraction using multi-objective genetic programming: a methodology and preliminary study on edge detection, in *Conference on Genetic and Evolutionary Computation*

Feature Selection Using Multi-Objective Optimization Technique for Supervised Cancer Classification

P. Agarwalla and S. Mukhopadhyay

1 Introduction

The importance of classifying cancer and appropriate diagnosis of advancement of disease has leveraged many research fields, from biomedical to the machine learning (ML) domains. For proper diagnosis of a disease and categorizing it into different classes, investigation in the changes of genetic expression level is needed. Gene expression data (Zhang et al. 2008) has a huge impact on the study of cancer classification and identification. The ability of machine learning approaches to detect key features from a huge complex dataset reveals their importance in the field of feature selection from microarray dataset. Modelling of cancer progression and classification of disease can be studied by employing learning-based approaches. The methodology that has been intimated here is based on supervised learning technique for different input feature genes and data samples. In the supervised learning process, a set of training data has been provided along with their class information. Based on the methodology, it will identify the relevant informative features which will further identify the class of an unknown test sample. Multi-objective optimization techniques are involved in this paper to select the required features which are efficient in the classification purpose as well as carry significant biological information related to disease. Then, those features are used to train the classifier and a new sample is diagnosed.

Different approaches are developed by the researchers for finding marker genes (Khunlertgit and Yoon 2013; Bandyopadhyay et al. 2014; Mukhopadhyay and Mandal 2014; Apolloni et al. 2016) related to different diseases. Various statistical filter

P. Agarwalla (✉)
Heritage Institute of Technology, Kolkata, India
e-mail: prativa.agarwalla87@gmail.com

S. Mukhopadhyay
Institute of Radio Physics & Electronics, Kolkata, India
e-mail: sumitra.mu@gmail.com

© Springer Nature Singapore Pte Ltd. 2018
J. K. Mandal et al. (eds.), *Multi-Objective Optimization*,
https://doi.org/10.1007/978-981-13-1471-1_9

approaches (Bandyopadhyay et al. 2014), clustering processes (Mukhopadhyay and Mandal 2014) and wrapper-based hybrid approaches (Apolloni et al. 2016) are utilized for this purpose. Many supervised and unsupervised classification techniques are used for classification or clustering of tissue samples. A family of bio-inspired algorithms has also been applied while formulating the problem as an optimization problem. The gene subset identification problem can be reduced to an optimization problem consisting of a number of objectives. However, identifying most relevant and non-redundant genes is the main goal that is to be achieved. Motivated by this, different multi-objective methodologies are proposed in the literature (Mukhopadhyay and Mandal 2014; Ushakov et al. 2016; Zheng et al. 2016; Mohamad et al. 2009; Hero and Fluery 2002; Chen et al. 2014). Recently, a number of literature involve multi-objective based methods for the feature selection from microarray datasets. To obtain a small subset of non-redundant disease-related genes by using the multi-objective criterions, different bio-inspired algorithms are applied. For example, a variable-length particle swarm optimization (Mukhopadhyay and Mandal 2014) is implemented. A bi-objective concept is implemented for clustering of cancer gene from microarray datasets (Ushakov et al. 2016). In work (Zheng et al. 2016), a numerical method is implemented with GA to extract informative features in the domain of bioinformatics. Multi-objective function is optimized by genetic algorithm (GA) (Mohamad et al. 2009) to obtain significant genes for cancer progression. Pareto-based analysis is performed for filtering the relevant genes (Hero and Fluery 2002). In this chapter, the problem is formulated as a multi-objective optimization problem, and multi-objective blended particle swarm optimization (MOBPSO), multi-objective blended differential evolution (MOBDE), multi-objective blended artificial bee colony (MOBABC) and multi-objective blended genetic algorithm (MOBGA) are proposed for this purpose. Here, the stochastic algorithms are modified using Laplacian blended operator to incorporate diversity in the search process. It helps to get more diversified and promising result. This has been established theoretically and experimentally in the subsequent sections. The modified multi-objective algorithms are searching for Pareto-front solution which represents the feature genes for cancer classification. Then, the comparative analysis is performed based on the efficiency of finding relevant marker genes which are significantly associated with the disease.

For the reliable classification of a disease, multiple objectives play an important role. In the context of gene selection from the microarray data, two objectives are considered. One of them allows selection of the most differentially expressed genes which help in identifying the separation between classes. Another consideration is given to the accurate classification of the disease. T-score is used for the job of selecting differentially expressed feature. Those selected genes may not be efficient to provide good classification result due to the heterogeneous nature of gene expression. Our mission is to choose the combination of feature which is providing high accuracy also. Again, if entire differential features are used, it can cause over-fitting of the classifier. So, it is necessary to eliminate redundant features for the task of classification. Here, our aim is to obtain high accuracy of classification. As well as the selected features should have good value of t-score and this in turn indicates differentiability in expression level. If the proper combination of genes for the determination of disease can be identified, then it will be a significant contribution for the diagnosis of disease and the treatment will be more effective and precise.

Experiments are performed using different types of microarray datasets which include Child_ALL (Cheok et al. 2003), gastric (Hippo et al. 2002), colon (Alon et al. 1999) and leukemia (Golub et al. 1999) cancer data. Initially, the performance of the proposed methodologies for the job of classification of disease through supervised learning process is evaluated. As mentioned, in this chapter, differential evolution (DE) (Price et al. 2006), artificial bee colony (ABC) (Karaboga and Basturk 2007), genetic algorithm (Yang and Honavar 1998) and particle swarm optimization (PSO) (Kennedy 2011) algorithms are involved for solving multi-objective feature selection problem along with Laplacian operator. Then, the results of classification are compared with other established methods for four real-life cancer datasets. The proposed Laplacian operator integrated with multi-objective swarm and evolutionary algorithms establishes good results in all respect. The results ascertain the ability of multi-objective blended particle swarm optimization (MOBPSO), multi-objective blended differential evolution (MOBDE), multi-objective blended artificial bee colony (MOBABC) and multi-objective blended genetic algorithm (MOBGA) to produce more robust gene selection activity. At the end of the chapter, the biological relevance of the resultant genes is also validated and demonstrated.

The remaining of the chapter is presented as follows: First, a description of experimental datasets is presented. Next, the proposed technique is presented for marker gene selection. In the next section, the result of the proposed technique is demonstrated and a comparative analysis is provided. Finally, the biological relevance of the result is given.

2 Experimental Datasets

Two classes of raw microarray data for different types of cancers are collected. In microarray data, the expressions of genes are arranged column-wise, whereas the samples, collected from different sources, are arranged in row. The changes at molecular level of genes can be visualized from the microarray technology. Here, gene expressions from different samples are analysed in a single microscopic slide. Samples from cancerous and non-cancerous tissues are taken and dyed using fluorescent colours. Then, through hybridization procedure, the combined colours are analysed. The intensity of different areas of microarray slide reveals the informative content and subsequently, conclusion can be made by investigating the expression level. Authors have collected microarray datasets of different variants of cancers from reliable sources such as **National Centre for Biotechnology Information** (**NCBI**). A brief description of the microarray datasets used for the experimental purpose is given below.

Child-ALL (GSE412) (Cheok et al. 2003): 110 samples of childhood acute lymphoblastic leukemia are collected. Among them, 50 and 60 examples are of before and after therapy, respectively. The samples are having expression level of 8280 genes. So, the dimension of the dataset matrix is $D_{110 \times 8280}$.

Gastric Cancer (GSE2685) (Hippo et al. 2002): Gastric cancer occurs due to the growth of cancerous cells in the lining of stomach. This experimental dataset is having expression of 4522 genes from total 30 number of different tissue samples. Two classes of tumour such as diffuse and intestinal advanced gastric tumour samples are considered. 22 samples are present in the first class and another class is having 8 samples.

Colon cancer (Alon et al. 1999): The colon dataset contains expression values of 6,000 genes column-wise. Totally, 62 cell samples are present row-wise, among which first 40 biopsies are from tumour cells and next 22 samples are from healthy parts of the colon. The data is collected from a public available website: www.bico nductor.com/datadet.

Lymphoma and Leukemia (GSE1577) (Golub et al. 1999): The leukemia dataset consists of 72 microarray experiments including two types of leukemia, namely AML (25 samples) and ALL (47 samples). Expressions of 5147 genes are present in the dataset. The data is collected from a public website: www.biolab.si/supp/bi-cancer/projections/info/.

Preprocessing Microarray Data: The microarray data generally consists of noisy and irrelevant genes which may mislead the computation. So, to extract most informative and significant gene subset which is relevant for the diagnosis of the disease, first the noisy and irrelevant genes are to be eliminated. To analyse the noise content, signal-to-noise ratio is calculated for each gene and based on the SNR value, the top 1000 genes are selected for the next level of computation. The formula for SNR value calculation is given in Eq. (1). Here, μ_1, μ_2 are the means of gene expression of a particular gene over the samples of first class and second class, respectively. sd_1, sd_2 are the standard deviations of gene expression of a particular gene over the samples of first and second class, respectively.

$$SNR = \frac{\mu_1 - \mu_2}{sd_1 + sd_2} \tag{1}$$

Next, using min-max normalization process (Bandyopadhyay et al. 2014), those 1000 genes are normalized. If the expression of a gene over the samples is represented by the variable g, then the min-max normalization formula for a data point g_i is described by Eq. (2). Thus, a data matrix D_{mx1000} is formed where m is the number of samples. This generated data matrix is used for the next level of computation.

$$x_i(normalized) = \frac{g_i - \min(g)}{\max(g) - \min(g)} \tag{2}$$

3   Objectives

A multi-objective optimization problem (MOP) can be represented as follows:

$$\text{maximize } F(x) = (f(x), \ldots, fm(x))^T \tag{3}$$

subject to $x \in \Omega$, where Ω is the search space and x is the decision variable vector. $F: \Omega \to R^m$, where m is the number of objective functions, and R^m is the objective space. The heterogeneity in the expression level of genes must be high from one patient to another and an optimal combination of feature set through learning process is to choose which will perform well for the classification of new sample. It has been noticed that a particular combination of gene set which is highly differentiable from one class to another sometimes fails to achieve good classification result. It sometimes causes over-fitting of the classifier. So, to keep a balance between them, a multi-objective problem is constructed. It gives rise to a set of trade-off between Pareto-optimal (P-O) solutions (Srinivas and Deb 1994). Here, two objectives are considered in this chapter which are described below:

$$t - score = \frac{\mu_1 - \mu_2}{\sqrt{\left(\frac{sd_1^2}{n_1} + \frac{sd_2^2}{n_2}\right)}} \tag{4}$$

$$Accuracy = \frac{t_n + t_p}{t_n + t_p + f_p + f_n} \tag{5}$$

For t-score calculation, the mean expression of the selected genes over the samples for both the classes is calculated. Then, the difference between the two mean expressions is computed. For fitness function for PSO computation t-score is utilized which is described in Eq. (4) where μ and sd represent the mean and the standard deviation value of the two classes, respectively. n_1 and n_2 are the number of samples present in the two classes, respectively. Higher fitness function indicates the better selectivity of genes. As another objective function, accuracy is estimated using the number of false positive (fp), true negative (tn), false negative (fn) and true positive (tp) for class prediction. The objective used for formulating multi-objective problem and the proposed methodology is discussed in brief in the following sections.

4   Proposed Methodology

The problem has been modelled as a multi-objective optimization problem, and different multi-objective evolutionary algorithms are employed. In the multi-objective optimization problem, a set of solutions called Pareto-optimal has to be achieved. Here, based on two objective functions, optimal Pareto solution is generated and for this purpose non-dominated sorting technique (Srinivas and Deb 1994) has been used. A new version of optimization algorithm is proposed and developed, entitled as multi-objective blended particle swarm optimization (MOBPSO) algorithm for finding gene subsets in cancer progression. PSO algorithm is modified, and Laplacian blended operator is integrated to provide better diversity in searching procedure. So, the multi-objective blended PSO based concept is implemented along with GA, DE and ABC algorithms and MOBABC, MOBGA and MOBDE are developed. Subsequently, a comparative study is performed using the proposed multi-objective stochastic computational methods. For the selection of genes from microarray data, supervised learning method is employed where the total experimental dataset is partitioned into two subsets. One is used for training purpose of the proposed model and the other one is used for evaluation of the model. The schematic diagram of the proposed methodology is shown for MOBPSO in Fig. 1, and the process selection of bio-markers from gene expression profile is described below.

4.1 Multi-Objective Blended Particle Swarm Optimization (MOBPSO)

4.1.1 Concept of Particle Swarm Optimization (PSO)

Particle swarm optimization, proposed by Eberhart and Kennedy in 1995, is a simple, well-established and widely used bio-inspired algorithm in the field of optimization. The technique is developed based on the social behaviour of a bird flock, as the flock searches for food location in a multidimensional search space. Location of a particle represents the possible solutions for the optimization function, $f(x)$. Velocity and the direction of a particle are affected by its own past experience as well as other particles in the swarm have an effect on the performance. The velocity and position update rule for ith particle at tth generation are given in Eqs. (6) and (7) where the values of two random weights, c_1 and c_2, represent the attraction of a particle towards its own success p_{best} and the attraction of a particle towards the swarm's best position g_{best} respectively. w is the inertia weight. After a predetermined number of iterations, the best solution of the swarm is the solution of the problem.

$$v^i(t) = w * v^i(t-1) + c_1 * rand * \left(p_{best}^i - x^i\right) + c_2 * rand * \left(g_{best}^i - x^i\right) \quad (6)$$

$$x^i(t) = x^i(t-1) + v^i(t) \quad (7)$$

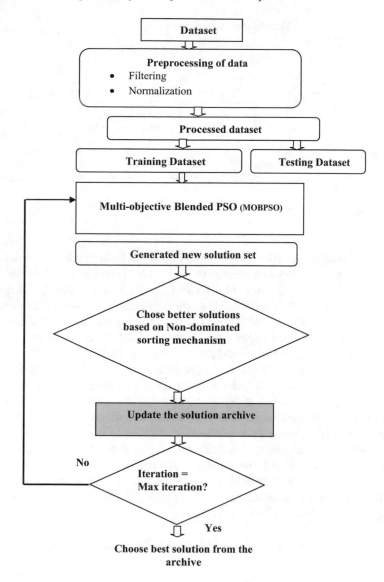

Fig. 1 Computational methods using MOBPSO-based approach

4.1.2 Concept of Multi-Objective Blended Particle Swarm Optimization (MOBPSO) for the Selection of Genes

The main drawback of PSO algorithm is that it is easily trapped to local optima due to scarcity of divergence which leads to premature convergence. To get rid of the issue, a diversity mechanism should be applied to get rid of any local optima. So,

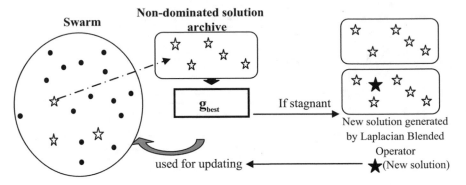

Fig. 2 Schematic diagram of the proposed methodology

in course of searching, better results can be achieved by introducing some sort of diversity technique. Being motivated by this, a blended operator is implemented with PSO and MOBPSO is proposed. In MOBPSO, particles are searching for optima, following the rules of PSO and if any particle is stuck at local optima, then it comes out of the situation by using the new probable solution generated through Laplacian blended operator. Blended Laplacian operator works very effectively to generate new probable solutions in the search space. The whole swarm is directed to the new solution which helps to discover new area of searching. As a whole the performance of the algorithm is accelerated and a better optimized result of the problem can be obtained. The mechanism is discussed below and shown in Fig. 2. MOBPSO is applied in the domain of multi-objective problem where the aim is to choose the Pareto-optimal solution. For the selection of feature genes, it is optimizing two objective functions and after each iteration non-dominated solutions are selected. As a single fitness value cannot be assigned, a modification is performed in the updating rule of the particles. In MOBPSO, the particles of the swarm are updating their velocity, and position towards the food using Eqs. (8), (9) and during updating the effect of g_{best} is only taken into consideration.

$$v^i(t) = w \cdot v^i(t-1) + c_1 \cdot rand \cdot (g^i_{best} - x^i) \tag{8}$$

$$x^i(t) = x^i(t-1) + v^i(t) \tag{9}$$

The g_{best} is the best solution chosen among the non-dominated solutions obtained so far. For the problem of identifying significant genes, the differentially expressed genes in different classes are important to be identified. As mentioned, t-score is used as one of the objective functions for the purpose. Accuracy is chosen as another objective function where the aim is to maximize the value of the accuracy. Now for each iteration, new subset of genes is generated by MOBPSO. The position of each particle represents a possible gene subset of the problem. Then, the fitness value of each particle is calculated and based on the non-dominated sorting, better solutions are sorted. The non-dominated solutions are stored in an archive. As, for multi-objective problem, a number of Pareto-optimal solutions can be achieved, one of the Pareto-optimal solutions is randomly chosen as g_{best}. The g_{best} is used for the updating of velocity and position of a particle of MOBPSO. In the next iteration, again new subsets of genes are selected by the particles and the non-dominated solutions are loaded into the archive. Now, the archive is updated with the non-dominated solutions among the solutions obtained so far. A binary version of the optimization algorithm is used to select feature genes which are to be presented in the computation. The selection of genes is done based on Algorithm 1.

Algorithm-1 (Implementation of Binary concept)

```
for j=1:dimension of particle
if x (j) > 0.5
    x(j)=1;
else x (j)=0;
end;
end;
```

During the search process, it may happen that generation of new better solution is stuck after few iteration due to the lack of diversity. So, the algorithm needs some mechanism which can direct the particles to a new probable region. Blended Laplacian operator works very efficiently to provide diversity to the swarm. If no better solution is generated, blended operator produces a new g_{best} at that point to provide diversification to the swarm. The mechanism is shown schematically in Fig. 2. First, two random solutions, sol_1 and sol_2, are chosen from the archive. Then, using a random coefficient termed as beta, two new solutions y_1 and y_2 are produced. The new g_{best}, g_{best_new}, is a combination of these two new solutions y_1 and y_2 having a weightage factor gamma. Blended Laplacian operator used for g_{best_new} generation is described below. g_{best_new} is completely a new solution generated from the old best non-dominated solutions, achieved so far. The new solution works to direct the entire swarm to a new possible direction.

$$gamma = 0.1 + (1-0.1)^{0.95*iter}$$
$$beta = 0.5 * log(rand)$$
$$y_1 = sol_1 + beta * (sol_1 - sol_2)$$
$$y_2 = sol_2 + beta * (sol_1 - sol_2)$$
$$g_{best\_new} = gamma * y_1 + (1 - gamma) * y_2$$

(10)

The new g_{best_new} provides a momentum in the velocity of the particles. The position of the particles consequently changes. So, the stagnancy in the movement of the particles can be overcome. To establish the effectiveness of blended Laplacian operator, few plots are provided in Fig. 3. The experimental analysis is performed for gastric cancer data, and the fitness values of searching particles for the two objectives t-score and accuracy are plotted for different iterations. After a 100 iteration when

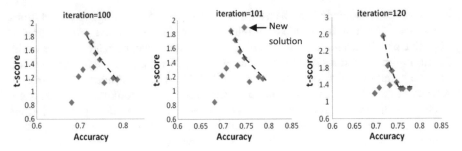

Fig. 3 Plot of fitness values for gastric cancer at different iterations

swam is unable to generate new better solution, a new solution is produced using blended operator. As a result, the swarm updates themselves and overcomes the stagnancy. The effect is shown for iteration number 120. New better solutions are generated by the searching particles. The overall MOBPSO technique is described in Algorithm 2.

Algorithm-2 (MOBPSO)

1. **Initialization.**
   Total Number of particle=N
   (a) Randomly initialize the position of particles, X_i (1=1, 2,..., N)
   (b)Initialize archive1 with few randomly chosen solutions
2. **Termination check.**
   (a) If the termination criterion holds go to step 8.
   (b) Else go to step 3.
3. **Set** t=1(iteration counter)
   For i= 1,2...N **Do**
      (a) **If** stagnancy occurs,
         Choose g_{best} randomly from the archive1
        Else choose g_{best} randomly from the archive2
        End If
      (b) Update the position according to Equations (8),(9)
      (c) Evaluate the fitness of the ith particle $f_1(X_i)$ and $f_2(X_i)$
        for two objectives
   End For
      (d) Choose the non-dominated solutions among N particles
      (e) Update the archive1 with non-dominated solutions
      (f) Check for stagnancy
        If stagnancy occurs
          i) Generate few new solutions (g_{best_new}) using blended laplacian
           operator as equation (10)
          ii) Construct a new archive2 using those g_{best_new}
        End If
4. **Set** t=t+1.
5. **Go to** step 2
6. Solution is the solution from archive1

4.2 Other Comparative Methods for the Selection of Genes

Most of the evolutionary and swarm intelligence algorithms such as genetic algorithm (GA), differential evolution (DE) and artificial bee colony (ABC) suffer from local trapping which results in premature convergence. The Laplacian blended operator can be implemented when such situation occurs as it produces few new solutions blending some previously generated solutions. So, the use of operator makes the optimization algorithms more efficient in the process of stochastic searching. Here, authors have integrated the blended operator with the above-mentioned algorithms and introduced multi-objective blended differential evolution (MOBDE), multi-objective blended artificial bee colony (MOBABC) and multi-objective blended genetic algorithm (MOBGA) in the similar fashion as it is done for MOBPSO. MOBDE, MOBABA and MOBGA are now applied to the four cancer datasets for marker gene selection. In the next subsection, the methodologies are discussed in brief.

4.2.1 Multi-Objective Blended Genetic Algorithm (MOBGA)

GA is a metaheuristic algorithm which is being inspired by the natural selection. It constitutes of few steps like parent selection, crossover and mutation (Yang and Honavar 1998). Initially, the algorithm starts with few solution termed as chromosome. Now, fitted chromosomes are considered as parents who are used to generate new child solutions. To create new solutions, a set of genetic operations like crossover and mutation are used. In MOBGA, initially non-dominated solutions are stored in an archive. Parents are chosen randomly from the archive to create next generation of solutions. Next, based on Pareto-optimal concept, fitted chromosomes survive and the archive is updated accordingly. Similar to the MOBPSO, when stagnancy occurs, blended Laplacian operator is utilized to overcome it. New parents are generated using blended Laplacian operator. For MOBGA, the process of gene selection is kept same as shown in Fig. 1, and only the MOBPSO block is replaced by the MOBGA.

4.2.2 Multi-Objective Blended Differential Evolution (MOBDE)

DE is a population-based stochastic optimization technique which adopts mutation and crossover operators to search for new promising areas in the search space (Price et al. 2006). The algorithm starts with a number of solutions based on non-dominated sorting and more promising solutions are kept in an archive. From the archive, fitted solutions are selected for mutation purpose and new solutions are produced. Similar to previously mentioned algorithms, when no further improvement is found, Laplacian blended operator is used. The binary format is implemented as done using Algorithm-1, and the process of gene selection is same as described in Fig. 1 except that the MOBPSO block is replaced by MOBDE.

Table 1 Parameters used in different swarm and evolutionary algorithms

Algorithms	Parameters	Explanation	Value
MOBPSO	N	Number of particle(s) in one swarm	20
	c_1, c_2	Acceleration constants	1.49
	w	Inertia	0.7
	r_1, r_2	Random numbers	[0, 1]
MOBGA	N	Number of genetic(s) in one group	20
	Ps	Selection ratio	0.8
	Pc	Crossover ratio	0.9
	Pm	Mutation ratio	0.01
MOBDE	N	Number of individual(s) in one group	20
	fm	Mutation factor	0.6
	CR	Crossover rate	0.9
MOBABC	N	Number of bee(s) in one swarm	20
	L	Limit for scout phase	100

4.2.3 Multi-Objective Blended Artificial Bee Colony (MOBABC)

Artificial bee colony (ABC) algorithm is inspired by the foraging behaviour of honey bees (Karaboga and Basturk 2007). Three groups of bees, employee bees, onlooker bees and scout bees, are involved in the searching process. The employee bee produces a modification on the position (solution) and depending on the non-dominated sorting procedure best positions are memorized. Here, those positions are stored in an archive. Onlooker bee chooses a food source from the archive and searches thoroughly across it. When stagnancy occurs, the archive is updated with new solutions, produced through Laplacian blended operator. MOBABC is applied to cancer datasets similar to the process as described in Fig. 1 just replacing the block of MOBPSO by MOBABC. The parameter settings of all other stochastic algorithms are given in Table 1.

5   Experimental Results

The experimental datasets consist of microarray data of Child_ALL, leukemia, colon and gastric cancer. Multi-objective blended GA (MOBGA), multi-objective blended DE (MOBDE), multi-objective blended ABC (MOBABC) and multi-objective blended PSO (MOBPSO) are employed for the task of feature gene selection using supervised learning process in the field of cancer classification. The performance of the proposed multi-objective gene selection techniques is analysed and compared for four real-life cancers. The evaluation is performed based on classification results such as sensitivity, specificity, accuracy and F-score (Agarwalla and Mukhopadhyay 2016) using 10-fold cross-validation. Different classifiers are involved for the classification such as support vector machine (SVM), decision tree (DT), K-nearest neighbour (KNN) classifier and naive Bayes (NB) classifier (Kotsiantis et al. 2007). Experiments are carried out 10 times, and the average results are reported. The performance of MOBPSO is given in the subsequent sections utilizing different classifiers. Then, a comparative study is performed involving all the algorithms. Next, the results are compared with other existing methods reported in different research articles. The proposed methodologies establish promising results, indicating the capability to produce more effective gene selection activity.

5.1 *Classification Results*

In this chapter, the aim is to identify top differentially expressed genes (DEGs) which are performing well in the process of classification. By involving MOBPSO, MOBGA, MOBDE and MOBABC algorithms, the optimized gene subset is obtained. The gene subset which is identified is validated by analysing the classification results. The proposed methodology is implemented using four different well-known classifiers (SVM, DT, KNN and NB). The experimental result ascertains that the proposed methodology is able to extract important features from the huge dataset. The classification results of MOBPSO algorithm for different cancer datasets are given in Table 2. For leukemia cancer, NB classifier shows better performance compared to others classifiers. Here, 100% accuracy is achieved which indicates the perfect classification of disease. For colon cancer, decision tree classifier is working efficiently in terms of providing good specificity of the result. Highest accuracy is achieved by the SVM classifier which is equal to 87%.

For gastric cancer, SVM achieves 89% accuracy which establishes its superiority over the other classifiers, used for the experiment. KNN classifier is providing 79% accuracy and 89% sensitivity as the classification result of Child_ALL data.

The accuracy of classification obtained using different classifiers is also given in the form of bar chart in Fig. 4 for better interpretability of the results. The comparative result shows that for leukemia data, NB and decision tree both work effectively to classify the cancer. In case of colon cancer, SVM classifier is producing reliable result. For gastric cancer, SVM and NB classifiers are able to find out relevant genes for disease classification. KNN classifier is performing top for Child_ALL cancer compared to all other classifier techniques. Similar to MOBPSO, the other methodologies like MOBGA, MOBABC and MOBDE are applied on the cancer datasets and a comparative study is performed in the next subsection.

Table 2 Results of classification using MOBPSO for different cancer datasets

Dataset	Algorithms	Sensitivity	Specificity	Accuracy	F-score
Leukemia	*MOBPSO-*SVM	1.00	0.78	0.89	0.87
	*MOBPSO-*KNN	0.89	0.87	0.89	0.91
	*MOBPSO-*tree	0.96	1.00	0.99	0.98
	MOBPSO-NB	**1.00**	**1.00**	**1.00**	**1.00**
Colon	*MOBPSO-*SVM	**0.91**	0.72	**0.87**	**0.81**
	*MOBPSO-*KNN	0.75	0.71	0.73	0.70
	*MOBPSO-*tree	0.72	**0.77**	0.77	0.75
	MOBPSO-NB	0.78	0.76	0.76	0.75
Gastric	*MOBPSO-*SVM	**1.00**	**0.89**	**0.89**	**0.92**
	*MOBPSO-*KNN	0.81	0.87	0.83	0.89
	*MOBPSO-*tree	0.78	0.83	0.82	0.86
	MOBPSO-NB	0.91	0.87	0.89	0.90
Child_ALL	*MOBPSO-*SVM	0.71	0.76	0.70	0.71
	*MOBPSO-*KNN	**0.89**	**0.73**	**0.78**	**0.81**
	*MOBPSO-*tree	0.78	0.77	0.74	0.72
	MOBPSO-NB	0.72	0.73	0.71	0.69

5.2   Comparative Analysis

To estimate the effectiveness of the proposed method, experiments are conducted on the four real-time cancer datasets. Here, authors have provided the results of classification of disease after applying MOBPSO, MOBDE, MOBGA and MOBABC on the datasets. SVM classifier is used for each classification purpose. Average result of 10 times 10-fold cross-validation is reported for the comparative study in Table 3. Best results are marked in bold. For leukemia, good results are obtained using MOBGA. For colon cancer, MOBPSO is the best performing feature selection technique and MOBABC is able to obtain second position. MOBDE has achieved promising result for gastric cancer, whereas all the algorithms are able to achieve 100% sensitivity for the data. For Child_ALL data, MOBPSO is able to estimate the proper genes for the classification of disease with an accuracy of 75%. The

Fig. 4 Accuracy of classification using different classifiers for **a** leukemia, **b** colon, **c** gastric, **d** Child_ALL cancer datasets

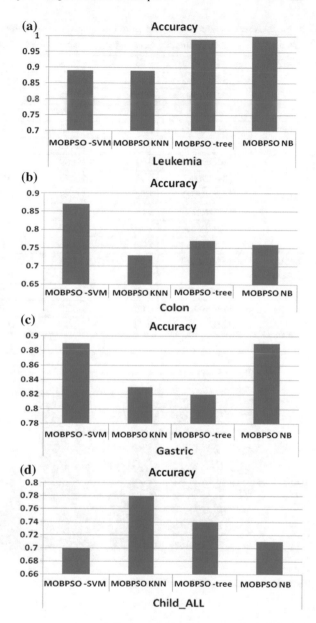

comparison of accuracy obtained from different proposed algorithms is also shown in Fig. 5.

In Table 4, results are again compared with other approaches, reported in different literature for gene selection methodology (Mukhopadhyay and Mandal 2014; Apolloni et al. 2016; Salem et al. 2017; Luo et al. 2011). NSGA-II (Deb et al. 2002) and

Table 3 Result of comparison with different swarm and evolutionary algorithms

Dataset	Algorithms	Sensitivity	Specificity	Accuracy	F-score
Leukemia	*MOBPSO*	**1.00**	0.78	0.89	0.87
	MOBABC	0.84	0.86	0.83	0.80
	MOBDE	0.71	0.83	0.81	0.79
	MOBGA	0.90	**0.91**	**0.90**	**0.89**
Colon	*MOBPSO*	**0.91**	0.72	**0.87**	**0.81**
	MOBABC	0.90	**0.76**	0.81	0.83
	MOBDE	0.71	0.69	0.67	0.68
	MOBGA	0.78	0.73	0.77	0.80
Gastric	*MOBPSO*	**1.00**	0.89	0.89	0.92
	MOBABC	**1.00**	0.90	0.90	0.92
	MOBDE	**1.00**	**0.94**	**0.91**	**0.93**
	MOBGA	**1.00**	0.86	0.87	0.89
Child _ALL	*MOBPSO*	**0.71**	**0.76**	**0.75**	**0.71**
	MOBABC	0.60	0.64	0.61	0.62
	MOBDE	0.65	0.70	0.67	0.63
	MOBGA	0.50	0.70	0.66	0.68

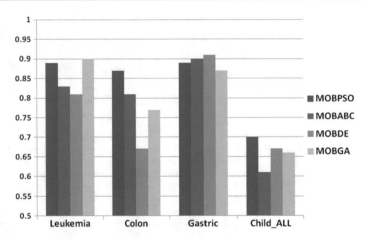

Fig. 5 Accuracy of classification using different algorithms

MOEA/D (Zhang and Li 2007) are also applied on the cancer datasets to obtain Pareto solutions for the objectives. For colon and Child_ALL datasets, MOBPSO is able to gain the best result of accuracy in classification of cancer among the techniques, used for comparison. For other two datasets, the results are also quite promising. The comparative result signifies the efficiency of the proposed methodology for supervised cancer classification.

Table 4 Comparison of accuracy of classification with other results reported in the literature

Reference	Year	Leukemia	Colon	Gastric	Child_ALL
(Salem et al. 2017)	2017	0.97	0.85	–	–
(Luo et al. 2011)	2011	0.71	0.80	–	–
(Apolloni et al. 2016)	2016	0.82	0.75	–	–
(Mukhopadhyay and Mandal 2014)	2014	–	–	0.96	0.74
NSGA-II (Deb et al. 2002)	2002	0.78	0.75	0.93	0.68
MOEA/D (Zhang and Li 2007)	2007	0.92	0.81	0.91	0.72
MOBPSO	2017	0.89	0.87	0.89	0.75
MOBDE	2017	0.83	0.81	0.90	0.61
MOBGA	2017	0.81	0.67	0.91	0.67
MOBABC	2017	0.90	0.77	0.87	0.66

Table 5 Biological significance for gene–disease association

Dataset	Associated diseases	Gene symbol
Leukemia	Leukemia	RAG1(3), MSH(61), CD36(2)
	Lymphomas	CCND3(7), LYN(4)
Colon	Colorectal cancer	MAPK3(11), EGR1(1)
	Malignant tumour of colon	IGF1(67), KLK3(781)
Gastric	Malignant neoplasm of stomach	CYP2C9(1), SPP1(20)
	Stomach carcinoma	SPP1(21), NOS2(2),
Child_ALL	Tumour progression	SMAD3(1), ITGA6(1)

5.3 Biological Relevance

Biological relevance of the experimentally selected genes is analysed by gathering the information about those genes from disease–gene association database. Also, the information of number of Pubmed citations against those genes is collected. In Table 5, disease information related to those top genes is given. For example, MSH gene has 61 Pubmed citations as evidence that the gene is related to leukemia cancer. Similarly, for colon cancer KLK3 is the most cited gene related to the disease. The information proves the biological significance of the proposed work. As a whole, it can be concluded that the proposed gene selection methodologies are more efficient in detection of the relevant genes for all the different types of datasets.

6 Conclusion

Classification of disease through supervised learning method leads to the investigation on feature selection technique. So, for the feature reduction and extraction from the huge dimension of data, authors involve new multi-objective blended particle swarm optimization (MOBPSO) technique. The methodology uses a new concept of integrating blended Laplacian operator in the algorithmic portion, and it generates a subset of genes based on two objectives. The multi-objective concept along with the proposed methodology is proved to be very useful in the context of diagnosis of disease as it identifies biologically significant genes related to the disease. Similarly, authors have implemented the concept with other swarm and evolutionary algorithms and developed multi-objective blended differential evolution (MOBDE), multi-objective blended artificial bee colony (MOBABC) and multi-objective blended genetic algorithm (MOBGA). The experimental result establishes that the proposed technique is able to provide promising result in the context of classification of disease which reflects its effectiveness of selecting relevant feature genes.

References

P. Agarwalla, S. Mukhopadhyay, Selection of relevant genes for pediatric leukemia using cooperative Multiswarm. Mater. Today Proc. **3**(10), 3328–3336 (2016)

U. Alon, N. Barkai, D.A. Notterman, K. Gish, S. Ybarra, D. Mack, A.J. Levine, Broad patterns of gene expression revealed by clustering analysis of tumor and normal colon tissues probed by oligonucleotide arrays. Proc. Natl. Acad. Sci. **96**(12), 6745–6750 (1999)

J. Apolloni, G. Leguizamón, E. Alba, Two hybrid wrapper-filter feature selection algorithms applied to high-dimensional microarray experiments. Appl. Soft Comput. **38**, 922–932 (2016)

S. Bandyopadhyay, S. Mallik, A. Mukhopadhyay, A survey and comparative study of statistical tests for identifying differential expression from microarray data. IEEE/ACM Trans. Comput. Biol. Bioinform. **11**(1), 95–115 (2014)

K.H. Chen, K.J. Wang, M.L. Tsai, K.M. Wang, A.M. Adrian, W.C. Cheng, K.S. Chang, Gene selection for cancer identification: a decision tree model empowered by particle swarm optimization algorithm. BMC Bioinform. **15**(1), 49 (2014)

M.H. Cheok, W. Yang, C.H. Pui, J.R. Downing, C. Cheng, C.W. Naeve, W.E. Evans, Treatment-specific changes in gene expression discriminate in vivo drug response in human leukemia cells. Nat. Genet. **34**(1), 85–90 (2003)

K. Deb, A. Pratap, S. Agarwal, T.A.M.T. Meyarivan, A fast and elitist multiobjective genetic algorithm: NSGA-II. IEEE Trans. Evol. Comput. **6**(2), 182–197 (2002)

T.R. Golub, D.K. Slonim, P. Tamayo, C. Huard, M. Gaasenbeek, J.P. Mesirov, C.D. Bloomfield, Molecular classification of cancer: class discovery and class prediction by gene expression monitoring. Science **286**(5439), 531–537 (1999)

A.O. Hero, G. Fluery, Pareto-optimal methods for gene filtering. J. Am. Stat. Assoc. (JASA) (2002)

Y. Hippo, H. Taniguchi, S. Tsutsumi, N. Machida, J.M. Chong, M. Fukayama, H. Aburatani, Global gene expression analysis of gastric cancer by oligonucleotide microarrays. Cancer Res. **62**(1), 233–240 (2002)

D. Karaboga, B. Basturk, A powerful and efficient algorithm for numerical function optimization: artificial bee colony (ABC) algorithm. J. Glob. Optim. **39**(3), 459–471 (2007)

J. Kennedy, Particle swarm optimization. *Encyclopedia of Machine Learning* (Springer, US, 2011), pp. 760–766

N. Khunlertgit, B.J. Yoon, Identification of robust pathway markers for cancer through rank-based pathway activity inference. Adv. Bioinform. (2013)

S.B. Kotsiantis, I. Zaharakis, P. Pintelas, Supervised machine learning: a review of classification techniques (2007)

L.K. Luo, D.F. Huang, L.J. Ye, Q.F. Zhou, G.F. Shao, H. Peng, Improving the computational efficiency of recursive cluster elimination for gene selection. IEEE/ACM Trans. Comput. Biol. Bioinform. **8**(1), 122–129 (2011)

M.S. Mohamad, S. Omatu, S. Deris, M.F. Misman, M. Yoshioka, A multi-objective strategy in genetic algorithms for gene selection of gene expression data. Artif. Life Robot. **13**(2), 410–413 (2009)

A. Mukhopadhyay, M. Mandal, Identifying non-redundant gene markers from microarray data: a multiobjective variable length PSO-based approach. IEEE/ACM Trans. Comput. Biol. Bioinform. (TCBB) **11**(6), 1170–1183 (2014)

K. Price, R.M. Storn, J.A. Lampinen, *Differential Evolution: A Practical Approach to Global Optimization* (Springer Science & Business Media, 2006)

H. Salem, G. Attiya, N. El-Fishawy, Classification of human cancer diseases by gene expression profiles. Appl. Soft Comput. **50**, 124–134 (2017)

N. Srinivas, K. Deb, Muiltiobjective optimization using nondominated sorting in genetic algorithms. Evol. Comput. **2**(3), 221–248 (1994)

A.V. Ushakov, X. Klimentova, I. Vasilyev, Bi-level and bi-objective p-median type problems for integrative clustering: application to analysis of cancer gene-expression and drug-response data. IEEE/ACM Trans. Comput. Biol. Bioinform. (2016)

J. Yang, V. Honavar, Feature subset selection using a genetic algorithm. IEEE Intell. Syst. Appl. **13**(2), 44–49 (1998)

L. Zhang, J. Kuljis, X. Liu, Information visualization for DNA microarray data analysis: a critical review. IEEE Trans. Syst. Man Cybern. Part C Appl. Rev. **38**(1), 42–54 (2008)

Q. Zhang, H. Li, MOEA/D: a multiobjective evolutionary algorithm based on decomposition. IEEE Trans. Evol. Comput. **11**(6), 712–731 (2007)

C.H. Zheng, W. Yang, Y.W. Chong, J.F. Xia, Identification of mutated driver pathways in cancer using a multi-objective optimization model. Comput. Biol. Med. **72**, 22–29 (2016)

Extended Nondominated Sorting Genetic Algorithm (ENSGA-II) for Multi-Objective Optimization Problem in Interval Environment

Asoke Kumar Bhunia, Amiya Biswas and Ali Akbar Shaikh

1 Introduction

In the existing literature of optimization, most of the works have been reported for optimization of single objective. However, in real-world design or decision-making problems, it is often required to simultaneously optimize more than one objective functions which are conflicting in nature. The goal of these problems is to maximize or minimize several conflicting objectives simultaneously. These types of problems are known as multi-objective optimization problems. In this subarea, most of the works have been done in crisp environment. However, in reality, due to uncertainty or ambiguity, the parameters of the problems are not always precise and should be considered as imprecise. To represent the impreciseness of a parameter, several approaches like stochastic, fuzzy, fuzzy stochastic and interval approaches have been reported in the existing literature. Among these approaches, interval approach is more significant. Due to this representation, either all the objectives or some of the objectives would be interval valued.

Thus, the general form of multi-objective optimization problem with interval objectives can be written as

$$\text{Minimize } \{A_1(x), A_2(x), \ldots, A_k(x)\}$$

A. K. Bhunia (✉) · A. A. Shaikh
Department of Mathematics, The University of Burdwan, Burdwan 713104, West Bengal, India
e-mail: akbhunia@math.buruniv.ac.in

A. A. Shaikh
e-mail: aliashaikh@math.buruniv.ac.in

A. Biswas
Department of Mathematics, A. B. N. Seal College, Cooch Behar 736101, India
e-mail: amiya2al@gmail.com

© Springer Nature Singapore Pte Ltd. 2018
J. K. Mandal et al. (eds.), *Multi-Objective Optimization*,
https://doi.org/10.1007/978-981-13-1471-1_10

$$\text{subject to } x \in S \subset \mathbb{R}^n$$

$$\text{where } A_i(x) = [f_{iL}(x), f_{iU}(x)], i = 1, 2, \ldots, k$$

$$\text{and } S = \left\{ x : g_j(x) \leq 0, j = 1, 2, \ldots, m \right\}$$

For solving multi-objective optimization problems with crisp objectives, several approaches have been reported in the existing literature. Among these approaches, NSGA-II (elitist nondominated sorting genetic algorithm) is very well known and widely used efficient algorithm. The idea of nondominated sorting genetic algorithm (NSGA) was proposed by Srinivas and Deb (1995). Thereafter, Deb et al. (2000) modified NSGA and proposed a computationally fast elitist genetic algorithm (called NSGA-II). After Deb et al. (2000), several researchers applied this algorithm to solve different types of application-oriented multi-objective optimization problems. To improve the performance of NSGA-II, Murugan et al. (2009) modified the algorithm by introducing virtual mapping procedure and controlled elitism. Then they applied the modified algorithm to solve the multi-objective generation expansion planning problem with two objectives (by minimizing investment cost and maximizing reliability). Again, introducing dynamic crowding distance and controlled elitism, Dhanalakshmi et al. (2011) proposed modified NSGA-II and applied this algorithm to solve the economic emission dispatch problem. Using the same algorithm, Jeyadevi et al. (2011) solved multi-objective optimal reactive power dispatch problem by minimizing real power loss and maximizing the system voltage stability. Deb and Jain (2012) suggested a reference point based many-objective NSGA-II that emphasizes population members which are nondominated close to a set of well-distributed reference points. Kannan et al. (2009) discussed the use of multi-objective optimization method, elitist nondominated sorting genetic algorithm version II (NSGA-II), for solving the generation expansion planning (GEP) problem. They formulated and solved two different bi-objective optimization problems. The first problem has been solved by minimizing the cost as well as the sum of normalized constraint violations. On the other hand, the second one has been solved by minimizing the investment cost and maximizing the reliability.

To the best of our knowledge, very few works have been reported in the existing literature regarding the solution methodology of multi-objective optimization problem with interval objectives. Wu (2009) first proposed an approach for solving this type of problem. For this purpose, he derived Kurush–Kuhn–Tucker optimality conditions by defining Pareto optimal solution based on the interval ranking proposed by Ishibuchi and Tanaka (1990). After Wu (2009), Sahoo et al. (2012) proposed different techniques based on interval distance between two interval numbers in the area of reliability optimization. In their paper, they also proposed new definitions of interval ranking by modifying earlier definitions. However, using this new definition of interval ranking, Pareto optimality cannot be proved for their proposed interval distance. Very recently, Bhunia and Samanta (2014) proposed new definitions of interval ranking and interval metric modifying earlier definitions. Using these definitions, they solved multi-objective optimization problem by converting it into a single objective optimization problem with interval objective. For solving the transformed problem, they proposed hybrid tournament genetic algorithm with interval fitness, whole arithmetical crossover and double mutation. Till now, none has developed any method/technique based on NSGA-II for solving multi-objective optimization problem with interval objectives.

In this book chapter, for the first time, we have proposed an extended version of NSGA-II (ENSGA-II) for solving the multi-objective optimization problem with interval objectives using recently developed interval ranking and interval metric (Bhunia and Samanta 2014), along with whole arithmetical crossover and double mutation. For this purpose, we have proposed crowding distance and crowded tournament selection along with nondominated sorting of solutions with interval fitness. Then, the efficiency and performance of the proposed algorithm have been tested by considering and solving four numerical examples. Finally, another set of four numerical examples with the same lower and upper bounds has been solved by the proposed algorithm for comparing the computational results with the existing one.

2 Interval Mathematics and Order Relations Between Intervals

2.1 Interval Mathematics

An interval number A can be expressed in two different forms:

 (i) Lower and upper bounds form, i.e., $A = [a_L, a_U]$
(ii) Center and radius form, i.e., $A = \langle a_c, a_w \rangle$

where a_L and a_U are the lower and upper bounds of the interval A, respectively $a_c = (a_L + a_U)/2$ and $a_w = (a_U - a_L)/2$.

Again, a real number p can also be expressed in interval form as $[p, p]$ with center p and radius zero.

According to Moore (1966), the definitions of addition, subtraction, multiplication, and division of interval numbers are as follows:

Definition 2.1.1 If $A = [a_L, a_R]$ and $B = [b_L, b_R]$

$$A + B = [a_L, a_R] + [b_L, b_R] = [a_L + b_L, a_R + b_R]$$

$$A - B = [a_L, a_R] - [b_L, b_R] = [a_L - b_R, a_R - b_L]$$

For any real number λ,

$$\lambda A = \lambda[a_L, a_R] = \begin{cases} [\lambda a_L, \lambda a_R] \ if \ \lambda \geq 0 \\ [\lambda a_R, \lambda a_L] \ if \ \lambda < 0, \end{cases}$$

$$A \times B = [\min\{a_L b_L, a_L b_R, a_R b_L, a_R b_R\}, \max\{a_L b_L, a_L b_R, a_R b_L, a_R b_R\}]$$

$$\frac{A}{B} = A \times \frac{1}{B} = [a_L, a_R] \times \left[\frac{1}{b_R}, \frac{1}{b_L}\right], \text{ provided } 0 \notin [b_L, b_R].$$

Definition 2.1.2 In center and radius form of interval, addition, subtraction, and multiplication by a scalar of interval numbers are defined as follows:

Now if $A = \langle a_c, a_w \rangle$ and $B = \langle b_c, b_w \rangle$

$$A + B = \langle a_c, a_w \rangle + \langle b_c, b_w \rangle = \langle a_c + b_c, a_w + b_w \rangle$$

$$A - B = \langle a_c, a_w \rangle - \langle b_c, b_w \rangle = \langle a_c - b_c, a_w + b_w \rangle$$

For any real number λ,

$$\lambda A = \lambda \langle a_c, a_w \rangle = \langle \lambda a_c, |\lambda| a_w \rangle$$

According to Hansen and Walster (2004), the definition of n-th power of an interval number is as follows:

Definition 2.1.3 Let $A = [a_L, a_R]$ be an interval and n be any nonnegative integer, then

$$A^n = \begin{cases} [1, 1] & \text{if } n = 0 \\ [a_L^n, a_R^n] & \text{if } a_L \geq 0 \text{ or if } n \text{ is odd} \\ [a_R^n, a_L^n] & \text{if } a_R \leq 0 \text{ and if } n \text{ is even} \\ [0, \max(a_L^n, a_R^n)] & \text{if } a_L \leq 0 \leq a_R \text{ and if } n(> 0) \text{ is even}. \end{cases}$$

In this definition, when $n=0$ and $0 \in A$ then $A^n \neq [1, 1]$.

Hence the revised definition is as follows:

Definition 2.1.4 Let $A = [a_L, a_R]$ be an interval and n be any nonnegative integer, then

$$A^n = \begin{cases} [1, 1] & \text{if } n = 0 \text{ and } 0 \notin [a_L, a_R] \\ [a_L^n, a_R^n] & \text{if } a_L \geq 0 \text{ or if } n \text{ is odd} \\ [a_R^n, a_L^n] & \text{if } a_R \leq 0 \text{ and if } n \text{ is even} \\ [0, \max(a_L^n, a_R^n)] & \text{if } a_L \leq 0 \leq a_R \text{ and if } n(> 0) \text{ is even}. \end{cases}$$

In the existing literature, there is no definition regarding the negative integral power of an interval. We have defined the same which is as follows:

Definition 2.1.5 Let $A = [a_L, a_R]$ be an interval and n be any positive integer, then

$$(A)^{-n} = \frac{1}{A^n} = \left(\frac{1}{A}\right)^n \text{ provided } 0 \notin [a_L, a_R].$$

According to Karmakar et al. (2009), the n-th root of an interval number is defined as follows:

Definition 2.1.6 Let $A = [a_L, a_R]$ be an interval and n be any positive integer, then

$$(A)^{\frac{1}{n}} = [a_L, a_R]^{\frac{1}{n}} = \sqrt[n]{[a_L, a_R]} = \left[\sqrt[n]{a_L}, \sqrt[n]{a_R}\right] \text{if } a_L \geq 0 \text{ or if } n \text{ is odd}$$
$$= \left[0, \sqrt[n]{a_R}\right] \text{if } a_L \leq 0, a_R \geq 0 \text{ and } n \text{ is even}$$
$$= \phi \text{ if } a_R < 0 \text{ and } n \text{ is even}$$

where ϕ is the empty interval.

However, for $a_L < 0$ and even value of n, $(A)^{1/n}$ is not a real interval. Moreover, it is a complex interval (For complex interval, see Kearfott 1996).

Hence, the revised definition of n-th root of an interval number will be as follows:

Definition 2.1.7 Let $A = [a_L, a_R]$ be an interval and n be any positive integer

$$(A)^{\frac{1}{n}} = [a_L, a_R]^{\frac{1}{n}} = \sqrt[n]{[a_L, a_R]} = \left[\sqrt[n]{a_L}, \sqrt[n]{a_R}\right] \text{ if } a_L \geq 0 \text{ or if } n \text{ is odd.}$$

Again, by applying the definition of power and different roots of an interval, we can find any rational power of an interval. Suppose we have to find $A^{\frac{p}{q}}$, where $A = [a_L, a_R]$, then it can be found by defining $A^{\frac{p}{q}}$ as $(A^p)^{\frac{1}{q}}$.

According to Sahoo et al. (2012), the interval power of an interval is defined as follows:

Definition 2.1.8 Let $A = [a_L, a_R]$ and $B = [b_L, b_R]$ be two intervals, then

$(A)^B = [a_L, a_R]^{[b_L, b_R]}$
$$= \begin{cases} \left[e^{\min(b_L \log a_L, b_L \log a_R, b_R \log a_L, b_R \log a_R)}, e^{\max(b_L \log a_L, b_L \log a_R, b_R \log a_L, b_R \log a_R)}\right] & \text{if } a_L \geq 0 \\ \text{a complex interval} & \text{if } a_L < 0 \end{cases}$$

However when $a_L = 0, a_R > 0$ and $0 \notin B = [b_L, b_R]$ or $a_L = 0, a_R = 0$ and $0 \notin B = [b_L, b_R]$, the result of $(A)^B$ will be different from the above definition. On the other hand, when $a_L < 0$ and either $0 \notin B = [b_L, b_R]$ or when $0 \notin A = [a_L, a_R]$, $(A)^B$ is not a real interval, it must be a complex interval.

Hence, the revised definition of interval power of an interval will be as follows:

Definition 2.1.9 Let $A = [a_L, a_R]$ and $B = [b_L, b_R]$ be two intervals, then

$$(A)^B = [a_L, a_R]^{[b_L, b_R]}$$

$$= \begin{cases} \begin{bmatrix} e^{\min(b_L \log a_L, b_L \log a_R, b_R \log a_L, b_R \log a_R)}, \\ e^{\max(b_L \log a_L, b_L \log a_R, b_R \log a_L, b_R \log a_R)} \end{bmatrix} & \text{if } a_L > 0 \\ \left[0, e^{\max(b_L \log a_R, b_R \log a_R)} \right] & \text{if } a_L = 0, a_R > 0 \text{ and } 0 \notin B = [b_L, b_R] \\ [0, \ 0] & \text{if } a_L = 0, a_R = 0 \text{ and } 0 \notin B = [b_L, b_R] \end{cases}$$

2.2   Order Relations of Interval Numbers

In solving optimization problems with interval objective, decision regarding the order relation between two arbitrary intervals is an important task. In this section, we shall discuss the earlier developments of order relations between interval numbers. Any two closed intervals $A = [a_L, a_R]$ and $B = [b_L, b_R]$ may be one of the following three types:

Type I: Nonoverlapping intervals, i.e., when either $a_L > b_R$ or $b_L > a_R$.
Type II: Partially overlapping intervals, i.e., when either $b_L < a_L < b_R < a_R$ or $a_L < b_L < a_R < b_R$.
Type III: Fully overlapping intervals, i.e., when either $a_L \leq b_L \leq b_R \leq a_R$ or $b_L \leq a_L \leq a_R \leq b_R$.

From the existing literature, it is observed that several researchers proposed the definitions of order relations between two interval numbers based on either set properties or fuzzy applications or probabilistic approaches or value-based approaches or depending upon some specific indices/functions. Most of the definitions are incomplete. The detailed comparison is available in Karmakar and Bhunia (2012). Here, we shall discuss some significant definitions only.

2.2.1   Mahato and Bhunia's Definition

Mahato and Bhunia (2006) proposed two types of definitions of interval order relations with respect to optimistic and pessimistic decision-making. In optimistic decision-making, the decision-maker ignores the uncertainty whereas the pessimistic decision-maker prefers the interval according to the principle "less uncertainty is better than more uncertainty". The order relations for these types of decision makers are as follows:

Optimistic Decision-Making

Definition For minimization problems, the order relation $\leq_{\text{omin}}$ between the intervals $A = [a_L, a_R]$ and $B = [b_L, b_R]$ is

$$A \leq_{\text{omin}} B \Leftrightarrow a_L \leq b_L$$

$$A <_{\text{omin}} B \Leftrightarrow A \leq_{\text{omin}} B \text{ and } A \neq B.$$

Definition For maximization problems, the order relation $\geq_{\text{omax}}$ between the intervals $A = [a_L, a_R]$ and $B = [b_L, b_R]$ is

$$A \geq_{\text{omax}} B \Leftrightarrow a_R \geq b_R$$

$$A >_{\text{omax}} B \Leftrightarrow A \geq_{\text{omax}} B \text{ and } A \neq B.$$

Pessimistic Decision-Making

Definition For minimization problems, the order relation $<_{\text{pmin}}$ between two intervals $A = [a_L, a_R] = \langle a_c, a_w \rangle$ and $B = [b_L, b_R] = \langle b_c, b_w \rangle$ for a pessimistic decision-maker is

(i) $A <_{\text{pmin}} B \Leftrightarrow a_c < b_c$, for Type I and Type II intervals
(ii) $A <_{\text{pmin}} B \Leftrightarrow (a_c \leq b_c) \wedge (a_w < b_w)$ for Type III intervals.

However, for Type III intervals with $(a_c \leq b_c) \wedge (a_w > b_w)$, a pessimistic decision cannot be taken. In this case, the optimistic decision is to be taken.

Definition For maximization problems, the order relation $>_{\text{pmax}}$ between two intervals $A = [a_L, a_R] = \langle a_c, a_w \rangle$ and $B = [b_L, b_R] = \langle b_c, b_w \rangle$ for a pessimistic decision-maker is

(i) $A >_{\text{pmax}} B \Leftrightarrow a_c > b_c$, for Type I and Type II intervals
(ii) $A >_{\text{pmax}} B \Leftrightarrow a_c \geq b_c \wedge a_w < b_w$ for Type III intervals.

However, for Type III intervals with $(a_c \geq b_c) \wedge (a_w > b_w)$, a pessimistic decision cannot be taken. In this case, the optimistic decision is to be considered.

2.2.2 Hu and Wang's Definition

Hu and Wang (2006) proposed a modified version of order relations for interval numbers based on center and radius of the intervals. In their work, they first studied the incompleteness of interval ranking techniques developed earlier. Then introducing new approach, they tried to fulfill the shortcomings of the previous definitions. Their interval ranking relation "$\prec_=$" is defined as follows:

Definition For any two intervals, $A = [a_L, a_R] = \langle a_c, a_w \rangle$ and $B = [b_L, b_R] = \langle b_c, b_w \rangle$

$$A \prec_= B \Leftrightarrow \begin{cases} a_c < b_c \; if \; a_c \neq b_c \\ a_w \geq b_w \; if \; a_c = b_c \end{cases}$$

and $A \prec B \Leftrightarrow A \prec_= B$ and $A \neq B$.

The relation "$\prec_=$" satisfies the following relational properties:

(i) $A \prec_= A$ for any interval A (*Reflexivity*).
(ii) If $A \prec_= B$ and $B \prec_= A$, then $A = B$ for any two intervals A and B (*Antisymmetry*).
(iii) If $A \prec_= B$ and $B \prec_= C$, then $A \prec_= C$ for any three intervals A, B and C (*Transitivity*).
(iv) One of $A \prec_= B$ and $B \prec_= A$ holds for any two intervals A and B (*Comparability*).

In this definition, for intervals with the same center, the interval with more uncertainty is considered as the lesser interval. However, it is not true in case of pessimistic decision-making for minimization problems.

2.2.3   Sahoo et al. Definition

In the definition of Mahato and Bhunia (2006) of pessimistic decision-making of Type III intervals, it is observed that sometimes optimistic decisions are to be taken. To overcome this situation, Sahoo et al. (2012) proposed two new definitions of order relations irrespective of optimistic as well as pessimistic decision-maker's point of view for maximization and minimization problems separately.

Definition For maximization problems, the order relation $>_{max}$ between two intervals $A = [a_L, a_R] = \langle a_c, a_w \rangle$ and $B = [b_L, b_R] = \langle b_c, b_w \rangle$ is

(i) $A >_{max} B \Leftrightarrow a_c > b_c$ for Type I and Type II intervals
(ii) $A >_{max} B \Leftrightarrow$ either $a_c \geq b_c \wedge a_w < b_w$ or $a_c > b_c \wedge a_R > b_R$ for Type III intervals.

Definition For minimization problems, the order relation $<_{min}$ between two intervals $A = [a_L, a_R] = \langle a_c, a_w \rangle$ and $B = [b_L, b_R] = \langle b_c, b_w \rangle$ is

(i) $A <_{min} B \Leftrightarrow a_c < b_c$ for Type I and Type II intervals
(ii) $A <_{min} B \Leftrightarrow$ either $a_c \leq b_c \wedge a_w < b_w$ or $a_c < b_c \wedge a_L < b_L$ for Type III intervals.

2.2.4   Bhunia and Samanta's Definition

In the definition of Sahoo et al. (2012), it is observed that for different types of intervals, different conditions for interval order relations have been considered which can be put in a much simpler way. Accordingly, Bhunia and Samanta (2014) proposed two order relations $\geq^{max}$ and $\leq^{min}$ separately for maximization and minimization problems, respectively.

Definition The order relation $\geq^{max}$ between two intervals $A = [a_L, a_R] = \langle a_c, a_w \rangle$ and $B = [b_L, b_R] = \langle b_c, b_w \rangle$ for maximization problems is as follows:

$$A \geq^{max} B \Leftrightarrow \begin{cases} a_c > b_c \ \ if \ a_c \neq b_c \\ a_w \leq b_w \ \ if \ a_c = b_c \end{cases}$$

and $A >^{max} B \Leftrightarrow A \geq^{max} B$ and $A \neq B$.

Definition The order relation $\leq^{min}$ between two intervals $A = [a_L, a_R] = \langle a_c, a_w \rangle$ and $B = [b_L, b_R] = \langle b_c, b_w \rangle$ for minimization problems is as follows:

$$A \leq^{min} B \Leftrightarrow \begin{cases} a_c < b_c \ \ if \ a_c \neq b_c \\ a_w \leq b_w \ \ if \ a_c = b_c \end{cases}$$

and $A <^{min} B \Leftrightarrow A \leq^{min} B$ and $A \neq B$.

Theorem 1 (i) $A \geq^{max} A$ *for any interval A (Reflexivity).*
(ii) *If* $A \geq^{max} B$ *and* $B \geq^{max} A$, *then* $A = B$ *for any two intervals A and B (Antisymmetry).*
(iii) *If* $A \geq^{max} B$ *and* $B \geq^{max} C$, *then* $A \geq^{max} C$ *for any three intervals A, B, and C (Transitivity).*
(iv) *One of* $A \geq^{max} B$ *and* $B \geq^{max} A$ *holds for any two intervals A and B (Comparability).*

Proof (i) Let $A = [a_L, a_R] = \langle a_c, a_w \rangle$ be an interval. Then $a_c = a_c$ and $a_w \leq a_w$. Therefore $A \geq^{max} A$. Since A is an arbitrary interval, this is true for all intervals. Hence $\geq^{max}$ is reflexive.

(ii) Let $A = [a_L, a_R] = \langle a_c, a_w \rangle$ and $B = [b_L, b_R] = \langle b_c, b_w \rangle$ be two intervals such that $A \geq^{max} B$ *and* $B \geq^{max} A$ holds, then $A \geq^{max} B \Rightarrow a_c \geq b_c$ and $B \geq^{max} A \Rightarrow b_c \geq a_c$. Therefore $a_c \geq b_c$ and $b_c \geq a_c \Rightarrow a_c = b_c$. Now if $a_c = b_c$, $A \geq^{max} B \Rightarrow a_w \leq b_w$ and $B \geq^{max} A \Rightarrow b_w \leq a_w$. Again $a_c = b_c, a_w \leq b_w$ and $b_w \leq a_w \Rightarrow a_c = b_c$ and $a_w = b_w \Rightarrow A = B$. Since A and B are arbitrary intervals, this is true for all intervals A, B. Hence $\geq^{max}$ is antisymmetric.

(iii) Let $A = [a_L, a_R] = \langle a_c, a_w \rangle$, $B = [b_L, b_R] = \langle b_c, b_w \rangle$ and $C = [c_L, c_R] = \langle c_c, c_w \rangle$ be three intervals such that $A \geq^{max} B$ and $B \geq^{max} C$ holds, then $A \geq^{max} B \Rightarrow a_c \geq b_c$ and $B \geq^{max} C \Rightarrow b_c \geq c_c$. Therefore, $a_c \geq b_c \geq c_c \Rightarrow$ either $a_c > c_c$ or $a_c = b_c = c_c$. Now if $a_c > c_c$, then $A \geq^{max} C$. On the other hand, if $a_c = b_c = c_c$ then $A \geq^{max} B \Rightarrow a_w \leq b_w$ and $B \geq^{max} C \Rightarrow b_w \leq c_w$. Again $a_c = b_c = c_c, a_w \leq b_w$ and $b_w \leq c_w \Rightarrow a_c = c_c$ and $a_w \leq c_w \Rightarrow A \geq^{max} C$. Since A, B, and C are arbitrary intervals, this is true for all intervals A, B, C. Hence, $\geq^{max}$ is transitive.

(iv) Let $A = [a_L, a_R] = \langle a_c, a_w \rangle$ and $B = [b_L, b_R] = \langle b_c, b_w \rangle$ be any two intervals. Then $a_c > b_c$ or $b_c > a_c$ or $b_c = a_c$ holds obviously. Now $a_c > b_c \Rightarrow A \geq^{max} B$, $b_c > a_c \Rightarrow B \geq^{max} A$ and if $b_c = a_c$ then either $a_w \leq b_w$ or $b_w \leq a_w$. Hence $b_c = a_c$ and $a_w \leq b_w \Rightarrow A \geq^{max} B$; $b_c = a_c$ and $b_w \leq a_w \Rightarrow B \geq^{max} A$. Since A and B are arbitrary intervals, this is true for all intervals A, B. Hence any two intervals are comparable.

Theorem 2 *(i)* $A \leq^{min} A$ *for any interval A (Reflexivity).*

(ii) $A \leq^{min} B$ *and* $B \leq^{min} A$ *then* $A = B$ *for any two intervals A and B (Antisymmetry).*

(iii) $A \leq^{min} B$ *and* $B \leq^{min} C$ *then* $A \leq^{min} C$ *for any three intervals A, B, and C (Transitivity).*

(iv) $A \leq^{min} B$ *or* $B \leq^{min} A$ *holds for any two intervals A and B (Comparability).*

Proof The proof is similar to *Theorem* 1.

Theorem 3 *Let A and B be two intervals then*

(i) $A >_{max} B \Leftrightarrow A >^{max} B$

(ii) $A <_{min} B \Leftrightarrow A <^{min} B.$

Proof It can easily be proved.

Definition Let M be any non-empty set. A function $d : M \times M \to$ IR (set of all real intervals) is said to be an interval metric if it satisfies the following properties:

(i) Reflexivity: $d(X, X)_c = 0 \in d(X, X), \forall X \in M$
where $d(X, X)_c$ is the center of the interval $d(X, X)$.
(ii) Triangular inequality $d(X, Y) \leq^{min} d(X, Z) + d(Z, Y), \forall X, Y, Z \in M$
(iii) Symmetry: $d(X, Y) = d(Y, X), \forall X, Y \in M$
(iv) Indiscernible identity: if $d(X, Y) = d(X, X)$, then $X = Y$ $(\forall X, Y \in M)$.

Definition Let X and $Y \in$ IR (set of all real intervals). An interval distance between X and Y, denoted by $d_I(X, Y)$, is defined by

$$d_I(X, Y) = \tilde{|}X - Y\tilde{|} = \tilde{|}\langle x_c, x_w \rangle - \langle y_c, y_w \rangle \tilde{|}$$
$$= \tilde{|}\langle x_c - y_c, x_w + y_w \rangle \tilde{|} = \langle |x_c - y_c|, x_w + y_w \rangle$$

Definition Let $X = (X_1, X_2, \ldots, X_k)$ and $Y = (Y_1, Y_2, \ldots, Y_k) \in$ IR^k. An interval distance between X and Y is denoted by $d_I(X, Y)$ and is defined by

$$d_I(X, Y) = \sum_{i=1}^{k} \tilde{|}X_i - Y_i\tilde{|}$$

3 Multi-Objective Optimization Problem with Interval Objectives

The general form of multi-objective optimization problem with interval objectives can be written as

$$\text{Minimize } \{A_1(x), A_2(x), \ldots, A_k(x)\}$$

$$\text{subject to } x \in S \subset \mathbb{R}^n$$

$$\text{where } A_i(x) = [f_{iL}(x), f_{iU}(x)], i = 1, 2, \ldots, k$$

$$S = \left\{ x : g_j(x) \leq 0, j = 1, 2, \ldots, m \right\}$$

and $f_{iL}, f_{iU} : \mathbb{R}^n \to \mathbb{R}, g_j : \mathbb{R}^n \to \mathbb{R}, i = 1, 2, \ldots, k, j = 1, 2, \ldots, m$.

Here, $x = (x_1, x_2, \ldots, x_n) \in S$ is the solution vector and S is the feasible space. As all the objectives are interval valued, so we have used the existing extended definitions of Pareto optimality and weak Pareto optimality (Bhunia and Samanta (2014)) for multi-objective optimization with interval valued objectives. According to them, the extended definitions are as follows.

Definition 3.1 A decision vector $x^* \in S$ is Pareto optimal if there does not exist another decision vector $x \in S$ such that $A_i(x) \leq^{\min} A_i(x^*)$ for all $i = 1, 2, \ldots, k$ and $A_j(x) <^{\min} A_j(x^*)$ for at least one index j.

Definition 3.2 A decision vector $x^* \in S$ is weakly Pareto optimal if there does not exist another decision vector $x \in S$ such that $A_i(x) <^{\min} A_i(x^*)$ for all $i = 1, 2, \ldots, k$.

4 Nondominated Sorting Genetic Algorithm for Interval Objectives

The nondominated sorting genetic algorithm-II (NSGA-II) is well known and widely used algorithm proposed by Deb et al. (2000) based on its preceding algorithm NSGA which was initiated by Srinivas and Deb (1995) as a computationally fast elitist genetic algorithm. In this method, the idea of ranking is involved to reflect the performance of an individual and the rank of an individual is obtained by the dominance relation. Again the idea of crowding distance is used to reflect the density of solutions surrounding a particular solution in the population and the individuals with a lower rank and a larger crowding distance have good performance in diversity; therefore they have more opportunities to be preserved. This NSGA-II algorithm was proposed for solving multi-objective optimization problem with crisp objectives, i.e., there is no uncertainty in these objectives. However, in reality, the parameters of decision-making problems need not be precise and may be imprecise in nature due to uncertainty. In this work, our objective is to solve the multi-objective optimization problem with interval objectives. For this purpose, we have extended the existing NSGA-II (proposed by Deb et al. 2000) algorithm in interval environment and we call it as ENSGA-II. As a result, we have to redefine the following components of ENSGA-II.

(i)   Constraint handling techniques
(ii) Nondominated sorting
(iii) Interval crowding distance
(iv) Crowded Tournament Selection.

4.1   Constraint Handling Techniques

In ENSGA-II, the constraint handling technique towards the feasible region is an important task. As all the constraints are crisp in nature, so we have used the following constraints handling technique.

Definition 4.1.1 A solution $x = (x_1, x_2, \ldots, x_n)$ is said to constraint dominate a solution $y = (y_1, y_2, \ldots, y_n)$, if any one of the following conditions is true:

(i)   Solution x is feasible and solution y is not.

(ii) Solutions x and y are both infeasible, but solution x has a smaller constraint violation.

(iii) Solutions x and y are both feasible and solution x dominates solution y.

4.2   Nondominated Sorting

The dominance relation is very important for solving multi-objective optimization problem. Here, as the objective functions are interval valued, so the existing dominance relation for fixed objectives in NSGA-II is not applicable for a Pareto optimal set of solutions. As a result, it is necessary to define a new dominance relation for interval objectives to get a Pareto optimal set of solutions for a multi-objective optimization problem with interval objectives. This dominance relation depends on the type (maximization/minimization/both) of objective functions, and also on the order relations between interval numbers. Accordingly, we have proposed the following dominance relation.

Definition 4.2.1: Dominance relation For two solutions $\underset{\sim}{x}$ and $\underset{\sim}{y}$, $\underset{\sim}{x}$ dominates $\underset{\sim}{y}$ if any one of the following conditions is satisfied:

(i) $A_i(\underset{\sim}{x}) \geq^{\max} A_i(\underset{\sim}{y})\,(i = 1, 2, \ldots, k)$ in case of maximization of all the objective functions

(ii) $A_i(\underset{\sim}{x}) \leq^{\min} A_i(\underset{\sim}{y})\,(i = 1, 2, \ldots, k)$ in case of minimization of all the objective functions

(iii) $A_i(\underset{\sim}{x}) \geq^{\max} A_i(\underset{\sim}{y})$ for all $i = 1, 2, \ldots, p$ and $A_i(\underset{\sim}{x}) \leq^{\min} A_i(\underset{\sim}{y})$ for all $i = p + 1, \ldots, k,$ when p number of objectives are to be maximized and the rest are to be minimized.

We know that in the objective function space, as the number of objectives is two, three, or even more, the product of the objective values gives an ordinary rectangle, an ordinary cuboid or generalized hyper-cuboid respectively. Usually, the solutions in a particular front (front 1 or 2 ...) of a multi-objective optimization problem with more than three objectives describe a hyper-surface, whereas the solutions in a Pareto front (front 1 or 2 ...) that we discuss here describe a hyper-cuboid. As a result, it is more complicated than the former, so it is a greater challenge to obtain the Pareto front in this case. In Fig. 1, the objective function values of a minimization problem with two objectives are marked. Here, the alphabets denote the solution name and the numerical digits denote its front. Clearly, the solutions a, b, c, d and e are nondominated because the centers of the corresponding objective function values are nondominated. However, for the solutions g and j, the corresponding objective function values have the same center but as g have lower uncertainty than j, so g dominates j.

Now in order to sort a population according to the level of nondomination, each solution must be compared with each other solution in the population to check whether it is dominated or not. This process is continued to select the members of the first nondominated class (we call it as first front) from the population. In order to find the solutions for the next front, the solutions of the previous fronts (in this case only the solutions of first front) are temporarily rejected and the above procedure is repeated to find the solutions of the subsequent fronts.

4.3   Interval Crowding Distance

For a multi-objective optimization problem, it is an important task to estimate the density of solutions in the objective space. In order to estimate the density of solutions near a particular solution in the population, we calculate the average distance of two solutions on either side of this solution along each of the objectives.

For a multi-objective optimization problem with interval objectives, the crowding distance guides the search process towards a uniformly spread-out Pareto optimal

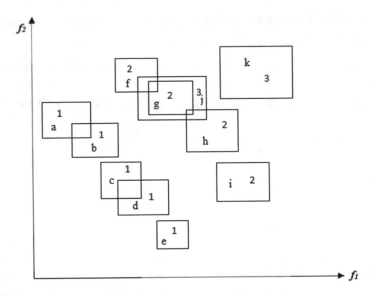

Fig. 1 Dominance relationships among solutions

front. However, the idea of existing crowded distance is suitable only for crisp objectives, not for interval objectives, so we have to redefine the crowding distance operator for interval objectives.

To compute the crowding distance of a solution, first we have to choose a front and then the population in that front is sorted in ascending order according to the values of each objective function values using interval ranking. Then for each objective function, the boundary solutions are assigned to an infinite distance values. All other intermediate solutions are assigned to a distance which is equal to the interval distance of the function values of two adjacent solutions. The crowding distance of intermediate solutions are as follows:

$$dI_j = dI_j + d_I\left(f_m^{(I_{j-1}^m)}, f_m^{(I_{j+1}^m)}\right), \ j = 2, 3, \ldots, l-1$$

where I_j denotes the solution index of the j-th member in the sorted list, dI_j is the crowding distance of the solution index I_j, $f_m^{(I_j)}$ denotes the m-th objective function value of the solution index I_j, $d_I\left(f_m^{(I_{j-1}^m)}, f_m^{(I_{j+1}^m)}\right)$ is the interval distance between objective function values $f_m^{(I_{j-1})}$ and $f_m^{(I_{j+1})}$, l is the number of solutions in that list, I_1, and I_l is the lowest and highest solution indices of f_m among all solution index in the list. This calculation is continued with other objective functions also. Finally, the overall crowding distance is computed as the sum of the individual distance values corresponds to each objective.

Now we have to compare the solutions according to the crowding distance using interval order relations (Bhunia and Samanta 2014). In Fig. 2, the crowding distance

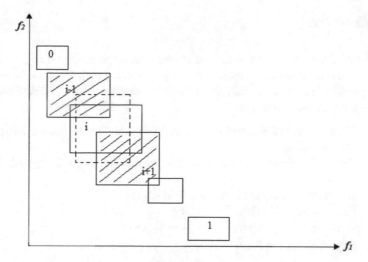

Fig. 2 Crowding distance of solutions for a particular front

of i-th solution in its front is an interval whose center is the average side length of the dotted rectangle and radius, i.e., the uncertainty is the sum of the average uncertainty of the objective function values of the $(i-1)$-th and $(i+1)$-th solutions.

4.4 Crowded Tournament Selection

This selection process is dependent on the following attributes of each solution:

  (i) A nondomination rank of a solution in the population.
 (ii) Local crowding distance of a solution in the population.

By tournament selection, i-th solution with nondomination rank r_i and local crowding distance d_i will be selected from all the solutions not yet selected if any one of the following conditions is true:

  (i) If the rank of i-th solution is less than that of all other solutions not yet selected.
 (ii) If there exist any other solution say, j-th solution with nondomination rank r_j and crowding distance d_j, not yet selected, have the same rank but solution i has a better crowding distance than j-th solution, i.e., if $r_i = r_j$ and $d_i > d_j$.

From the first condition, it is ensured that the selected solution lies in a better nondominated front. On the other hand, when both the solutions are taken from the same nondominated front, the selection of a better solution is done according to their crowding distances.

4.5 Crossover

The goal of crossover operation is to exchange the information between randomly selected parent chromosomes (individuals) by combining their features. In our work, we have used whole arithmetical crossover (Michalewicz 1996). The different steps of this operation are as follows:

Step 1: Find the integral value of the product of population size and crossover rate and store it in M.
Step 2: Select the chromosomes v_i and v_j randomly from the population for crossover.
Step 3: Generate a random real number λ in [0, 1].
Step 4: Produce two offspring v'_i and v'_j by
$$v'_i = \lambda v_i + (1 - \lambda)v_j \text{ and } v'_j = (1 - \lambda)v_i + \lambda v_j.$$
Step 5: Repeat Steps 2–4 for $\frac{M}{2}$ times.

4.6 Mutation

The aim of mutation operation is to introduce random variations into the population. Generally, it performs with lower probability. In this work, we have used double mutation which is a combination of two different mutation operations, viz., nonuniform mutation and boundary mutation (Michalewicz 1996). The action of nonuniform mutation is dependent on the age of population. If the gene v_{ij} of chromosome v_i is selected for this operation and if the domain of v_{ij} is an interval $[l_j, u_j]$ then the reduced value of v_{ij} is given by

$$v'_{ij} = \begin{cases} v_{ij} + \Delta(t, u_j - v_{ij}) \ if \ r \leq 0.5 \\ v_{ij} - \Delta(t, v_{ij} - l_j) \ if \ r > 0.5 \end{cases}$$

where $j \in \{1, 2, \ldots, n\}$, $\Delta(t, y)$ returns a value in the range [0,y] and r is a uniformly distributed random number in [0,1].
   In our work, we have considered

$$\Delta(t, y) = y r_1 \left(1 - \frac{t}{T} \right)^b$$

where r_1 is a uniformly distributed random number in [0,1], T and t represent the maximum and current generation numbers respectively and b is called the nonuniform mutation parameter which is constant (in this work, we have considered $b=2$).
   In boundary mutation, the new value of v_{ij} is generated as follows:

$$v'_{ij} = \begin{cases} l_j & if\ r_2 \leq 0.5 \\ u_j & if\ r_2 > 0.5 \end{cases}$$

where the domain of v_{ij} is an interval $[l_j, u_j]$ and r_2 is a uniformly distributed random number in $[0,1]$.

The computational steps of mutation operation are as follows:

Step 1: Find the integral value of the product of population size and mutation rate and store it in M'.

Step 2: Select a non-mutated gene v_{ij} of individual v_i for mutation.

Step 3: Create new gene v'_{ij} by the following process:

If $r \leq 0.5$ (r being a uniformly distributed random number), create the new gene by nonuniform mutation process, otherwise create the same by boundary mutation process.

Step 4: Repeat Step 2 and Step 3 for M' times.

Step 5: Stop.

4.7 Algorithm

The proposed algorithm corresponding to ENSGA-II is as follows:

---

begin

    initialize different parameters of GA, bounds of variables of the problem;

    set $t \leftarrow 0$; [t represents the generation number]

    initialize $p(t)$; [$p(t)$ represents the population at t-th generation]

    evaluate the fitness function of each chromosome of $p(t)$.

    while (not termination condition) do

    begin

      set $t \leftarrow t+1$;

      create a population $p'(t)$ of offspring by crossover and mutation on $p(t)$.

      evaluate the fitness function of each offspring chromosome of $p'(t)$.

      create $p''(t)$ by combining $p(t)$ and $p'(t)$.

      obtain front for each chromosome of $p''(t)$.

      select the population $p(t+1)$ from $p''(t)$ for next generation by Crowded Tournament Selection

      Operation.

    end

    print the best found result.

end

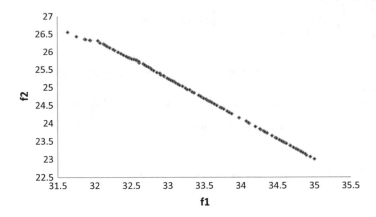

Fig. 3 Center of interval valued objectives corresponding to nondominated solutions algorithm for Example 1

5 Numerical Simulation

To test the performance of the proposed extended nondominated sorting genetic algorithm-II (ENSGA-II) for solving multi-objective nonlinear programming problems with interval objectives, we have considered and solved eight examples (Examples 1–8). These examples are given in the Appendix. Among these examples, the objective functions of first four examples are interval valued whereas for the others, the objective functions are crisp. The computational results of examples with crisp objectives have been compared with the solutions available in the existing literature. In these cases, where crisp objectives are given, crisp objectives have been converted into interval objectives considering lower and upper bounds to be the same. Then the transformed problems have been solved by the proposed algorithm.

In solving all the examples, we have used the following values of different GA parameters: population size = 100, probability of crossover = 0.8, probability of mutation = 0.1, and maximum number of generations = 500. The proposed algorithm has been coded in C programming language and the numerical experiments have been carried out in a PC with Dual Core Processor in LINUX environment.

In case of Example 1, the problem is linear bi-objective optimization problem with two constraints and two decision variables. Using the proposed algorithm (i.e., ENSGA-II), Example 1 has been solved and the values of center of both interval valued objective functions corresponding to 100 nondominated solutions after 500 generations have been considered. These results have been shown in Fig. 3. On the other hand, the values of radius of the interval valued objective functions corresponding to 100 nondominated solutions have been shown in Fig. 4.

In case of Example 2 (Wu 2009), there are two nonlinear interval objectives whereas constraints are linear with nonnegativity restrictions of the variables. Solv-

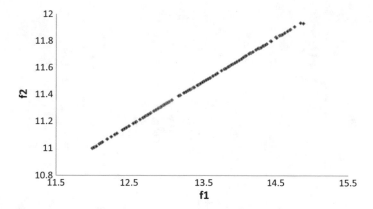

Fig. 4 Uncertainty of interval valued objectives corresponding to nondominated solutions for Example 1

ing this example, we obtain a Pareto optimal set which contains only one solution and this solution is given by $x^* = (1.8, 0.6)$, $z_1 = [4.6, 5.6]$ and $z_2 = [10.2, 11.2]$.

In Example 3, the given problem is a three-variable multi-objective optimization problem with three objectives in which the first objective is interval valued and the others are crisp. Using the proposed algorithm (i.e., ENSGA-II), Example 3 has been solved and the center values of interval valued objective functions corresponding to nondominated solutions after 500 generations have been considered. These results have been shown in Fig. 5. On the other hand, the values of radius of the interval valued objective function lie between 0.1444 and 0.8818 whereas the values of radius corresponding to other objectives are zero, as these are crisp objectives.

In Example 4, there are two interval valued objectives with 10 constraints and 7 decision variables. Solving this example, the values of center of the two interval valued objective functions corresponding to 100 nondominated solutions after 500 generations have been considered. These results have been shown in Fig. 6. On the other hand, the values of radius of these interval valued objective functions corresponding to 100 nondominated solutions have been shown in Fig. 7.

Example 5 has been considered from Deb et al. (2000). This example contains two objective functions, and two constraints with two decision variables. Using the proposed algorithm, we have solved the problem and plotted the objective function values corresponding to 100 nondominated solutions after 500 generations in Fig. 8. On the other hand, the problem has been solved by the existing NSGA-II and after 500 generations, the computational results have been considered corresponding to 100 nondominated solutions. These results have been plotted in Fig. 9.

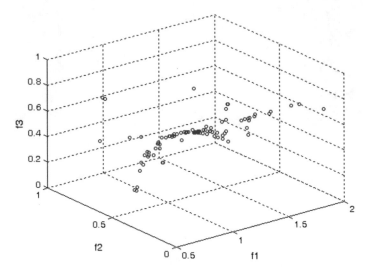

Fig. 5 Center values of interval valued objectives corresponding to nondominated solutions for Example 3

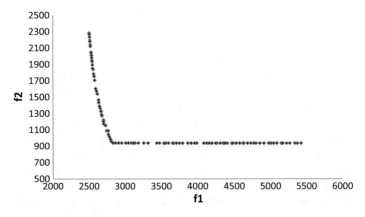

Fig. 6 Center of interval valued objectives corresponding to nondominated solutions for Example 4

In the same way, Examples 6, 7, and 8 have been solved by the proposed algorithm and existing NSGA-II and the computational results have been plotted in Figs. 10, 11, 12, 13, 14 and 15.

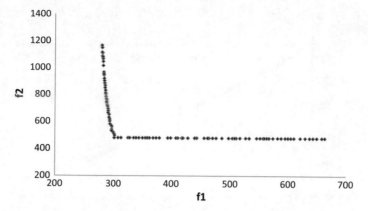

Fig. 7 Uncertainty of interval valued objectives corresponding to nondominated solutions for Example 4

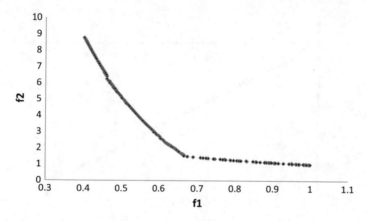

Fig. 8 Nondominated solutions of Example 5 by the proposed algorithm (ENSGA-II)

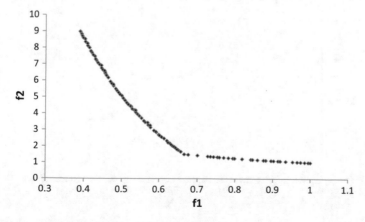

Fig. 9 Nondominated solutions of Example 5 by NSGA-II

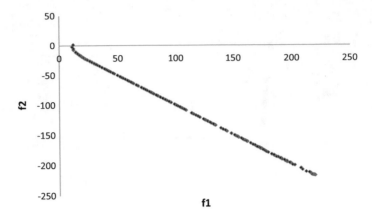

Fig. 10 Nondominated solutions of Example 6 by the proposed algorithm (ENSGA-II)

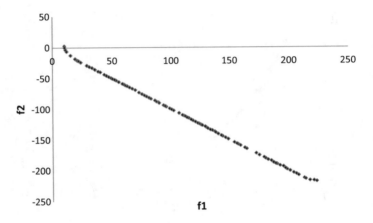

Fig. 11 Nondominated solutions of Example 6 by NSGA-II

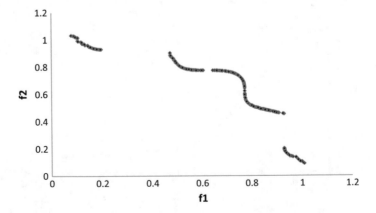

Fig. 12 Nondominated solutions of Example 7 by the proposed algorithm (ENSGA-II)

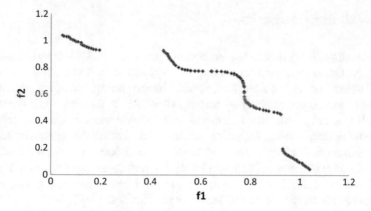

Fig. 13 Nondominated solutions of Example 7 by NSGA-II

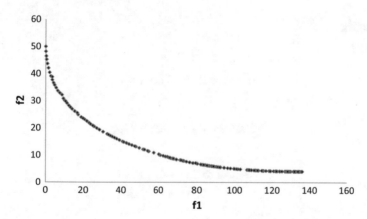

Fig. 14 Nondominated solutions of Example 8 by the proposed algorithm (ENSGA-II)

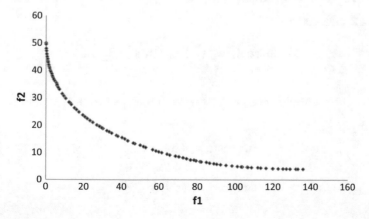

Fig. 15 Nondominated solutions of Example 8 by NSGA-II

6 Concluding Remarks

For the first time, in this book chapter, we have proposed ENSGA-II by extending the existing NSGA-II algorithm for solving multi-objective optimization problem with interval objectives. In this algorithm, nondominated sorting, crowding distance, and crowded tournament selection play an important role. To suitably adapt the existing NSGA-II (for crisp objective functions) with interval valued objective functions, the nondominated sorting, crowding distance, and crowded tournament selection have been modified using the recently developed definitions of interval ranking and interval distance (Bhunia and Samanta 2014). For further research, one may apply this algorithm to solve different types of application problems in the areas of Engineering Design, Operations Research, and Management Science.

Appendix

Example 1

$$\text{Maximize } Z_1 = C_1 x_1 + C_2 x_2$$

$$\text{Maximize } Z_2 = C_3 x_1 + C_4 x_2$$

$$\text{subject to } 3x_1 + 4x_2 \leq 42$$
$$3x_1 + x_2 \leq 24$$

$C_1 = [1, 2.5], C_2 = [3, 4], C_3 = [2, 3], C_4 = [1.5, 2.5]$
  search region $x_1 \in [0, 10], x_2 \in [0, 9]$.

Example 2 (Wu 2009)

$$\text{Minimize } Z_1 = \left[x_1^2 + x_2^2 + 1, \ x_1^2 + x_2^2 + 2 \right]$$

$$\text{Minimize } Z_2 = \left[2x_1^2 + 2x_2^2 + 3, \ 2x_1^2 + 2x_2^2 + 4 \right]$$

$$\text{subject to } x_1 + x_2 \geq 1$$
$$6x_1 + 2x_2 \geq 12,$$
$$\text{and} \qquad x_1 \geq 0, \ x_2 \geq 0$$

Example 3

$$\text{Minimize } Z_1 = C_1 + \{x_1 - C_2\}^{C_3}$$

$$\text{Minimize } Z_2 = x_2$$

$$\text{Minimize } Z_3 = x_3$$

$$\text{subject to } x_3^2 + \{1 + (x_1 - 1)^5\}^2 \geq 0.5$$
$$x_2^2 + x_3^2 \geq 0.5$$

where $C_1 = [0.82, 1.1]$, $C_2 = [0.9, 1.25]$, $C_3 = [4.5, 5.5]$
  search region: $x_1 \in [1.25, 2]$, x_2, $x_3 \in [0, 1]$.

Example 4

$$\text{Minimize } Z_1 = C_1 x_1 x_2^2 \left(C_2 x_3^2 + C_3 x_3 - C_4 \right) - C_5 x_1 \left(x_6^2 + x_7^2 \right)$$
$$+ C_6 \left(x_6^3 + x_7^3 \right) + C_1 2 x_4 x_6^2$$

$$\text{Minimize } Z_2 = \frac{\sqrt{C_7 \frac{x_4^2}{x_2^2 x_3^2} + C_8}}{C_9 x_6^3}$$

$$\text{subject to } \frac{1}{x_1 x_2^2 x_3} - \frac{1}{27} \leq 0$$

$$\frac{1}{x_1 x_2^2 x_3^2} - \frac{1}{397.5} \leq 0$$

$$\frac{x_4^3}{x_1 x_3 x_6^4} - \frac{1}{1.93} \leq 0$$

$$\frac{x_5^3}{x_1 x_3 x_7^4} - \frac{1}{1.93} \leq 0$$

$$x_2 x_3 \leq 40$$

$$\frac{x_1}{x_2} - 12 \leq 0$$

$$5 - \frac{x_1}{x_2} \leq 0$$

$$10.9 - x_4 + 10.5 x_6 \leq 0$$
$$10.9 - x_5 + 10.1 x_7 \leq 0$$

$$\frac{\sqrt{555025 \frac{x_4^2}{x_2^2 x_3^2} + 1.575 \times 10^8}}{0.1 x_7^3} \leq 1100$$

where $C_1 = [0.7, 0.8]$, $C_2 = [3, 3.5]$, $C_3 = [14.5, 15.5]$, $C_4 = [43, 44]$, $C_5 = [1.5, 1.6]$, $C_6 = [7, 7.75]$, $C_7 = [555020, 555030]$, $C_8 = [1.65, 1.75] \times 10^7$, $C_9 = [0.05, 0.15]$.

search region: $x_1 \in [2.6, 3.6]$, $x_2 \in [0.7, 0.8]$, $x_3 \in [17, 28]$, x_4, $x_5 \in [7.3, 8.3]$, $x_6 \in [2.9, 3.9]$ $x_7 \in [5, 5.5]$.

Example 5 (Deb et al. 2000)

$$\begin{aligned}
&\text{Minimize} && f_1(x) = x_1 \\
&\text{Minimize} && f_2(x) = (1 + x_2)/x_1 \\
&\text{subject to} && g_1(x) = x_2 + 9x_1 \geq 6 \\
&&& g_2(x) = -x_2 + 9x_1 \geq 1
\end{aligned}$$

search region: $x_1 \in [0.1, 1.0]$, $x_2 \in [0, 5]$

Example 6 (Srinivas and Deb 1995)

$$\begin{aligned}
&\text{Minimize } f_1(x) = (x_1 - 2)^2 + (x_2 - 1)^2 + 2 \\
&\text{Minimize } f_2(x) = 9x_1 - (x_2 - 1)^2 \\
&\text{subject to} \\
&g_1(x) = x_1^2 + x_2^2 \leq 225 \\
&g_2(x) = x_1 - 3x_2 \leq -10
\end{aligned}$$

search region: $x_i \in [-20, 20]$, $i = 1, 2$

Example 7 (Tanaka 1995)

$$\begin{aligned}
&\text{Minimize} && f_1(x) = x_1 \\
&\text{Minimize} && f_2(x) = x_2 \\
&\text{subject to} && g_1(x) = -x_1^2 - x_2^2 + 1 + 0.1 \cos(16 \arctan x_1/x_2) \leq 0 \\
&&& g_2(x) = (x_1 - 0.5)^2 + (x_2 - 0.5)^2 \leq 0.5
\end{aligned}$$

search region: $x_i \in [0, \pi]$, $i = 1, 2$

Example 8 (Binh and Korn 1997)

$$\begin{aligned}
&\text{Minimize} && f_1(x) = 4(x_1^2 + x_2^2) \\
&\text{Minimize} && f_2(x) = (x_1 - 5)^2 + (x_2 - 5)^2 \\
&\text{subject to} && g_1(x) = (x_1 - 5)^2 + x_2^2 \leq 25 \\
&&& g_2(x) = (x_1 - 8)^2 + (x_2 + 3)^2 \geq 7.7
\end{aligned}$$

search region: $x_1 \in [0, 5]$ $x_2 \in [0, 3]$

References

A.K. Bhunia, S.S. Samanta, A study of interval metric and its application in Multi-objective optimization with interval objective. Comput. Ind. Eng. **74**, 169–178 (2014)

T. Binh, U. Korn, A. Mobes, Multi-objective evolution strategy for constrained optimization problems, in *Proceedings of the Third International Conference on Genetic Algorithms*, Czech Republic, 1997, 176–182

K. Deb, H. Jain, Handling many-objective problems using an improved NSGA-II procedure, in *Evolutionary Computation, IEEE Congress*, 2012, pp. 1–8

K. Deb, A. Pratap, S. Agrawal, T. Meyarivan, A fast and elitist multi-objective genetic algorithm: NSGA-II, Technical report No. 2000001, Kanpur: Indian Institute of Technology Kanpur, India, 2000

S. Dhanalakshmi, S. Kannan, K. Mahadeyan, S. Baskar, Application of modified NSGA-II algorithm to combined economic and emission dispatch problem. Int. J. Electr. Power Energy Syst. **33**(4), 992–1002 (2011)

E. Hansen, G.W. Walster, *Global Optimization using Interval Analysis* (Marcel Dekker Inc, New York, 2004)

B.Q. Hu, S. Wang, A novel approach in uncertain programming Part I: New arithmetic and order relation for interval numbers. J. Ind. Manag. Optim. **2**(4), 351–371 (2006)

H. Ishibuchi, H. Tanaka, Multi-objective programming in optimization of the interval objective function. Eur. J. Oper. Res. **48**, 219–225 (1990)

S. Jeyadevi, S. Baskar, C.K. Babulal, M.W. Iruthayarajan, Solving multi-objective optimal reactive power dispatch using modified NSGA-II. Int. J. Electr. Power Energy Syst. **33**(2), 219–228 (2011)

S. Kannan, S. Baskar, J.D. McCalley, P. Murugan, Application of NSGA-II algorithm to generation expansion planning. IEEE Power Syst. **24**(1), 454–461 (2009)

S. Karmakar, S.K. Mahato, A.K. Bhunia, Interval oriented multi-section techniques for global optimization. J. Comput. Appl. Math. **224**(2), 476–491 (2009)

S. Karmakar, A.K. Bhunia, A comparative study of different order relations of intervals. Reliab. Comput. **16**, 38–72 (2012)

B.R. Kearfott, *Rigorous Global Search: Continuous Problems* (Kluwer Academic Publishers, Netherlands, 1996)

S.K. Mahato, A.K. Bhunia, Interval-arithmetic-oriented interval computing technique for global optimization. Appl. Math. Res. Express 1–19 (2006)

Z. Michalewicz, *Genetic Algorithm + Data structures = Evolution Programs* (Springer Verlag, Berlin, 1996)

R.E. Moore, *Interval Analysis* (Prentice-Hall, Englewood Cliffs, NJ, 1996)

P. Murugan, S. Kannan, S. Baskar, NSGA-II algorithm for multi-objective generation expansion planning problem. Electr. Power Syst. Res. **79**(4), 622–628 (2009)

L. Sahoo, A.K. Bhunia, P.K. Kapur, Genetic algorithm based multi-objective reliability optimization in interval environment. Comput. Ind. Eng. **62**, 152–160 (2012)

N. Srinivas, K. Deb, Multi-objective function optimization using non-dominated sorting genetic algorithms. Evol. Comput. **2**, 221–248 (1995)

M. Tanaka, GA-based decision support system for multi-criteria optimization, in *Proceedings of the International Conference on Systems, Man and Cybernetics-2*, 1995, pp. 1556–1561

H.C. Wu, The Karush-Kuhn-Tucker optimality conditions in multi-objective programming problem with interval-valued objective functions. Eur. J. Oper. Res. **196**, 49–60 (2009)

A Comparative Study on Different Versions of Multi-Objective Genetic Algorithm for Simultaneous Gene Selection and Sample Categorization

Asit Kumar Das and Sunanda Das

1 Introduction

Gene selection and sample categorization is one of the vital tasks to identify and monitor the target diseases. Genetic algorithm (GA) in gene selection and clustering has been proved beneficial in a variety of contexts to solve for a globally optimal solution of several complex problems. Genetic Algorithm makes an effort to swallow up the ideas of natural evolution. The basic idea behind genetic algorithm is to replicate natural evolution of evolving solutions generation wise by applying genetic operators. So, for the clustering task, the evolutionary algorithm, in particular, the GA is chosen frequently as the appropriate algorithm. A good GA always explores the search space in appropriate manner to search the globally optimal solution. The main objective of MOGA is to find out the possible pareto optimal solutions and diverse set of solutions in non-dominated fronts. In the last decade, many pareto-based algorithms have been proposed like pareto-based ranking procedure (FFGA), niched pareto genetic algorithm (NPGA), non-dominated sorting genetic algorithm (NSGA), and NSGA II, the strength pareto evolutionary algorithm (SPEA), and SPEA2. Fonseca and Fleming Genetic Algorithm (FFGA) is a multi-objective GA, which is now an interesting topic on multi-objective evolutionary algorithms (MOEAs). This method is a pareto-based ranking scheme, which is highly incorporated on the sharing factor. A Niched Pareto Genetic Algorithm (NPGA) is a kind of pareto dominance based tournament selection with multiple objectives. The NSGA and SPEA able to handle any number of

A. K. Das
Department of Computer Science and Technology, Indian Institute of Engineering, Science & Technology, Shibpur, Howrah, West Bengal, India
e-mail: akdas@cs.iiests.ac.in

S. Das (✉)
Department of Computer Science and Engineering, The Neotia University, South 24 Paragana, West Bengal, India
e-mail: das.sunanda2012@gmail.com

© Springer Nature Singapore Pte Ltd. 2018
J. K. Mandal et al. (eds.), *Multi-Objective Optimization*,
https://doi.org/10.1007/978-981-13-1471-1_11

objectives. SPEA2, an advance version of Strength Pareto Evolutionary Algorithm, is a density estimation technique, which uses a fine-grained fitness assignment strategy and an enhanced archive truncation method. In this section, several versions of genetic algorithm are discussed with their merits and demerits.

In this chapter, different versions of MOGA such as NSGA, SPEA and SPEA2 are explored for gene selection and sample categorization. The objective functions used in the methods are: (i) external clustered validation index between two set of clusters, one based on sample type and the other based on applied clustering technique on data subset associated to the chromosome, (ii) intra-correlation among the genes of the associated data subset, and (iii) overlapped cluster validation index.

Based on these three objective functions, the GA is run and after the convergence, the best non-dominated solution is identified which gives the important genes and the clusters of sample. Sample clustering is very helpful and widely used data mining techniques mainly to deal with finding the natural structures in collected experimental data, which is essential for data analysis to reveal interesting patterns in the underlying data. Till now none of the clustering algorithm is proved to be the best for all possible solutions even different configurations of the same algorithm. Cluster validation is a process of estimating how well a partition fits the structure underlying the data. Many researchers have proposed several internal and external cluster validation indices that quantize the goodness of a partition. The experimental results of all three methods are compared and observed that SPEA2 outperforms the others.

The chapter is organized into four sections. In Sect. 2, a brief overview of existing GA based optimization techniques is mentioned. Section 3 describes the proposed simultaneous gene selection and sample categorization method based on above three mentioned multi-objective genetic algorithms. Experimental results pertaining to the performance evaluation of the proposed methods with respect to the existing state-of-the-art methods are presented in Sect. 4 using various microarray cancer datasets. Finally, the chapter is summarized in Sect. 5.

2 Brief Overview of State-of-the-Art Methods

Genetic algorithm (GA) is a heuristic search technique for providing a global optimal solution. It imitates both the genetic and the evolutionary process of natural evolution. Professor Holland first proposed this technique in Michigan University of the United States. It is very useful for finding the solutions of the optimization problems because of its robustness. The method is also highly benefited for searching a solution in a high dimensional search space. The method starts with a randomly generated population and finds the fitness of each chromosome in the population iteratively until it converges, providing the chromosomes or candidate solutions with the optimum fitness values. All the versions of GA perform three basic operations such as selection, crossover and mutation but they may use one or more objective functions. Different versions of GA are briefly discussed below.

A. The Vector Evaluated Genetic Algorithm (VEGA)

The Simple Genetic Algorithm (SGA) was extended by David Schaffer in the year 1984 to propose Vector Evaluated Genetic Algorithm (VEGA) (Schaffer 1987). VEGA is almost same as SGA except the selection operation. At each generation its operators are modified till the number of sub populations are generated. So a problem having k objectives and n size of population generates k sub populations each of size n/k. Finally, the sub populations are mixed together to get a new population of size n and crossover and mutation operations are performed on the new population in a usual way.

B. Multi-Objective Genetic Algorithm proposed by Fonseca and Fleming (FFGA)

Using the concept of Goldberg's suggestion, the FFGA is implemented in a different way than the traditional GA. Firstly, the entire population is searched and all non-dominated individuals are identified as top ranked individuals in FFGA (Fonseca and Fleming 1993). The remaining individuals are ranked based on their non-dominant factors considering the rest of the population. If an individual ch_i of t-th generation is dominated by $n_i^{(t)}$ number of solutions in the current generation, then its present rank is defined by Eq. (1).

$$Rank(ch_i, t) = 1 + n_i^{(t)} \tag{1}$$

The fitness to each individual is assigned after completion of the ranking procedure. FFGA assigns fitness value to the individuals by any one of the following two fitness-assigning methods: (i) rank based method and (ii) Niche formation method. The rank based fitness assignment method suffers from a large selection pressure, which may result into an early convergence. To avoid this, FFGA mainly uses the second method of fitness assignment or the Niche formation method. In this method, sharing on the fitness values is used and the population is distributed over Pareto optimal region.

C. The Niched Pareto based Genetic Algorithm (NPGA)

Tournament selection is one of the most widely used selection techniques in GA. Horn, Nafloitis and Goldberg proposed the NPGA (Horn et al. 1994), which works using the concepts of Pareto dominance tournament selection and equivalence class sharing. This kind of tournament selection method selects some subsets of individuals randomly from the current population and the best one is retained in the subsequent population to determine the dominant individuals. The convergence speed of the method can be increased by adjusting the size of the tournament, which controls over the quantity of selection difficulty. When individuals become in a tie, then fitness sharing helps to choose the winner.

The equivalence class sharing procedure with an aim to select the most fitted individuals from the population is shown in Fig. 1. Here, the niche radius (σ_{share}) is chosen and according to the radius, candidates that have the least number of individuals are chosen as the 'best fit'.

Fig. 1 Equivalence class
sharing

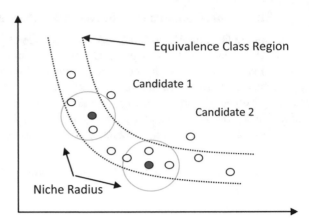

D. The Non-dominated Sorting based Genetic Algorithm (NSGA)

NSGA is different from a simple GA in terms of selection operator with keeping the crossover and mutation operations the same. NSGA (Srinivas and Deb 1995) eliminates the bias in VEGA with an objective to circulate the population in the area of pareto optimal front. It includes two steps selection process namely (i) the rank of solutions computed using individual's non-domination level and (ii) fitness sharing to each individual in the population. In the first step, all the dominated individuals are detected and in the second step strength is given to each solution based on its non-domination level. Lower fitness assignment to higher level individual gives a result towards Pareto optimal region.

E. Non-dominated sorting GA II (NSGA II)

In multi-objective GA, the three main difficulties are in: (i) computational complexity which is O (mN^3), where m is the number of fitness functions and N is the size of the population (ii) implementing non-elitism approach and (iii) specifying a sharing parameter. All these 3 difficulties are eliminated by NSGA II (Deb et al. 2002). NSGA II adopts a fast non-dominated sorting methodology with computational complexity O(mN^2). The selection operator generates a mating pool from parent and child populations and selects the best N solutions.

F. Strength Pareto Evolutionary Algorithm (SPEA)

Zitzler and Thiele in the year 1999 (Zitzler and Thiele 1999) proposed the SPEA. It combined the elitism and non-domination approach. At every generation of this algorithm, an external population is maintained which takes part in genetic operations. The external population is generated considering a set of non-dominated solutions from the current population. The fitness of each individual in the current and external population is defined using the dominated solutions in the population. After generating population for the next generation, it is required to update the external population. Each individual in the current population is checked whether both current population and external population dominate it. If both the populations do not dominate the individual then it is added to the external population. Finally, all individuals of external populations, which are dominated by the added individual, are eliminated. This is the updating procedure followed by SPEA.

G. Stronger Version of SPEA (SPEA 2)

Zitzler and Thiele proposed a modified version of SPEA known as Strength Pareto Evolutionary Algorithm 2 (SPEA2) (Zitzler et al. 2002). There are three main differences of this algorithm with SPEA; these are as follows: (i) it defines a fitness function for each individual using both the number of solutions that dominates it and the number of solutions dominated by it (ii) it computes the nearest neighbor density of each solution for efficient searching and (iii) it has an improved archive truncation method to retain the boundary solutions.

3 Proposed Methodology

Simultaneous gene selection and sample clustering help to increase understandability, scalability and accuracy by identifying relevant genes for sample categorization. Here, sample clustering and gene subset selection is done together using genetic algorithm (GA). Basically, GA is used to select the optimal subset of genes which automatically finds optimal set of clusters of samples when the GA converges. Clustering is a data mining technique used in different fields to discover the overall distribution pattern and correlation among the data. Till now many clustering algorithms have been devised by the researchers for extracting hidden patterns from the dataset. In this section, multi-objective genetic algorithm (MOGA) based clustering algorithm is proposed for selecting relevant gene subset and obtaining optimal set of clusters of samples. Almost all real world optimization problems have multiple objectives that is why the MOGA is so useful for searching optimal solutions. This technique provides an appropriate distribution of individuals with multiple equivalent solutions. Optimality of the clusters is measured using various cluster validation indices. Here, *NSGA* (Srinivas and Deb 1995), SPEA (Zitzler and Thiele 1999), and *SPEA2* (Zitzler et al. 2002) have been explored for finding both optimal gene subset and clusters of samples from microarray datasets.

The methods use the concepts of cellular automata to generate initial population of binary chromosomes as input. Assume that the size of the population is m and length of each chromosome in the population is n, which is the number of genes in each sample. The fitness of each chromosome is calculated using the subset of the gene set. This subset is generated considering some genes of the original set. Which genes will be selected that depends on in which position of the chromosome the bit '1' occurs. Since the sample size and chromosome size are same, so all the genes of the gene set are selected for which in corresponding positions of the chromosome have bit '1's. Thus, a chromosome provides a collection of genes. At the end of the process, the chromosomes of the final population are placed into pareto fronts so that each non-dominated chromosomes are in the same front. Thus, the chromosomes in

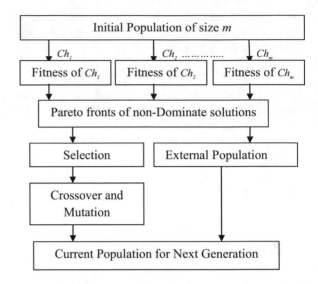

Fig. 2 Basic workflow of proposed methods between two successive iterations

the population are partitioned into non-dominated sets, which regulate flow of the genetic algorithm. A chromosome ch_1 is said to dominate the other solution ch_2, if both the following conditions are true.

(i) For all objectives, ch_1 will never be worse than ch_2.
(ii) For some objective functions, ch_1 is strictly better than ch_2.

Two individuals of same front are independent to each other and considered as similar solutions. The best solution of first front (either it is determined by measuring some strength excluding the fitness value or it is selected randomly) of the final population gives the optimal gene subset and clusters of samples. Figure 2 describes the overall work flow of the proposed work.

3.1 Initial Population and External Population

Initial population generation plays an important role in evolutionary algorithms (Gong et al. 2015; Price et al. 2005; Gu et al. 2015). When there is no idea about the solution of the problem, then the obvious choice is to generate the initial population randomly. But the GA takes long computational time to converge when the searching space is unpredictable. At the same time, the initial population generated using quasi-random sequences is very complex and not appropriate in high dimension (Maaranen et al. 2004). As a result, a suitable cellular automata based technique is applied for making a simple and faster initial population in high dimensional space.

Abstract Cellular Automata (*CA*) (Waters 2012), a pseudorandom pattern generator, plays an important role for any population based stochastic search method. Here,

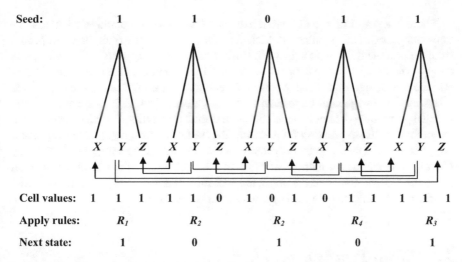

Fig. 3 Next-state population generation using cellular automata

the nonlinear hybrid uniform cellular automata have been used for generating the initial population so that the initial solutions are situated over the major portion in the searching space. As a result, optimization becomes more efficient using the proposed population generation approach. The nonlinear hybrid uniform CA is modeled as a large number of cells organized like a lattice. It is a very powerful automaton where each cell has the ability of self-breeding. The proposed *GA* based methods generate next state (*NS*) of each cell c using its present state (*PS*) and states of its neighbor cells based on the four basic rules R_1, R_2, R_3, and R_4 as defined below. The method considers only 3-neighbors (namely; left neighbor (*LNB*), self or current (*SLF*) and right neighbor (*RNB*)) one-dimensional cellular automata and each cell only have any one of two states (0 or 1).

$R_1 : NS(c) = (LNB(c) \wedge SLF(c)) \vee (\sim LNB(c) \wedge RNB(c))$
$R_2 : NS(c) = (LNB(c) \wedge RNB(c)) \vee (SLF(c) \wedge \sim RNB(c))$
$R_3 : NS(c) = LNB(c) \oplus SLF(c) \oplus RNB(c)$
$R_4 : NS(c) = SLF(c) \oplus (LNB(c) \vee \sim RNB(c))$

Where, LNB(c) denotes the left cell value of current cell c, *SLF(c)* denotes the c-th current cell value and *RNB(c)* denotes the right cell value of current cell c. For every cell to generate the next state, a feasible rule is chosen dynamically. Among the rules, R_1, R_2, and R_4 are nonlinear while R_3 is linear and therefore, it is named as nonlinear hybrid *CA*. Population generation uses these rules as mentioned in Fig. 3 with an example, where *X*, *Y* and *Z* stand for Right (*RNB*), Current (*SLF*) and Left (*LNB*) cell respectively.

Thus, the initial population of chromosomes is created and used as current population for the first iteration and after every iteration a new current population is generated.

There is a possibility in *GA* based heuristic that some comparatively better solutions can get rid of the population which may slower the process of finding the optimal solution or quite difficult to achieve it. Elitism is the property used in *GA* to keep the copy of the better chromosomes in an external population and move them directly into the next generation. Thus, a portion of external population is chosen to produce offspring for new generation in each successive generation. Here, together with the internal population or simply population P, external population P' is also generated to maintain the elitism property for allowing the best chromosome(s) from the present population to the next population directly. Thus, quality of the solutions obtained by *GA* will not degrade from one generation to other. Generally the size of the external population is one-fourth of the original population. The external population is filled by the non-dominated solutions of the optimal pareto front.

3.2 Fitness Function

The fitness function is defined to determine the quality of a solutions in the population. It gives the direction of looking for the optimal solution of the problem that helps to quick convergence of the process. These functions for each solution in the population are defined using (i) external cluster validation index between the two set of clusters, one based on sample type and the other based on applied clustering technique on data subset associated to the chromosome, which needs to be maximized, (ii) intra-correlation among the genes of the associated data subset, which is to be minimized, and (iii) overlapped cluster validation index, needs to be minimized.

3.2.1 External Cluster Validation Index

To define the external cluster index, the samples are clustered using the Enhanced Cluster Affinity Search Technique (E-CAST) (Bellaachia et al. 2000). It determines clique graphs, which are basically a set of disjoint undirected complete graph. Each complete graph inside the clique is a cluster, where every node (i.e., sample) in the complete graph is similar to each other and dissimilar to node (i.e., sample) of other complete graph inside the clique. The similarity is measured using cosine similarity of the samples. In E-CAST algorithm, each cluster has a *connectivity threshold* and each member/sample has certain *affinity* to a particular cluster. A node/sample is said to be a high connectivity node to a cluster only when the affinity of that node to that cluster is larger than or equal to the connectivity threshold of that cluster; else the node/sample is called low connectivity node to that cluster.

The external cluster index is computed using the following process:

(a) The data subset associated to each chromosome is first classified into different groups based on the sample types. Thus, a set of clusters $CL_1 = \{Cl_{11}, Cl_{12}, \ldots, Cl_{1s}\}$ are obtained, where s is the number of sample types exist in the dataset.

The following two steps give the second set of clusters CL_2.

(b) For every pair of samples in the data subset, Cosine similarity is calculated using Eq. (2) and a Cosine similarity matrix is obtained.

$$S(P, Q) = \frac{\sum_{i=1}^{n} P_i Q_i}{\sqrt{\sum_{i=1}^{n} P_i^2}\sqrt{\sum_{i=1}^{n} Q_i^2}} \qquad (2)$$

where, P_i and Q_i are i-th gene of samples P and Q respectively. It is a measure of orientation and not magnitude; two vectors of equal orientation have a cosine similarity value 1, so higher the similarity value implies more similar the samples are and vice versa. Instead of cosine similarity, several other similarity measures can be used like Euclidean distance, Pearson correlation, and so on. The E-$CAST$ clustering algorithm uses cliques of the graph and divisive hierarchical clustering algorithm to partition the samples into different groups in a top down approach. The algorithm relies on input called affinity threshold μ, on which the size and cluster quality depends. A hardcoded predetermined value for affinity threshold μ is used in Ben-Dor et al. but better option is the dynamic calculation of its value. When affinity threshold value is calculated dynamically, only the similarity values of samples not yet clustered are considered. By using the similarity matrix S defined in Eq. (2), the affinity of a sample p is found to a particular cluster C_i as $a(p) = \sum_{u \in C_i} S(p, u)$. Each cluster have a connectivity threshold defined as $\chi = \mu |C_i|$, where $|C_i|$ is the number of elements in that cluster. A sample is known as a highly connected node to a cluster if its affinity to that cluster is higher than the connectivity threshold of that cluster; otherwise the sample is called low connectivity node to that cluster. Each Cluster is formed using following three operations:

(i) Node Addition: Add a sample to the cluster having highest affinity or connectivity.

(ii) Node Deletion: Remove a sample from a cluster having low connectivity or affinity.

(iii) Cluster Clearing: It is used to remove any node from its current assigned cluster to another cluster to which it has highest affinity. But sometimes this step may not be required to improve the cluster's performance. Thus, after applying E-$CAST$ algorithm, a set of t-clusters $CL_2 = \{Cl_{21}, Cl_{22}, ..., Cl_{2t}\}$ are generated.

(c) Now, the external cluster index namely, F-Measure between two sets of clusters CL_1 and CL_2 is measured. This index value ranges in [0–1]. Higher value indicates clusters obtained by both the clustering algorithms are more similar to each other.

3.2.2 Intra Correlation Among Genes

The intra correlation among genes is measure to define a fitness function of the chromosome *ch* to maximize non-redundancy i.e. minimize intra-correlation among the genes in the dataset, defined in Eqs. (3) and (4).

$$f3(ch) = \frac{1}{MC(ch)} \tag{3}$$

$$MC(ch) = \frac{2}{N(N-1)} \sum_{i=1}^{N} \sum_{j=i+1}^{N} \left| corr(ch_i, ch_j) \right| \tag{4}$$

where, *f3(ch)* is the non-redundancy, *MC(ch)* is the intra-correlation of genes in chromosome *ch*, and corr(ch_i, ch_j) is the correlation between two genes ch_i and ch_j of *ch*, *N* is the number of genes in chromosome *ch*.

3.2.3 Overlapped Cluster Validation Index

Now, the overlapped external cluster validation index between two set of clusters CL_1 and CL_2 are measured using the concept discussed in paper (Campo et al. 2016), briefly discussed as follows:

For a cluster C_i probability that every pair of samples belongs to the cluster is measured by Eq. (5).

$$\Pr((s_x, s_y) \in C_i) = \frac{\binom{M_i}{2}}{\binom{N}{2}} = \frac{M_i(M_i - 1)}{N(N - 1)} \tag{5}$$

where, M_i is the number of objects in cluster C_i. Here, the numerator represents the number of pairs that can be found in cluster C_i. In order to normalize it, the denominator represents a similar situation where all of the samples are grouped together in a single cluster; hence any possible pair could be found. The probability of finding a pair of elements in any cluster C_i for all of the existing clusters *k* is estimated using Eq. (6).

$$\tilde{P} = \frac{\sum_{i=1}^{k} \binom{|M_i|}{2}}{k \binom{N}{2}} \tag{6}$$

where, the numerator accumulates all of the pairs found in each cluster. The denominator represents a normalization factor, which acts as if all of the samples were grouped together. The k factor considers the situation where the overlapping is complete up to all k clusters. An identical reasoning could be applied to obtain a comparable expression for C' in Eq. (7).

$$\widetilde{P}' = \frac{\sum_{j=1}^{k'} \binom{|M_i|}{2}}{k \binom{N}{2}} \tag{7}$$

The same analysis described for $\widetilde{P}$ and $\widetilde{P}'$ can be performed for both solutions together which is defined in Eq. (8), where M_{ij} = number of objects common in C_i and C_j'.

$$\Pr((s_x, s_y) \in C_i \wedge (s_x, s_y) \in C_j') = \frac{\binom{M_{ij}}{2}}{\binom{N}{2}} = \frac{M_{ij}(M_{ij} - 1)}{N(N - 1)} \tag{8}$$

It is seen that, as an approximation to the probability, the pair of data points (S_x, S_y) is present in both solutions. The whole expression stands for the event of drawing two samples that are in both clusters C_i and C_j'. It is assumed that the same analysis is made for every possible pairing between clusters of C and C'. So, the probability of finding (S_x, S_y) in both solutions can be estimated using Eq. (9).

$$\tilde{t} = \frac{\sum_{i=1}^{k} \sum_{j=1}^{k'} \binom{M_{ij}}{2}}{\binom{N}{2} \frac{\max(n, n')}{N} \min(k, k')} \tag{9}$$

where n and n' are the number of samples that can be counted in solutions C and C', respectively. Similar to $\widetilde{P}$ and $\widetilde{P}'$, the numerator of Eq. (9) counts all of the effective pairs of samples that can be found in both solutions simultaneously. The denominator acts once again as a normalization term. It basically covers the extreme scenario where all of the samples are clustered together several times. Just as in Eqs. (6) and (7), $\binom{N}{2}$ counts the number of pairs that can be arranged given all N samples. Since there could be overlap in both solutions, the given number of pairs should be multiplied by a factor. On the one hand, there could be as many overlaps as k in C and k' in C'. Also, it was found that the matching between clusters of both solutions produces at most $\min(k, k')$ pairs of clusters in the comparison.

Finally, $\max(n, n')/N$ is the average number of samples that can be found considering overlaps.

With these elements in mind, the new index for overlapped clusters (OC) could be defined as the ratio between the probabilities of finding two items grouped together in both solutions and the maximum probability of finding them in one of the given solutions, which is defined in Eq. (10).

$$OC = \frac{\widetilde{T}}{\max\left(\widetilde{P}, \widetilde{P'}\right)} \tag{10}$$

This validation index value ranges in [0–1]. Higher value indicates clusters obtained by both the clustering algorithms are more similar to each other.

3.3 Tournament Selection

Selection operation is considered as the first genetic operator of *GA*. The solutions selected as parents are involved in crossover operation to produce offspring for the next operation. Selection operations popularly used are roulette wheel selection, boltzmann selection, tournament selection, rank selection, and steady state selection to select the best chromosomes for mating pool generation. In the proposed work, tournament selection method is used, where four competitors participate in the tournament each time. Chromosome with the best fitness value is chosen in each tournament and the selection process is continued until the mating pool is filled up.

3.4 Crossover Operation

The fittest chromosomes of the current population *P* are selected using the selection operation, which gives some direction of searching solutions but offsprings are created by crossover operation for searching the better solutions throughout the whole search space. Crossover is one of the basic operations of *GA* by which new chromosomes are generated to give the direction of searching solutions in the population towards local optima. As the main goal of *GA* is to make the population convergence, so crossover is basically happened more frequently, generally in every generation. New solutions are found from old solutions by the process of crossover and mutation, much like what happens in nature with chromosomes. So, new solutions are the offspring of the old solutions or the children of the old solutions. There are many cross over operations like single point crossover, uniform crossover, multi point crossover out of which two-point crossover operation is used in the work between two members present in the mating pool with crossover probability 0.8.

3.5   Mutation Operation

Mutation is a crucial operation to maintain the diversity in the population. In spite of popularity of the single bit mutation in *GA*, diversity could not be maintained all the time in the population. The reason behind this is no change of the first bit of binary string. Jumping gene mutation methodology (Pati et al. 2013) helps to overcome this demerit for mutating the genes with probability 0.02. The genes capable of jumping dynamically within the chromosome are known as jumping genes. These genes have the great potential to maintain diversity throughout the entire population and take very crucial role for evolutionary searching algorithm. Let, a chromosome in population is $(x_1, x_2, ..., x_n)$. Jumping genes of length p $(p \ll n)$ say, $(y_1, y_2, ..., y_p)$ and the initial position of the chromosome from which mutation occurs is selected randomly and the portion of the chromosome is substituted by the jumping gene. Let r be the starting position, so after mutation the muted chromosome is $(x_1, x_2, ..., x_{r-1}, y_1, y_2, ..., y_p, x_{r+p}, ..., x_n)$.

3.6   Multi-Objective Genetic Algorithm for Gene Selection and Sample Clustering

Clustering is an important data mining technique to find the actual clusters of data. Many clustering algorithms are devised for searching optimal set of actual clusters to identify the overall distribution patterns and that helps in prediction of future trend. In this section, a multi-objective GA based clustering algorithm is proposed to find the clusters of genes. Generally, real world optimization problems have multiple objectives so multi-objective genetic algorithm is a perfect treatment for finding multiple optimal solutions. Thus, the usefulness of application of this algorithm is to partition the samples into different clusters based on the defined objective functions. In this chapter, *NSGA*, SPEA, and *SPEA2* have been explored for finding both important gene subset and optimal set of clusters of samples from the dataset. The corresponding algorithms are named as *GNSGA*, *GSPEA* and *GSPEA2* respectively.

All the above methods need a population of chromosomes which are different individuals in the search space. Length of each chromosome is set as n, the number of genes present in the dataset and population size is set as m, a predefined constant. Initial population of binary chromosome is generated using cellular automata. The '1' in the i-th position of a chromosome corresponds to i-th gene in the dataset. Thus, a chromosome in the population represents a collection of genes corresponds to value '1' across its length. This collection of genes is considered as a cluster associated to that chromosome. After convergence, final population is divided into pareto optimal fronts based on their objective functions and genes and clusters of samples associated to the best chromosome of the first front are considered as the optimal gene subset and optimal clusters of samples of the gene datasets.

3.6.1 The Proposed *GNSGA* Method

The main difference between *NSGA* and Single Objective Genetic Algorithm (*SGA*) is only in selection operation but crossover and mutation operations are same for both the GA. In Vector Evaluation Genetic Algorithm (*VEGA*) (Schaffer 1987), there are some biasness which is removed by *NSGA* and the *NSGA* distributes the population over the entire Pareto optimal region. In *NSGA*, the selection operation functions using following steps: (i) the chromosomes are ranked by their non-domination level and (ii) sharing is used to assign fitness to each chromosome. Lower fitness is assigned to a chromosome in higher level. This fitness assignment provides a result towards Pareto optimal region.

(A) Parameter Setting

The sharing function, defined later, requires two parameters; α and σ_{share}. The parameter α does not have much influence on the performance of sharing function but σ_{share} need to be set perfectly in order to define the niche size. In our method, similar to paper (Deb 2001), α is considered as 2 and genotypic sharing is considered. For genotypic sharing, the Hamming distance is used instead of Euclidean distance. It takes an integer value between $[0, n]$, where n is the length of the string. For a given string, the number of strings having different bit differences with it in the entire search space is calculated. Let the number of allowed bit differences is K. Then all strings having K or less bit differences, occupy at least $1/q$th portion of the entire search space, are determined in Deb and Goldberg (1989). K is calculated using Eq. (11). If K is not an integer then the least value of K is chosen so that the left side of the equation greater than or equal to the right side. This minimum value K is assigned to σ_{share}.

$$\sum_{i=0}^{K} \binom{l}{i} = \frac{2^l}{q} \tag{11}$$

(B) Shared Fitness Assignment

NSGA provides solutions which are classified into mutually exclusive non-dominated fronts. The solutions in first front are the best solutions of the current population. Similarly, solutions in the second fronts are the next best solution of the current population and so on. Therefore, the highest fitness is assigned to the chromosomes in first front, and gradually worse fitness to solutions in higher level fronts. This is because, the best non-dominated solutions in a population are closest to the actual Pareto-optimal front compared to other solutions in the population. The fitness assignment to the chromosome starts from the first non-dominated front and successively fitness assignment is continued to other fronts in order. In any r-th solution of the first front, a fitness value $F_r = m$ is assigned; m is the population size. As the diversity among the solutions in a front need to be maintained, so shared function method is used that degrades the assigned fitness based on the number of neighboring solutions.

Sharing function value is computed between every pair of solutions in a front using Eq. (12), d in the distance between any two solutions in the population.

$$Sh(d) = \begin{cases} 1 - \left(\frac{d}{\sigma_{share}}\right)^{\alpha}, & \text{if } d \leq \sigma_{share} \\ 0, & otherwise \end{cases} \quad (12)$$

The function $Sh(d)$ takes a value in $[0, 1]$, depending on the value of d and σ_{share}. Both the solutions are same if d is zero and in this case $Sh(d) = 1$, which implies that the solution has full sharing effect on itself. Two solutions are at least at a distance of σ_{share} from each other if $d \geq \sigma_{share}$ and in this case $Sh(d) = 0$ indicating that they don't have any sharing effect on each other. The two solutions have a partial effect on each other for any other distance d. This sharing function values are computed for each individual considering all solutions in the same front. Thus, niche count nc_i for i-th solution is computed using Eq. (13). Here, d_{ij} is the distance between i-th and j-th solution. The niche count provides an estimation of crowding near a solution. If no other solutions exist within a radius σ_{share} of a solution in the same front, the niche count would be one for that solution. But, if all solutions in the front are very close to each other and within radius σ_{share}, the niche count of any solution in the group are closer to number of solutions in the front.

$$nc_i = \sum_{j=1}^{n} Sh\left(d_{ij}\right) \quad (13)$$

After computing niche count of a solution, the fitness F_i assigned to i-th solution is divided by nc_i to obtain its share fitness value. As all over-represented optima will have a larger nc_i value, the fitness of all representative solutions would be degraded more and vice versa. The procedure is performed for all solutions in the first front. Next, the fitness value slightly smaller than the least shared fitness in the first front, is assigned to all the solutions in the second front. The whole fitness sharing process is repeated for second front and so on. The chromosome with maximum share fitness value of each front is considered as the best chromosome of that front.

As already mentioned, NSGA used a mating restriction scheme (Deb 1989) based on genotypic distances between mating individuals. To select a mate for an individual, their hamming distance is computed. If the distance is smaller than the parameter σ_{share}, they are involved in crossover operation, otherwise another mate is selected at random.

3.6.2 The Proposed GSPEA Method

Here, instead of NSGA, SPEA (Zitzler and Thiele 1999) is applied for important gene subset selection. An external population is selected from the current population to breed a new generation. The method uses a binary tournament selection operation based on the strength assigned to the chromosomes. The strength $SE(ch_i)$ assigned to each chromosome ch_i in external population is computed using Eq. (14).

$$SE(ch_i) = \frac{n_i}{m+1} \tag{14}$$

where, n_i is the number of current population members dominated by chromosome ch_i and m is the population size. So, the strength of a chromosome is directly proportional to the number of chromosomes dominated by it. The strength of each chromosome ch_j in internal population is defined using Eq. (15).

$$SI(ch_j) = 1 + \sum_{i \in P' \wedge \mathbf{i} \neq \mathbf{j}} SE(ch_i) \tag{15}$$

In Eq. (15), $i \neq j$ signifies that the fitness of j-th chromosome in internal population is one more than the sum of the strength values of all chromosomes in external population which weakly dominate j. So, according to *SPEA* algorithm, the chromosome having less strength is considered as the best chromosome to be selected for the mating pool. This selection method is continued till the mating pool is filled.

3.6.3 The Proposed *GSPEA2* Method

In this method, improved version of strength pareto evolutionary algorithm (*SPEA2*) (Zitzler and Thiele 1999) is explored to identify gene subset sufficient for sample clustering. Like other methods, external population P' is generated together with the current population P to allow the best chromosome(s) from the current population to the next population. The method is differed from *SPEA* in two respects; (i) the number of external population member is constant and (ii) it uses truncation operation, which allow to considered the boundary solutions. The external population of current generation is made considering all the non-dominated solutions of the previous internal population and the external population. If the size of the external population exceeds than its predefined size then truncation operation is used. On the other hand, if the number of solutions in the external population is less than its size then the dominated solutions with more fitness value are selected from the previous internal and external population to fill up the size. All non-dominated members are identified from the combined internal population and external population whose fitness value $F(i) < 1$ i.e. $P'_{t+1} = \{ k | k \in P_t + P'_t \wedge F(k) < 1 \}$.

Thus, if $|P'_{t+1}|$ is the external Population Size, then nothing need to be done and if $|P'_{t+1}| <$ the external Population Size (N') then fill the remaining $N' - |P'_{t+1}|$ portion of the population, i.e., copy the first $N' - |P'_{t+1}|$ members of the internal and external populations having $F(k) \geq 1$. But if $|P'_{t+1}| >$ External Population Size then the size of the external population is reduced by truncation operator discussed below:

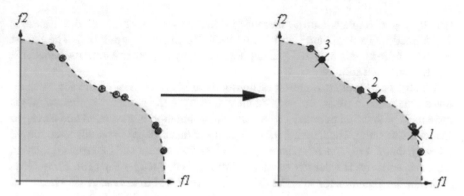

Fig. 4 Illustration of the truncation method used in *GSPEA2*

Truncation operator

The k-th solution of the external population P'_{t+1} is removed when $k \leq_d j$ for all $j \in P'_{t+1}$ with

$$k \leq_d j :\Leftrightarrow \forall 0 < l < |P'_{t+1}| : \sigma_k^l = \sigma_j^l \lor \exists 0 < l < |P'_{t+1}| : [\left(\forall 0 < m < l : \sigma_k^m = \sigma_j^m\right) \land \sigma_k^l < \sigma_j^l]$$

where, σ_k^l denotes the distance of k-th solution to its l-th nearest neighbor in P'_{t+1}. Thus, the individual with the minimum distance to other individuals is chosen at each stage. If there are multiple such individuals, then the tie is broken by considering the second smallest distance and so on. Figure 4 expresses the truncation operation used in the proposed *SPEA2* based method. The left side diagram shows the external population of eight solutions, where the population size is considered as five. Thus, after truncation operation, the external population contains a non-dominated set of five solutions as shown in the right side diagram. The figure also depicts which solutions are removed in which order by the truncate operator. Though the proposed method uses three objective functions, but for better understanding, the diagram is explained with first two objective functions.

Each individual or member of the external population and current population has the strength value $S(k)$. The strength $S(k)$ of a member k is equal to the number of members in both current and external population to which it dominates, as defined by Eq. (16).

$$S(k) = \{j | j \in P_t + P'_t \land k \neq j\} \tag{16}$$

where, '$k \neq j$' implies that the member k dominates member j.

Then the raw fitness value $R(k)$ is calculated for each member/solution by Eq. (17).

$$R(k) = \sum_{j \in P_t + P'_t \land j \neq k} S(k) \tag{17}$$

where, '$j \neq k$' implies that the member j dominates member k, i.e., raw fitness is determined by its dominators in both internal and external population. The lowest the value of $R(k)$ indicates that it is not dominated by any other solutions, hence it is a non-dominated solution.

Finally, the density estimation technique is used when most of the solutions do not dominate each other. So, the distance of every k-th solution to other solutions present in both initial population and external population is measured and arranged in increasing order. Then, taking the K-th entry of that ordered list as distance sought σ_k^K such that $K = \sqrt{N + N'}$, where N and N' are the current and external population sizes respectively. The density of each solution k is defined by Eq. (18). Here, '2' is added in the denominator to make the value of $D(k)$ within the range [0, 1].

$$D(k) = \frac{1}{\sigma_k^l + 2} \tag{18}$$

The Fitness function is defined by Eq. (19).

$$F(k) = R(k) + D(k) \tag{19}$$

The fitness value of non-dominated solution is less than 1.

3.6.4 Gene Selection and Sample Clustering

In all three multi-objective GA, one fitness function is the external cluster validation index which is measured using the Enhanced Cluster Affinity Search Technique (*E-CAST*) based clustering algorithm. This algorithm is applied for each member in the population for partitioning the samples into different groups. Thus, after convergence of each of the GA based method, the genes associated to the best solution in the population is the final gene subset and clusters corresponding to the best solution in the population is the final set of clusters.

3.6.5 Cluster Validation

Cluster validation, a very important issue in cluster analysis, is the measurement of goodness of the clusters relative to others created by clustering algorithms using different parameter values. The difficulty of determining the clusters number is known as "the fundamental problem of cluster validity" (Akogul and Erisoglu 2017). There are many validation indices like Calinski Harbasz index (*CH*) (Calinski and Harabasz 1974), Davies-Bouldin (*DB*) index (Davies and Bouldin 1979), *I*-index (Maulik and

Bandyopadhyay 2002) and Xie–Beni index (XB) (Xie and Beni 1991) for predicting quality of the clusters.

(A) Calinski Harbasz index

Calinski-Harabasz index (CH) (Calinski and Harabasz 1974), is defined by Eq. (20).

$$CH(K) = \frac{[trace\ B/K - 1]}{[trace\ W/N - K]} for\ K \in N \tag{20}$$

where, B is the sum of squares error between different clusters and W is the square of the differences of all objects in a cluster from their respective cluster center.

$$trace\ B = \sum_{k=1}^{K} |C_k| \|\overline{C_k} - \bar{x}\|^2$$

$$trace\ W = \sum_{k=1}^{K} \sum_{i=1}^{K} w_{k,i} \|x_i - \overline{C_k}\|^2$$

The maximum value of the CH-index (Calinski and Harabasz, 1974) gives the optimal number of clusters.

(B) Davies-Bouldin Validity Index

This index (Davies and Bouldin 1979) is a function defined as the ratio of the sum of intra-cluster scatter to inter-cluster separation. Let $C_1, C_2, \ldots, C_k$ be the k number of clusters, then the DB-index is defined using Eq. (21).

$$DB = \frac{1}{k} \sum_{i=1}^{k} \max_{1 \le j \le k\ and\ i \ne j} \left\{ \frac{\delta_i^2 + \delta_j^2}{d_{ij}^2} \right\} \tag{21}$$

where, δ_i^2 and δ_j^2 are the variance of clusters C_i and C_j, respectively and d_{ij}^2 is the distance between centers of clusters C_i and C_j. A low variance and high distance between clusters lead to low value of DB that corresponds to clusters that are compact and centers are far away from each other.

(C) I-Index

The I-Index index (II) (Maulik and Bandyopadhyay 2002) is defined in Eq. (22).

$$II(K) = \left(\frac{1}{K} \times \frac{E_1}{E_K} \times D_K \right)^P \tag{22}$$

where, $E_1 = \sum_j \|x_j - \bar{c}\|_2$ and $E_K = \sum_{k=1}^{K} \sum_{j \in c_k} \|x_j - c_k\|_2$, $D_K = max_{i,j}^K \|c_i - c_j\|_2$ and the power P is a constant, which normally is set to be two. The optimal K is the one that maximizes $II(K)$.

(D)   Xie–Beni index

The Xie-Beni index (*XB*) (Xie and Beni 1991) is an index of fuzzy clustering which exploits the compactness and separation. *XB* computes the ratio between the compactness and the separation. The *XB* for a given dataset X with a partition with K clusters is mathematically defined by Eq. (23).

$$XB(K) = \frac{V_C(K)/N}{V_S(K)} = \frac{\sum_{n=1}^{N} \sum_{k=1}^{K} u_{k,n}^m \|x_n - c_k\|^2}{N \times \min_{i,j} \|c_i - c_j\|} \tag{23}$$

$V_C(K)$ gives the compactness of K clusters, $c_{k=} \sum_{n=1}^{N} u_{k,n}^m x_n / \sum_{n=1}^{N} u_{k,n}^m$ is the centroid of the k-th cluster. $V_S(K)$ is the degree of separation between clusters. In general, an optimal K is found for $\min_{k \in [2, N-1]} XB(K)$ to produce the best clustering performance.

4   Experimental Results

Extensive experiments are done to evaluate the proposed methods using experimental microarray datasets ('Kent Ridge Bio-medical Data Set Repository') describes as follows:

4.1   *Microarray Dataset Description*

I.   **Leukemia Dataset**: It is a well-understood gene database published in (Baraldi and Blonda 1999). The raw data is available in (Steinbach et al. 2000). It consists of 27 *ALL* and 11 *AML* over 7129 human genes.

II.   **Lung Cancer Dataset**: It is a well-understood gene database published in (Pedrycz and Hirota 2007). The raw data is available in (Tetko et al. 1995). The dataset contains 32 samples of which 16 are Malignant Pleural Mesothelioma (*MPM*) and 16 are Adenocarcinoma (*ADCA*).

III.   **Prostate Cancer Dataset**: The dataset published in (Gowda and Krishna 1978). The raw dataset is available in Huang and Ng (1999). It contains 52 *prostate-tumor* samples and 50 *non-tumor* prostate samples with around 12600 genes.

IV.   **Breast Cancer Dataset**: The experimental dataset is published in (Liu et al. 2006). The raw data is available in (Bhat 2014). The dataset contains samples of 78 patients, 34 of which are from patients (labeled as "*relapse*") who had developed distance metastases within 5 years, the rest 44 samples (labeled as

Table 1 Parameters for GA control

Parameter	Value
Size of the population	200
Number of generations	100
Crossover probability	0.80
Mutation probability	0.02

"*non-relapse*") are from patients who remained healthy from the disease after their initial judgment for interval of at least 5 years. The number of genes is 24481.

V. **Diffuse Large B-cell Lymphoma (DLBCL) Dataset**: The DLBCL dataset is published by (Petrovskiy 2003). The raw data available in (Nag and Han 2002). The classification about data addressed in the publication is *DLBCL* against Follicular Lymphoma (*FL*) morphology. The dataset contains 58 *DLBCL* samples and 19 *FL* samples. The expression profile contains 6817 genes.

VI. **Colon Cancer Dataset**: This gene database is published in (Merz 2003). The raw data available in (Rousseeuw 1987). It Contains 62 samples collected from colon-cancer patients. Among them, 40 tumor (labeled as "*negative*") biopsies are from tumors and 22 normal (labeled as "*positive*") biopsies are from healthy parts of the colons of the same patients. The expression profile contains 2000 genes.

4.2 Parameter Setup and Preprocessing

The datasets contain high volume of unwanted genes with random noise and the samples are linearly inseparable. To obtain the optimal solution, parameter tuning for the GA is important. Crossover probability and mutation probability are two important parameters of the GA. Here, tuning of these two parameters are completed owing to the optimal performance of the GA. It is observed that varying values of the parameters tend to verify the convergence of the GA. Finally, these parameters are selected after several test evaluation of proposed methods on the listed datasets. The parameters used in GA environment are listed in Table 1. Here, the ChiMerge discretization (Kerber 1992) algorithm is used to convert continuous attributes into discretized values.

4.3 Performance Measurement

The experimental results have been provided for the mentioned microarray datasets. The proposed methods use cellular automata for initial population generation instead

Table 2 Comparison between *GNSGA*, *GSPEA* and *GSPEA2* regarding time (*minutes*)

Method	Dataset					
	Leukemia	Lung	Prostate	Breast	DLBCL	Colon
GNSGA	46.48	52.5	57.3	68.25	57.35	43.27
GSPEA	42.52	**40.03**	**50.21**	57.43	**52.27**	41.36
GSPEA2	**34.75**	40.36	53.17	**55.18**	54.04	**40.25**

Table 3 Comparison between different methods in terms of classification accuracy and number of selected genes

Dataset	Classification accuracy (# Number of genes)				
	GNSGA	*GSPEA*	*GSPEA2*	(Mundra and Rajapakse 2010)	(Alonso-Gonzalez et al. 2012)
Leukemia	**97.34 (146)**	96.41(154)	96.85 (148)	96.88 (88)	95.49 (74)
Lung	99.87 (220)	99.91 (216)	**99.92 (210)**	99.90 (29)	99.63 (5)
Prostate	91.27 (236)	94.31 (239)	**94.55 (234)**	93.41 (85)	90.26 (4)
Breast	**91.63 (211)**	90.26 (219)	90.77 (215)	87.65 (67)	86.93 (57)
DLBCL	93.49 (256)	94.32 (253)	**94.35 (250)**	91.23 (76)	90.84 (62)
Colon	**91.23 (197)**	90.62 (215)	91.21 (201)	88.18 (95)	88.41 (25)

of random initialization, tournament selection strategy, two-point crossover operation for new offspring generation instead of single point crossover and jumping gene mutation instead of multi-bit mutation to improve the efficiency of the generated system. The Table 2 shows the comparison between *GNSGA*, *GSPEA* and *GSPEA2*. It is observed that the *GSPEA* and *GSPEA2* are more efficient than *GNSGA* with respect to the execution time.

The experimental results prove the usefulness of all these methods. It is observed that all these simultaneous gene selection and clustering methods are very efficient for high dimensional microarray dataset. The experimental results for these three methods are described in Table 3 and Table 4 to make a comparative study among them. In Table 3, the classification performance by Support Vector Machine (SVM) and number of genes selected are listed considering related state-of-the-art methods for microarray datasets. It is observed that both GNSGA and GSPEA2 represent better classification accuracy. It is also noticed that though the state-of-the-art methods give less number of genes in compare to the proposed methods but the advantage of the proposed methods is its simultaneous gene selection and clustering.

Whereas, a comparison of *GNSGA*, *GSPEA* and *GSPEA2* based on validity indices is shown in Table 4. The result in Table 4 shows that, in most of the cases the index values are better in *GNSGA* and *GSPEA2* methods compare to *GSPEA* method. Thus, it is concluded that *GNSGA* and *GSPEA2* methods both are more efficient than *GSPEA* method for simultaneous gene selection and clustering of samples.

Table 4 Cluster analysis for *GNSGA*, *GSPEA* and *GSPEA2* based on validity indices

Dataset		Leukemia	Lung	Prostate	Breast	DLBCL	Colon
GNSGA	*CH*	0.424	0.226	0.217	0.473	0.314	0.253
	DB	**0.343**	0.271	**0.136**	0.346	0.223	0.219
	II	0.315	**0.352**	0.242	0.425	0.541	**0.438**
	XB	**6.283**	**2.214**	3.492	**4.168**	**2.174**	3.215
GSPEA	*CH*	0.427	0.362	0.256	0.513	0.337	0.336
	DB	0.347	0.253	0.215	**0.286**	0.219	0.212
	II	0.428	0.275	0.257	0.445	0.552	0.416
	XB	5.431	2.324	3.464	4.529	2.194	**3.151**
GSPEA2	*CH*	**0.429**	**0.392**	**0.268**	**0.515**	**0.341**	**0.345**
	DB	0.403	**0.234**	0.214	0.288	**0.217**	**0.211**
	II	**0.465**	0.277	**0.264**	**0.445**	0.558	0.424
	XB	5.356	2.303	**3.339**	4.453	2.187	3.172

5 Summary

The microarray technology, a salient tool for research in the field of Bio-Technology, leads to the global view of high dimensional gene dataset over different time-points of different biological experiments. The gene expression analysis of microarray data manifests the relationships among the patterns present in the data. Gene selection and sample clustering are two significant tasks for data analysis to locate the characteristics of genes. Instead of gene selection followed by sample clustering, finding clusters for samples along with gene selection for a specific gene expression data in a single process is seemed to be a better approach. In this chapter, three methods are demonstrated for optimal gene subset selection and sample clustering simultaneously. The optimality of the clusters is measured by some important cluster validation indices. A comparative study between *GNSGA*, *GSPEA* and *GSPEA2* is presented to show the goodness of proposed methods. The result proves the superiority of *GNSGA* and *GSPEA2* methods over the *GSPEA* method. The values of internal cluster validation indices are also satisfactory which imply that the methods may be useful for important gene selection and sample clustering simultaneously from a gene dataset without sample class.

References

C.J. Alonso-Gonzalez, Q.I. Moro-Sancho, A. Simon-Hurtado, R. Varela-Arrabal, Microarray gene expression classification with few genes: criteria to combine attribute selection and classification methods. Expert Syst. Appl. **39**(8), 7270–7280 (2012)

S. Akogul, M. Erisoglu, An approach for determining the number of clusters in a model-based cluster analysis. Entropy **19**(452), 1–15 (2017)

A. Baraldi, P. Blonda, A Survey of fuzzy clustering algorithms for pattern recognition—part I and II. IEEE Trans. Syst. Man Cybern. B, Cybern. **29**(6), 778–801 (1999)

A. Bellaachia, D. Portno, Y. Chen, A.G. Elkahloun, E-CAST: a data mining algorithm for gene expression data. J. Comput. Biol. **7**, 559–584 (2000)

A. Ben-Dor, R. Shamir, Z. Yakhini, Clustering gene expression patterns. J. Comput. Biol. **6**(3–4), 281–297 (1999)

A. Bhat, K-Medoids clustering using partitioning around mediods performing face recognition. Int. J. Soft Comput. Math. Control (IJSCMC) **3**(3), 1–12 (2014)

D.N. Campo, G. Stegmayer, D.H. Milone, A new index for clustering validation with overlapped clusters. Expert Syst. Appl. **64**, 549–556 (2016)

R.B. Calinski, J. Harabasz, A dendrite method for cluster analysis. Commun. Stat. **3**, 1–27 (1974)

D.L. Davies, D.W. Bouldin, A cluster separation measure. IEEE Trans. Pattern Recogn. Mach. Intell. **1**(2), 224–227 (1979)

K. Deb, *Multi-Objective Optimization Using Evolutionary Algorithms*. Wiley, vol. 16 (2001)

K. Deb, D. Goldberg, An investigation of niche and spices formation in genetic function optimization, in *Proceedings of the Third International Conference on Genetic Algorithms* (1989), pp. 42–50

K. Deb, Genetic Algorithm in Multi-Modal Function Optimization, Master's Thesis, Tuscaloosa, University of Alabama (1989)

K. Deb, A. Pratap, S. Agarwal, T. Meyarivan, A fast elitist multi-objective genetic algorithm: NSGA-II. IEEE Trans. Evol. Comput. **6**(2), 182–197 (2002)

C.M. Fonseca, P.J. Fleming, Genetic algorithms for multi-objective optimization: formulation, discussion and generalization, in *Proceedings of the Fifth International Conference on Genetic Algorithms*, ed. by S. Forrest (Morgan Kauffman, San Mateo, CA, 1993), pp. 416–423

D. Gong, G. Wang, X. Sun, Y. Han, A set-based genetic algorithm for solving the many-objective optimization problem. Soft Comput. **19**(6), 1477–1495 (2015)

K.C. Gowda, G. Krishna, Agglomerative clustering using the concept of mutual nearest neighborhood. Pattern Recogn. **10**, 105–112 (1978)

F. Gu, H.L. Liu, K.C. Tan, A hybrid evolutionary multi-objective optimization algorithm with adaptive multi-fitness assignment. Soft Comput. **19**(11), 3249–3259 (2015)

J. Horn, N. Nafploitis, D.E. Goldberg, A niched Pareto genetic algorithm for multi-objective optimization, in *Proceedings of the First IEEE Conference on Evolutionary Computation*, ed. by Z. Michalewicz (IEEE Press, Piscataway, NJ, 1994), pp. 82–87

Z. Huang, M.K. Ng, A fuzzy k-Modes algorithm for clustering categorical data. IEEE Trans. Fuzzy Syst. **7**(4), 446–452 (1999)

R. Kerber, ChiMerge: discretization of numeric attributes, in *Tenth National Conference on Artificial Intelligence* (1992), pp. 123–128

H. Liu, B. Dai, H. He, Y. Yan, The k-prototype algorithm of clustering high dimensional and large scale mixed data, in *Proceedings of the International computer Conference*, China (2006), pp. 738–743

H. Maaranen, K. Miettinen, M.M. Makela, A quasi-random initial population for genetic algorithms, in *Computers and Mathematics with Applications*, vol. 47(12) (Elsevier, 2004), pp. 1885–1895

U. Maulik, S. Bandyopadhyay, Performance evaluation of some clustering algorithms and validity indices. IEEE Trans. Pattern Anal. **24**(12), 1650–1654 (2002)

P. Merz, An Iterated Local Search Approach for Minimum Sum of Squares Clustering. IDA 2003 (2003), pp. 286–296

P.A. Mundra, J.C. Rajapakse, Gene and sample selection for cancer classification with support vectors based t-statistic. Neurocomputing **73**(13–15), 2353–2362 (2010)

R.T. Nag, J. Han, CLARANS: a method for clustering objects for spatial data mining. IEEE Trans. Knowl. Data Eng. **14**(5), 1003–1016 (2002)

S.K. Pati, A.K. Das, A. Ghosh, Gene selection using multi-objective genetic algorithm integrating cellular automata and rough set theory in *Swarm, Evolutionary, and Memetic Computing* (2013), pp. 144–155

W. Pedrycz, K. Hirota, Fuzzy vector quantization with the particle swarm optimization: a study in fuzzy granulation-degranulation information processing. Signal Process. **87**(9), 2061–2071 (2007)

M.I. Petrovskiy, Outlier detection algorithms in data mining systems. Program. Comput. Softw. **29**(4), 228–237 (2003)

K. Price, R.M. Storn, J.A. Lampinen, Differential Evolution: A Practical Approach to Global Optimization. Natural Computing Series (Springer, 2005). ISBN: 3540209506

P.J. Rousseeuw, Silhouettes: a graphical aid to the interpretation and validation of cluster analysis. J. Comput. Appl. Math. **20**, 53–65 (1987)

J.D. Schaffer, Multiple objective optimization with vector evaluated genetic algorithms, in *Proceedings of the First International Conference on Genetic Algorithms* ed. by J.J. Grefensttete (Lawrence Erlbaum, Hillsdale, NJ, 1987), pp. 93–100

N. Srinivas, K. Deb, Multi-objective function optimization using non dominated sorting genetic algorithms. Evol. Comput. **2**(3), 221–248 (1995)

M. Steinbach, G. Karypis, V. Kumar, A Comparison of document clustering technique, Technical Report number 00 - 034, University of Minnesota, Minneapolis (2000)

I.V. Tetko, D.J. Livingstone, A.I. Luik, Neural network studies. 1. Comparison of overfitting and overtraining. J. Chem. Inf. Comput. Sci. **35**, 826–833 (1995)

D.P. Waters, Von Neumann's theory of self-reproducing automata: a useful framework for biosemiotics? Biosemiotics **5**(1), 5–15 (2012)

X.L. Xie, G. Beni, A validity measure for fuzzy clustering. IEEE Trans. Pattern Anal. Mach. Intell. **13**(4), 841–846 (1991)

E. Zitzler, M. Laumanns, L. Thiele, SPEA2: improving the strength pareto evolutionary algorithm for multiobjective optimization, in *Evolutionary Methods for Design, Optimisation, and Control* (2002), pp. 95–100

E. Zitzler, L. Thiele, Multiobjective evolutionary algorithms: a comparative case study and the strength Pareto approach. IEEE Trans. Evol. Comput. **3**(4), 257–271 (1999)

A Survey on the Application of Multi-Objective Optimization Methods in Image Segmentation

Niladri Sekhar Datta, Himadri Sekhar Dutta, Koushik Majumder, Sumana Chatterjee and Najir Abdul Wasim

1 Introduction

Image segmentation scheme is a pixel clustering procedure, which involves a sequence of steps. It is the most significant and crucial step in imaging systems (Zaitoun and Aqel 2015). Segmentation finds the homogeneous regions in the image which are distinct in nature. Classical image segmentation procedure involves a sequence of steps where decisions are made prior to final output creation. In Fig. 1, the general framework of image analysis is given. It has three layers: Processing (Bottom), Analysis (Middle), and Understanding (Top). Segmentation of image would be the first stage of image analysis. From multiple perspectives, to achieve the goal of segmentation, multi-objective optimization plays a crucial role. Applying multiple objectives in problem formulation in image segmentation appears as a new trend in recent research work (Nakib et al. 2009; Ganesan et al. 2013). Generally, in case of MOO, objective functions would be conflicting and prevent concurrent

N. S. Datta
Department of Information Technology, Future Institute of Engineering and Management, Kolkata, West Bengal, India

H. S. Dutta (✉)
Department of Electronics and Communication Engineering, Kalyani Government Engineering College, Kalyani, West Bengal, India
e-mail: himadri.dutta@gmail.com

K. Majumder
Department of Computer Science and Engineering, West Bengal University of Technology, Salt Lake, Kolkata, West Bengal, India

S. Chatterjee
Department of Ophthalmology, Sri Aurobindo Seva Kendra, Kolkata, West Bengal, India

N. A. Wasim
Department of Pathology, Dr. B.C. Roy (PGIPS), 111 Narkel bagan main road, Kolkata, West Bengal, India

© Springer Nature Singapore Pte Ltd. 2018
J. K. Mandal et al. (eds.), *Multi-Objective Optimization*,
https://doi.org/10.1007/978-981-13-1471-1_12

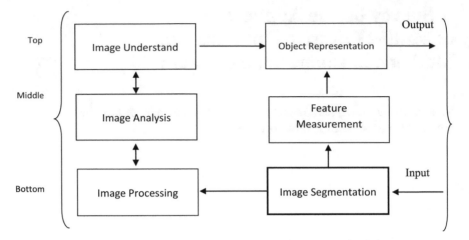

Fig. 1 Image segmentation framework

optimization for objectives. Image segmentations in the real world have multiple objectives. Here, the aim is to maximize the inter-cluster connectivity and minimize overall intra-cluster data deviation. As a result, it reduces the classifiers' error rate. These objectives formulation creates a gap in between the characteristics of segmentation problem with the solution of realistic problem. Here, in this concern, MOO is a suitable scheme for the said issue (Wong et al. 2017; Saha and Bandyopadhyay 2010). The aim of this research article is to explore the survey report in this concern. It will provide a general review of the following:

Multi-objective optimization in image segmentation scheme. The multi-objective schemes are classified according to their relevant features. In this research article, explanation of image segmentation is given in Sect. 2 and relates it with the problem of multiple objectives. Here, also MO is identified, which is related with the image segmentation problems. Section 3 will describe the design issues applying objective optimization concept. Section 4 will provide the application of MO optimization methods along with different classifications. Section 5 will provide the survey of image application with multi-objective optimization. Finally, conclusion is given.

2   Image Segmentation and MOO

Image segmentation involves several activities. Initially, pattern representation generates the classes for clustering operation. Thereafter, feature selection and extraction are the important activities. Feature selection is a method to select the best subset for clustering, where feature extraction is the transformation of input subset to generate new salient features. Either feature selection or extraction or both can be employed to achieve appropriate features for clustering. Hence, to select the suitable features,

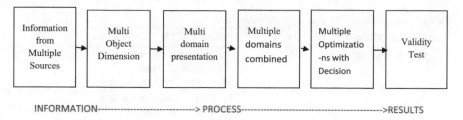

INFORMATION----------------------------> PROCESS--->RESULTS

Fig. 2 Selection of multiple objective criteria on image segmentation methods

there is a chance to select multiple features at the clustering process. For illustration, the detection of diabetic retinopathy, multiple features of the retinal image like shape, intensity may be considered. After that, pattern proximity is considered. It is generally measured by the distance function characterized by the group of patterns. In this contrast, Euclidean distance reflects variation in between two patterns. In this regard, inter-pattern similarity is a crucial issue for multiple objective criteria. As an illustration, in medical fundus image segmentation, special coherence through feature homogeneity may be considered. The feasible criteria may include here the inter-region connectors versus intra-area compactness. The classification may affect the feature extraction and similarity calculation. The grouping operation can be done in different manners. The resultant image (Fig. 1) can be presented in fuzzy or hard rules. Algorithms generate multiple numbers of groups based on a set of factors. For an instance, a hierarchical clustering method generates a sequence of partitions based on splitting or merging. Basically, partition clustering methods specify the partition that optimizes the clustering criterion. In case of validity test for clustering, validity index with multiple objectives may be formulated. The aim of these optimization schemes is to achieve the optimal set of clusters for image segmentation method. Multiple criteria or objectives begin from the realization of data to its preferred segmentation method and ends with the evaluation of the output. Figure 2 reflects multiple sources of information for the exact segmentation problem. Hence, multiple representations may be considered. In image segmentation method, there is also scope to combine multiple methods to achieve the desired output. In case of MO, select multiple optimizations as well as decision-making methods, where multiple validity testing methods should be applied (Chen et al. 2015; Nedjah and Mourelle 2015; Arulraj et al. 2014).

3 Image Segmentation Design Issue

Image segmentation is a decision-making process. In computer science, segmentation in the image is the procedure to partition a digital image into multiple segments. The aim of segmentation is to simplify or make changes in image representation into more meaningful for analysis. The stages are intelligence, problem design, and

Table 1 Three phases in image segmentation and MOO

Image segmentation (Decision-making)	Issues (MOO)
1. Intelligence phase	a. Understanding problem b. Aim identification c. Possible conflict may define
2. Design phase	a. Define objective function b. Identify weight values c. Customization of objective function
3. Choice phase	a. Verify the decision rules b. Interact with decisions-makers

choice. Initially, intelligent decision-making ability may be considered. This stage is concerned with decision environment. All the raw or input data are, therefore, collected, processed, and tested. So, from multi-objective optimization point of view, concerned issues would be understanding of the problem, goal definition, identify the possible conflict in between the existing objectives. The second stage is involved with analyzing and developing a set of solutions or alternative paths. In the optimization phase, the objective function is defined by selected attributes and weighted values. So, from MOO point of view, this stage would be: identification of objective function, realization of attributes, optimization of search strategies, etc. The final phase is to design the choice. Choice is the selection of alternatives from the available options. In this stage, each and every alternative is evaluated in terms of specific rules. Basically, the rules are applied for ranking different alternatives. So, in multi-objective point of view, this stage deals with different types of rules and creating links with decision makers (Table 1).

4   Image Segmentation Classification Using Multi-Objective Perspective

In case of image classification, chiefly two methods are available: unsupervised and supervised. In the unsupervised method, the classes will be unfamiliar and the said method begins by partitioning image dataset into clusters or by means of groups. To measure the similarity of the results, the outcome has to be compared with the reference data. These overall tasks are done by a data analyst. So, for that reason, classification using unsupervised method is also called clustering problem (Xu and Wunsch 2005) In case of the supervised method, the mean value and the variance number in image classes are familiar in advance. These datasets are applied in training phase, which is tagged on next stage, i.e., at classification (Bezdek et al. 1993). The applications of multi-objective optimization (MOO) in the classification technique, a number of objective criterions are defined. In the clustering method, generally, cluster validity is examined which is formulated as the objective function. Here, the objective functions are generally associated with a set of rules which are defined by the classifiers. Basically, objective functions are characterized by the tradeoff between sensitivity and the variety of classifier. In this concern, Cococcioni et al. (2007) have used an evolutionary 3-objective optimization method to produce the optimal solution approximately. It generates the trade-offs between exactness along with the complexity of classifiers. From the review work, it appears that the multi-objective optimization (MOO) schemes are most accepted for the design of multiple classifiers (Cococcioni et al. 2011; Oliveira et al. 2006; Ahmadian and Gavrilova 2012; Nojima 2006). Multiple classifiers have two levels in general. For an example, Ahmadian et al. (2007), initially generates a classifier which is based on aggregation error into every separate class. The subsequent level explores the best ensemble for classifiers with MO genetic algorithm. Pulkkinen and Koivisto (2008) have applied the multiple layer perceptron neural networks (i.e., MLP) as the classifiers to create classifiers. Here, the next level utilizes hidden Markov Model to select the best classifiers. Generally, the classifiers have been applied to reduce the classification ambiguity of the model, also increase the level of performance. This scheme is clearly described by Oliveira et al. (2006). In review of this specific area, it is observed that a superior quality ensemble would be the collection of individual classifiers which is highly accurate. To create any classifier, accuracy and diversity are two significant objective criteria or crucial issues that should be considered. For an example, Ahmadian et al. (2007) represent the simulation of classifiers using the concept of error that are created in each class. The researchers examined the level of performance of multi-objective classifier and realized the need to generate the classifiers including high diversity (Ishibuchi and Nojima 2015). Multi-objectivity with diversity and accuracy is still an open area for research work. Now, in image segmentation where ambiguities arise, fuzzy segmentation technique will be more suitable. Generally, fuzzy segmentation deals with inexact data (Datta et al. 2016). Rule-based fuzzy image segmentation methods are proficient to assimilate expert-based knowledge and also computationally less expensive (Karmakar and Dooleya 2002). Fuzzy segmentation is able to infer linguistic variables. Hence, the rule-based fuzzy classification system including multi-objective (MO) scheme is also a significant research area. In this concern, Cococcioni et al. (2007) used an evolutionary MOO algorithm considering rule-based fuzzy classifiers. To gain a better performance, MO fuzzy scheme has been merged with artificial neural networks. Kottathra and Attikiouzel (1996) presented the multi-objective setup using a neural network. Here, set up a branch and bound technique to find out the number of hidden neurons. Ghoggali et al. (2009) have been applied to support vector machine on multi-objective genetic algorithm for limited training dataset. Thus, a lot of facility was provided by genetic algorithms and the evolutionary algorithms for real-life difficult problems which involve features like multimodality, discontinuities, noisy function assessment, etc. Here, the population-oriented nature is well suited for optimizing multiple objectives (Bandyopadhyay and Pal 2007). Though a lot of efficient algorithms have been formulated for classifiers, but unable to show a single algorithm, this is theoretically or empirically better than in the mentioned scenarios. Multi-objective optimization scheme includes scatter search, particle swarm optimization (PSO), artificial immune systems, etc., which makes the classifier more effective.

Table 2 Image dataset where MO is applied

Natural images	Remote sensing images	Simulating images	Medical images
Nakib et al. (2009), Bhanu et al. (1993), Shirakawa and Nagao (2009), Omran et al. (2005), Zhang and Rockett (2005), Pulkkinen and Koivisto (2008), Bandyopadhyay et al. (2004)	Mukhopadhyay and Maulik (2009), Mukhopadhyay et al. (2007), Paoli et al. (2009), Saha and Bandyopadhyay 2008, Ghoggali et al. (2006)	Saha and Bandyopadhyay (2010), Guliashki et al. (2009), Matake et al. (2007), Zhang and Rockett (2005)	Cococcioni et al. (2011), Datta et al. (2016), Collins and Kopp (2008), Cocosco et al. (1997), Wen et al. (2017), Gupta (2017))

5 Survey on Image Application Including MOO

Today multi-objective scheme is very frequently used in image application. Bhanu et al. (Bhanu et al. 1993) recorded the first use of multi-objective scheme in image segmentation. Histogram thresholding, image clustering, and classification are the major areas, where multi-objective schemes have been applied. Table 2 represents a variety of image dataset that has been applied in image applications including multi-objective scheme.

Current review reflects that multi-objective schemes are becoming popular in medical and also remote-sensing image processing. As a result, a number of recent articles were found in which multi-objective schemes have been applied (Cococcioni et al. 2011; Datta et al. 2016; Collins and Kopp 2008; Cocosco et al. 1997; Wen et al. 2017; Gupta 2017; Nakib et al. 2007; Wu and Mahfauz 2016; Das and Puhan 2017; Alderliesten et al. 2012). For research purposes, there are a number of online datasets offered for experiments in multi-objective methods. For example, the renowned UCI repository (Newman et al. 1998) with images is related to iris, dermatology, wine yeast, a cancerous cell, etc. Other online available medical repositories are BrainWeb (Cocosco et al. 1997) and so on. The nature of modern imaging problem is associated with multi-spectra and the problem formulation with multi-objective is suitable for merging images with several spectral, spatial, and temporal resolutions. This concept is also applicable to generate new images. The multi-objective schemes have been applied in different areas of medical imaging as in dermatology, iris, breast cancer, and in BrainWeb (Cocosco et al. 1997). For an example, Carla and Luís (Carla and Luís 2015) have proposed a multi-objective scheme to detect exudates in retinal fundus images. The performance is evaluated by online dataset and the outcomes prove that the proposed scheme is better than traditional Kirsch filter for the identification of exudates.

Testing the retinal fundus images is useful to detect ocular disease like diabetic retinopathy (DR) and also able to interpret several problems in medical science. Usually, fundus camera produces low-quality noisy retinal images for DR and other

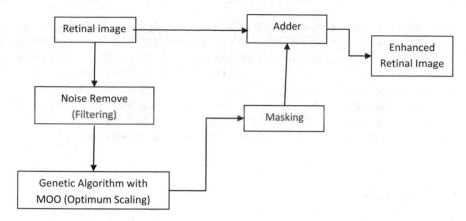

Fig. 3 Application of MOO in retinal image segmentation

ocular disease identification. In this scenario, contrast enhancement of retinal fundus image is a mandatory task. Recently, Punia and Kaur (2017) has presented a genetic algorithm-based multi-objective optimization method for the retinal image contrast enhancement and obtained a good result. Initially, the chromosomes are randomly distributed. In the next stage, initialize the random chromosome using heuristic methods. Then, adaptive crossover and mutation have been done. Here, MO is applied to get optimum scaling. Figure 3 represents the overall retinal image contrast enhancement scheme using MO.

Besides this, recently, multi-objective schemes have been used for medical resource allocation. Wen et al. (2017) have proposed a method for medical resource allocation applying multi-objective optimization method. The proposed algorithm provides a new searching technique for discrete value optimization. Generally, image segmentation is a pixel-based classification scheme. This scheme is conducted by calculating the features of each pixel and provides a decision surface at feature space. In case of classification, the partition of object space is done by training objects. So, multiple objective schemes have been applied for classifier design. Now, in case of segmentation in medical images, classifiers are "trained" to "learn" about the image. Here, input fundus images are a collection of sub-images with a variety of spatial locations in fundus image. In these classification schemes, a few are fully supervised whereas the rest of these are semi-supervised schemes. In this concern, clustering method is an important issue and unsupervised classification scheme is used to create a group in the clusters. Generally, the clustering scheme has been applied on image segmentation method especially for unsupervised partitioning of the object space using predefined lists like feature homogeneity, spatial coherence, etc. In this regard, to optimize different criteria, multi-objective schemes have been selected. Another image segmentation problem is to apply threshold method in image processing applications. Generally, thresholding is a region segmentation scheme. In this scheme, a threshold value is chosen so that the image is subdivided into pixels group with the value either lesser or greater than or equal to the selected threshold value. Comparing with multi-objective clustering application, a limited effort has been found in multi-objective thresholding method. Nakid et al. (2007) have applied multi-objective scheme to get the optimal threshold value. Here, the optimization scheme uses the simulated annealing for Gaussian Curve. Besides these, for unpopular applications like shape- and region-based image segmentation problems are associated with multi-objective scheme. Though in research, these are unusual but it is a probable research area in favor of multi-objective optimization method.

6 Conclusion

In recent years, there is an increase in the research interest on multi-objective optimization domain. Most of the realistic image segmentation involves multi-objective optimization method. For the past few years, evolutionary scheme and different types of heuristic methods have been useful to solve realistic MO methods. Currently, for problem formulation in image segmentation multi-objective optimization (MOO) methods have been applied. The MOO procedure is the collection of realistic complex optimization problems, where the objective functions are generally conflicting and have the ability to make decisions. In image processing, image segmentation is the clustering of pixels by applying definite criteria. It is one of the crucial parts in image processing. Here, in this article, the segmentation models are categorized by the problem formulation with relevant optimization scheme. The survey also provides the latest direction and challenges of MOO methods in image segmentation procedure. It is clearly observed that varieties of schemes are needed for optimizations. At the same time, it is very common to manage or customize the property of the problem to achieve the objective. In this concern, it is not essential to detect each pareto-optimal solution as researchers study MOO from different angles.

References

K. Ahmadian, M. Gavrilova, Chaotic neural network for biometric pattern recognition. Adv. Artif. Intell. **2012**, 1 (2012)

K. Ahmadian, A. Golestani, M. Analoui, M.R. Jahed, Evolving ensemble of classifiers in low-dimensional spaces using multi-objective evolutionary approach, in *6th IEEE/ACIS International Conference on Computer and Information Science* (ICIS, 2007), pp. 20–27

T. Alderliesten, J.J. Sonke, P. Bosman, Multi objective optimization for deformable image registration: proof of concept, in *Proceedings of the SPIE Medical Imaging 2012* (54), 32–43 (2012)

M. Arulraj, A. Nakib, Y. Cooren, P. Siarry, Multi criteria image thresholding based on multi objective particle swarm optimization. Appl. Math. Sci. **8**(4), 131–137 (2014)

S. Bandyopadhyay, S. Pal, Multiobjective VGA-classifier and quantitative indices of classification and learning using genetic algorithms, in *Applications in Bioinformatics and Web Intelligence* (Springer, Berlin, Heidelberg, 2007)

S. Bandyopadhyay, S.K. Pal, B. Aruna, Multi objective GAs, quantitative indices, and pattern classification. IEEE Trans. Syst. Man Cybern.-Part B Cybern. **34**, 2088–2099 (2004)

J.C. Bezdek, L.O. Hall, L.P. Clarke, Review of MR image segmentation techniques using pattern recognition. Med. Phys. **20**, 1033–1048 (1993)

B. Bhanu, S. Lee, S. Das, Adaptive image segmentation using multi objective evaluation and hybrid search methods. Mach. Learn. Comput. Vis. **3**(1993), 30–33 (1993)

P. Carla, G. Luís, Manuel Ferreira Exudate segmentation in fundus images using an ant colony optimization approach. Inf. Sci. **296**, 14–24 (2015)

L. Chen, F.P.T. Henning, A. Raith, Y.A. Shamseldin, Multiobjective optimization for maintenance decision making in infrastructure asset management. J. Manag. **31**(6), 1–12 (2015)

M. Cococcioni, P. Ducange, B. Lazzerini, F. Marcelloni, A Pareto-based multi objective evolutionary approach to the identification of Mamdani fuzzy systems. Soft Comput. **11**(11),1013–1031 (2007)

M. Cococcioni, P. Ducange, B. Lazzerini, F. Marcelloni, Evolutionary multi-objective optimization of fuzzy rule-based classifiers in the ROC space. FUZZ-IEEE 1–6 (2011)

C.A. Cocosco, V. Kollokian, R.K.S. Kwan, A.C. Evans, BrainWeb: online interface to a 3D MRI simulated brain database. Neuro Image **5** (1997)

M.J. Collins, E.B. Kopp, On the design and evaluation of multi objective single-channel SAR image segmentation. IEEE Trans. Geosci. Remote Sens. (46), 1836–1846 (2008)

V. Das, N. Puhan, Tsallis entropy and sparse reconstructive dictionary learning for exudates detection in diabetic retinopathy. J. Med. Imaging **4**(2), 1121–1129 (2017)

N.S. Datta, H.S. Dutta, K. Majumder, An effective contrast enhancement method for identification of microaneurysms at early stage. IETE J. Res. 1–10 (2016)

T. Ganesan, I. Elamvazuthi, K.Z.K. Shaari, P. Vasant, An algorithmic framework for multi objective optimization. Sci. World J. **2013**, 1–11 (2013)

N. Ghoggali, Y. Bazi, F. Melgani, A multi objective genetic data inflation methodology for support vector machine classification, in *IEEE International Conference on Geoscience and Remote Sensing Symposium* (2006), pp. 3910–3916

N. Ghoggali, F. Melgani, Y. Bazi, A multiobjective genetic SVM approach for classification problems with limited training samples. IEEE Trans. Geosci. Remote Sens. **47**, 1707–1718 (2009)

V. Guliashki, H. Toshev, C. Korsemov, Survey of evolutionary algorithms used in multi objective optimization. Probl. Eng. Cybern. Robot. Bulg. Acad. Sci. **2009**, 42–54 (2009)

P. Gupta, Contrast enhancement for retinal images using multi-objective genetic algorithm. Int. J. Emerg. Trends Eng. Dev. **6**, 7–10 (2017)

H. Ishibuchi, Y. Nojima, Performance evaluation of evolutionary multi objective approaches to the design of fuzzy rule-based ensemble classifiers, in *Fifth International Conference on Hybrid Intelligent Systems* (5) (2015), pp. 16–18

G.C. Karmakar, L.S. Dooleya, A Generic fuzzy rule based image segmentation algorithm. Pattern Recogn. Lett. **23**, 1215–1227 (2002)

K. Kottathra, Y. Attikiouzel, A novel multi criteria optimization algorithm for the structure determination of multilayer feed forward neural networks. J. Netw. Comput. Appl. **19**, 135–147 (1996)

A. Mukhopadhyay, U. Maulik, Unsupervised pixel classification in satellite imagery using multi objective fuzzy clustering combined with SVM classifier. IEEE Trans. Geosci. Remote Sens. **47**, 1132–1138 (2009)

A. Mukhopadhyay, S. Bandyopadhyay, U. Maulik, Clustering using multi-objective genetic algorithm and its application to image segmentation. IEEE Int. Conf. Syst. Man Cybern. **3**, 1–6 (2007)

N. Matake, T. Hiroyasu, M. Miki, T. Senda, Multi objective clustering with automatic k-determination for large-scale data, in *Genetic and Evolutionary Computation Conference, London, England* (2007), pp. 861–868

A. Nakib, H. Oulhadj, P. Siarry, Image histogram thresholding based on multi objective optimization. Signal Process. **87**, 2515–2534 (2007)

A. Nakib, H. Oulhadj, P. Siarry, Fractional differentiation and non-Pareto multi objective optimization for image thresholding. Eng. Appl. Artif. Intell. **22**, 236–249 (2009)

A. Nakid, H. Oulhadj, P. Siarry, Fast MRI segmentation based on two dimensional survival exponential entropy and particle swarm optimization, In *Proceedings of the IEEE EMBC'07 International Conference*, 22–26 August 2007.

N. Nedjah, LdM Mourelle, Evolutionary multi-objective optimisation: a survey. Int. J. Bio-Inspired Comput. **7**(1), 1–25 (2015)

D. Newman, S. Hettich, C. Blake, C. Merz, UCI repository of machine learning databases, University of California, Department of Information and Computer Sciences (1998)

Y. Nojima, Designing fuzzy ensemble classifiers by evolutionary multi objective optimization with an entropy-based diversity criterion, in *Sixth International Conference on Hybrid Intelligent Systems*, vol. 16(4) (IEEE, 2006), pp. 11–17

L.S. Oliveira, M. Morita, R. Sabourin, Feature selection for ensembles using the multi-objective optimization approach. Stud. Comput. Intell. (SCI) **16**, 49–74 (2006)

M.G.H. Omran, A.P. Engelbrecht, A. Salman, Differential evolution methods for unsupervised image classification. Congr. Evol. Comput. **3**(8), 331–371 (2005)

A. Paoli, F. Melgani, E. Pasolli, Clustering of hyper spectral images based on multi objective particle swarm optimization. IEEE Trans. Geosci. Remote Sens. **47**, 4179–4180 (2009)

P. Pulkkinen, H. Koivisto, Fuzzy classifier identification using decision tree and multi objective evolutionary algorithms. Int. J. Approx. Reason. **48**, 526–543 (2008)

P. Punia, M. Kaur, Various genetic approaches for solving single and multi objective optimization problems: a review, Int. J. Adv. Res. Comput. Sci. Softw. Eng. **3**(7), 1014–1020 (2017)

S. Saha, S. Bandyopadhyay, Unsupervised pixel classification in satellite imagery using a new multi objective symmetry based clustering approach, in *IEEE Region 10 Annual International Conference* (2008)

S. Shirakawa, T. Nagao, Evolutionary image segmentation based on multi objective clustering, in *Congress on Evolutionary Computation (CEC '09)*, Trondheim, Norway (2009), pp. 2466–2473

S. Saha, S. Bandyopadhyay, A symmetry based multi objective clustering technique for automatic evolution of clusters. Pattern Recogn. **43**(3), 738–751 (2010)

T. Wen, Z. Zhang, Q. Ming, W. Qingfeng, Li Chunfeng, A multi-objective optimization method for emergency medical resources allocation. J. Med. Imaging Health Inform. **7**, 393–399 (2017)

T.E. Wong, V. Srikrishnan, D. Hadka, K. Keller, A multi-objective decision-making approach to the journal submission problem. PLOS ONE **12**(6), 1–19 (2017)

J. Wu, M.R. Mahfauz, Robust X-ray image segmentation by spectral clustering and active shape model. J. Med. Imaging **3**(3), 1–9 (2016)

R. Xu, D. Wunsch, Survey of clustering algorithms. IEEE Trans. Neural Netw. **16**(2005), 645–678 (2005)

Y. Zhang, P.I. Rockett, Evolving optimal feature extraction using multi-objective genetic programming: a methodology and preliminary study on edge detection, in *Conference on Genetic and Evolutionary Computation* (2005), pp. 795–802

M.N. Zaitoun, J.M. Aqel, Survey on image segmentation techniques. Int. Conf. CCMIT **65**, 797–806 (2015)

Y. Zhang, P.I. Rockett, Evolving optimal feature extraction using multi-objective genetic programming, a methodology and preliminary study on edge. Artif. Intell. Rev. **27**, 149–163 (2005)

Bi-objective Genetic Algorithm with Rough Set Theory for Important Gene Selection in Disease Diagnosis

Asit Kumar Das and Soumen Kumar Pati

1 Introduction

In DNA microarray data analysis (Causton et al. 2003), usually biologists determine the gene expressions in the samples from patients and discover details regarding how the genes of samples relate to the specific varieties of disease. A lot of genes could robustly be correlated to a specific kind of disease; though biologists wish to indicate a small set of genes that controls the results before conducting expensive experiments on the high dimensional microarray data. Consequently, automated selection of the minimal subset of genes (Schaefer 2010; Leung and Hung 2010) is highly advantageous.

Gene selection is frequently performed in data mining and knowledge innovation to select an optimal set of genes from the high dimensional gene dataset based on some appropriate evaluation functions. Evolutionary Algorithms (EAs) (Price et al. 2005; Gu et al. 2015) are the optimization techniques used for finding an optimal subset of genes for disease prediction in an efficient way. A standard genetic algorithm deals with single fitness function, however, most of the real-life problems are intrinsically multi-objective in nature where concurrent satisfaction of two or more conflicting fitness functions is required. The purposes of this type of algorithm are to approximate a set of Pareto-optimal outcomes (Zitzler and Thiele 1999; Knowles and Corne 2000) instead of a single one; because the objectives regularly conflict with each other and enhancement of one objective may direct to deterioration of other. These methods

A. K. Das (✉)
Department of Computer Science and Technology, Indian Institute of Engineering Science and Technology, Shibpur, Shibpur, West Bengal, India
e-mail: akdas@cs.iiests.ac.in

S. K. Pati (✉)
Department of Computer Science and Engineering, St. Thomas' College of Engineering and Technology, Kolkata 700023, West Bengal, India
e-mail: soumenkrpati@gmail.com

© Springer Nature Singapore Pte Ltd. 2018
J. K. Mandal et al. (eds.), *Multi-Objective Optimization*,
https://doi.org/10.1007/978-981-13-1471-1_13

are optimized all objectives concurrently for a particular outcome. Therefore, the Pareto-optimal outcomes are important to a decision-maker instead of the best trade-off results.

The most popular multi-objective EAs, such as NSGA-II (Deb et al. 2002), provide much promising outcome by the modification of Pareto dominance relation and rank definition to increase the selection difficulty in identifying the nondominated Pareto optimality set. The main disadvantage of the technique is its high computational complexity, which is most expensive when the size of population is increased. The MOEA with decomposition (MOEA/D) (Zhang and Li 2007) optimizes scalar optimization problems instead of directly resolving a multi-objective problem (optimization) as a whole. The MOEA/D has lesser computational complexity at every generation compared to NSGA-II (Deb et al. 2002). The MOEA/D with a small amount of population size is capable to construct a smaller number of evenly distributed outcomes. A single and bi-objective evolutionary algorithm based filter approach for feature selection is proposed in Santana and Canuto (2014).

In the chapter, a new gene selection technique is proposed based on bi-objective GA (GSBOGA) to pick the small number of genes without sacrificing any information in the dataset. The proposed method uses nonlinear uniform hybrid Cellular Automata (CA) (Neumann 1996) which is highly acceptable for its capability as an exceptional random pattern creator for producing initial population of the binary strings. Generally, normal GA deals with single objective function but most of our real-life problems are multi-objective in nature where simultaneously two or more conflicting fitness functions are required for optimization purpose. Two objective functions for *GSBOGA* are defined as the one which uses the lower bound approximation with exploration of the boundary regions of the Rough Set Theory (RST) (Jing 2014) to incorporate some vagueness to achieve comparatively better solutions and the other uses Kullback–Leibler (KL) divergence method (Kullback and Leibler 1951) of Information Theory to select more precise and informative genes from the dataset. To create new individuals, most of the GA literature uses single-point crossover but the proposed method uses multipoint crossover. The motivation for using it is that the newly created offspring are mostly similar to one of their quality full parents than they are in single-point crossover. Consequently, convergence is accepted to arise earlier. In single-bit mutation, a gene is randomly selected to be mutated and its value is changed depending on the underlying encoding scheme although it lacks diversity in the population as the first bit of the binary string normally does not modify. In multi-bit mutation, multiple genes are randomly selected and their values are changed depending on the encoding type used. So, both of the mutations are depended on the random bit position generated with respect to mutation probability which is inefficient in high dimensional space. To overcome these demerits, a unique jumping gene mutation methodology (Chaconas et al. 1996) is used in the method for mutating the genes. Thus, the proposed method preserves the diversity of the population applying multipoint crossover and jumping gene mutation techniques. The replacement strategy for the creation of the next generation population is based on the Pareto-optimal concept (Zitzler and Thiele 1999; Knowles and Corne 2000) with respect to both objective functions.

The performance of the *GSBOGA* method is compared with some popular state-of-the-art feature selection techniques (like *CFS* (Hall 1999) and *CON* (Yang and Pedersen 1997)) is provided in the chapter. The *GSBOGA* also confirms most promising outcome with lower computational cost as there need not require of any global computation typical of other Pareto optimality based *MOEA*. The *MOEA* method employs a steady state selection procedure, no need for the fitness sharing parameter utilized in *NSGA* or crowding distance applied in *NSGA*-II (Deb et al. 2002) or converting the multi objectives problem into scalar objective problem and use weighted aggregation concept of the individual objectives in MOEA/D (Zhang and Li 2007).

The chapter is structured into four sections. The proposed gene selection methodology is described based on bi-objective GA in sufficient details in Sect. 2. Experimental outcomes pertaining to the performance assessment of the proposed method with respect to the existing state of the art are presented in Sect. 3 using various microarray cancer data. At last in Sect. 4, the chapter is concluded.

2 Bi-objective Gene Selection

Gene selection based on single criteria may not always yield the best result due to varied characteristics of the datasets used. If multiple criteria are combined for feature selection, an algorithm generally provides more important features compared to the algorithm relying on a single criterion. Here, two criteria are united and a novel bi-objective genetic algorithm (*GSBOGA*) is reported for gene selection, which effectively reduces the dimensionality of the dataset without sacrificing the classification accuracy. The method uses nonlinear hybrid uniform cellular automata (Waters 2012) for generating initial population, stable selection strategy, multipoint crossover operation for creating new offspring, and a unique jumping gene mechanism (Chaconas et al. 1996) for mutation in the population to maintain diversity.

2.1 Initial Population Generation

Initial Population generation is a fundamental task in evolutionary algorithms (Gu et al. 2015; Price et al. 2005; Gong et al. 2015). Random initialization is the most commonly used mechanism to initiate population when no information regarding the way out is given, but it takes long computational time, particularly when the solution space is difficult to investigate. The creation of quasi-random sequences is very complicated and their benefit disappears for high dimension space (Maaranen et al. 2004). So, a suitable cellular automata based technique is applied for making a simple and faster initial population in high dimensional space.

Abstract Cellular Automata (*CA*) (Waters 2012), a pseudorandom pattern generator, plays an important role in any population-based stochastic search method. Here, the nonlinear hybrid uniform cellular automata have been used for generating the initial population covering major portion of search space. Since most of search space can be explored, optimization becomes more efficient using the proposed population generation approach. The model is represented as a large number of cells organized in the form of a lattice where each cell has the potential of self-reproduction and is as influential as a general Turing machine. The *GSBOGA* methodology generates next state of a cell using its current state and states of its neighboring cells based on the rules $R1$, $R2$, $R3$, and $R4$ as defined below. The method considers only three neighborhoods namely, left neighbor, self or current, and right neighbor one-dimensional cellular automata and each cell only has any one of two states (0 or 1).

$$R_1 : Next\_state(i) = (L(i) \wedge C(i)) \vee (\sim L(i) \wedge R(i))$$
$$R_2 : Next\_state(i) = (L(i) \wedge R(i)) \vee (C(i) \wedge \sim R(i))$$
$$R_3 : Next\_state(i) = L(i) \oplus C(i) \oplus R(i)$$
$$R_4 : Next\_state(i) = C(i) \oplus (L(i) \vee \sim R(i))$$

where $L(i)$, $C(i)$, and $R(i)$ are the values of left cell, current cell (i), and right cell, respectively. For each of the cell to generate the next state, a feasible rule is chosen dynamically. Among the rules, $R1$, $R2$, and $R4$ are nonlinear while $R3$ is linear and therefore it is named as nonlinear hybrid *CA*. These rules are utilized for population generation as illustrated with an example.

Example 1 The binary chromosomes are randomly generated having a length equivalent to the number of genes in the experimental data. Suppose, there are five genes in the data and a randomly generated chromosome is 11011 (called seed). To every cell, one of the rule from $R1$ to $R4$ is randomly assigned and the value of next state for the corresponding cell is achieved, as shown in Fig. 1, where $R(i)$ and $L(i)$ for any cell are obtained from $C(i)$ value of right and left cell, respectively. The same procedure is continued for a definite number of times to obtain all chromosomes in the population.

2.2 Bi-objective Objective Function

The objective function determines excellence of a solution in the population; therefore a strong objective function is imperative for getting a good outcome. Contrary to single objective *GA* (Goldberg and Holland 1988), multi-objective *GA* (Gong et al. 2015) handles with simultaneous optimization of various incommensurable and regularly opposite objectives in nature. The objectives often conflict with one another. The enhancement of one objective may direct to deterioration of the other. The method uses a bi-objective fitness function with two parameters based on attribute dependency value with exploring boundary region (*LBA*) (Pawlak 1998) in *RST* and Kullback–Leibler Divergence (*KLD*) method (Kullback and Leibler 1951) in information theory. These two objectives are conflicting (i.e., minimum for *LBA* and

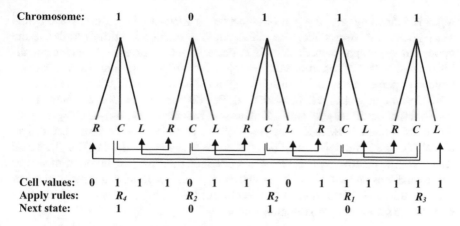

Chromosome:

| | 1 | 0 | 1 | 1 | 1 |

| R | C | L | R | C | L | R | C | L | R | C | L | R | C | L |

Cell values:	0	1	1	1	0	1	1	1	0	1	1	1	1	1	1
Apply rules:	R_4			R_2			R_2			R_1			R_3		
Next state:	1			0			1			0			1		

Fig. 1 Generation of next state chromosome

maximum for *KLD*) in nature and used to approximate a number of Pareto optimality based solutions instead of only one.

2.2.1 Modified Attribute Dependency Exploring Boundary Region (*LBA*)

The objective function concludes the excellence of a chromosome. Consequently, a quality full objective function is imperative for producing a good outcome. Here, the objective function measures a chromosome by its dependency value by exploring boundary region for the gene set presented by a chromosome, as explained below.

(A) Attribute dependency value

Let $\mathbb{I} = (\mathbb{U}, \mathbb{A})$ be an information system where $\mathbb{U}$ is the non-empty, finite set of objects (called *universe*) and $\mathbb{A}$ is a finite, non-empty set of *attributes or features*. Each attribute $a \in \mathbb{A}$ can be defined mathematically by Eq. (1).

$$f_a: \mathbb{U} \to \mathcal{V}_a, \forall a \in \mathbb{A} \tag{1}$$

where $\mathcal{V}_a$ is the collection of values of attribute a, that denoted the *domain* of a. . For any $\mathcal{P} \subseteq \mathbb{A}$, there exists a binary relation $IND(\mathcal{P})$, called indiscernibility relation as defined in Eq. (2).

$$IND(\mathcal{P}) = \{(x, y) \in (\mathbb{U} \times \mathbb{U}) | f_a(x) = f_a(y), \forall a \in \mathcal{P}\} \tag{2}$$

where $f_a(x)$ denotes the value of attribute a for object $x \in \mathbb{U}$. Clearly, $IND(\mathcal{P})$ is an equivalence relation which provides equivalence classes. All equivalence classes of $IND(\mathcal{P})$ are defined by $\mathbb{U}/IND(\mathcal{P})$ (or $\mathbb{U}/\mathcal{P}$ and an equivalence class of $\mathbb{U}/\mathcal{P}$, containing x is defined by $[x]_p$. If a pair of object $(x, y) \in IND(\mathcal{P})$, then x and y are called $\mathcal{P}$ -*indiscernible.-indiscernible*. The indiscernibility relation is utilized to define the upper and lower approximations in *RST*. For each set of attributes $\mathcal{P}$, an indiscernibility relation $IND(\mathcal{P})$ partitions the m objects into number of equivalence classes, defined as partition $\mathbb{U}/IND(\mathcal{P})$ (or $\mathbb{U}/\mathcal{P}$), equal to $\{[x]_p\}$, where $|\mathbb{U}/\mathcal{P}| = m$. Similarly, equivalence classes $[x]_D$ are formed using Eq. (2) for the decision attribute. Thus, two different partitions $\mathbb{U}/\mathcal{P}$ and $\mathbb{U}/\mathcal{D}$ of the equivalence classes $[x]_p$ and $[x]_D$ respectively, are formed. Now each class $[x]_D$ ($\mathbb{U}/\mathcal{D}$) is considered as the target set $\mathcal{X}$ (i.e., $\mathcal{X} \in \mathbb{U}/\mathcal{D}$). The lower approximation of $\mathcal{X}$ under $\mathcal{P}$ is computed using Eq. (3), for all $\mathcal{X} \in \mathbb{U}/\mathcal{D}$. The positive region $POS_\mathcal{P}(\mathcal{D})$ is obtained by taking the union of the lower approximations under $\mathcal{P}$ for all $\mathcal{X}$ in $\mathbb{U}/\mathcal{D}$, using Eq. (4). Then, dependency value of decision attribute $\mathcal{D}$ on $\mathcal{P}$ (i.e., $\gamma_\mathcal{P}(\mathcal{D})$) is obtained using Eq. (5), which is dependent only on the set lower approximation value.

$$\underline{\mathcal{P}}\mathcal{X} = \{ x \,|[x]_p \subseteq \mathcal{X} \} \tag{3}$$

$$POS_\mathcal{P}(\mathcal{D}) = \cup_{\mathcal{X} \in \mathbb{U}/\mathcal{D}} \underline{\mathcal{P}}\mathcal{X} \tag{4}$$

$$\gamma_\mathcal{P}(\mathcal{D}) = \frac{|POS_\mathcal{P}(\mathcal{D})|}{|\mathbb{U}|} \tag{5}$$

If, $\gamma_\mathcal{P}(\mathcal{D}) = 1$, then $\mathcal{P}$ is totally reliable with respect to $\mathcal{D}$. Feature selection in *RST* is achieved by identifying only a necessary and sufficient subset of features (called *reduct*) of a given set of features. A set $\mathcal{P} \subseteq \mathbb{A}$ is a reduct of $\mathbb{A}$, if (i) $\gamma_\mathcal{P}(\mathcal{D}) = \gamma_\mathbb{A}(\mathcal{D})$ and (ii) there does not exist $\mathcal{Q} \subset \mathcal{P}$ such that $\gamma_\mathcal{Q}(\mathcal{D}) = \gamma_\mathbb{A}(\mathcal{D})$. Each reduct has the property that a feature cannot be eliminated from it without altering the dependency of the decision attribute on it.

(B) Exploring Boundary region

The upper approximation $\bar{\mathcal{P}}\mathcal{X}$ of target set $\mathcal{X}$,, for all $\mathcal{X} \in \mathbb{U}/\mathcal{D}$ under attribute subset $\mathcal{P}$, is computed using Eq. (6) which contains the set of attributes possibly belong to the target set $\mathcal{X}$ and the boundary region, as shown in Fig. 2, for the decision system is obtained using Eq. (7) which possesses the degree of uncertainty as the objects in this region may or may not belong to the target set.

$$\bar{\mathcal{P}}\mathcal{X} = \{ x \,|[x]_p \cap \mathcal{X} \neq \emptyset \} \tag{6}$$

$$BND_\mathcal{P}(\mathcal{D}) = \cup_{x \in \frac{\mathbb{U}}{\mathcal{D}}} (\bar{\mathcal{P}}\mathcal{X}) - \cup_{x \in \frac{\mathbb{U}}{\mathcal{D}}} (\underline{\mathcal{P}}\mathcal{X}) \tag{7}$$

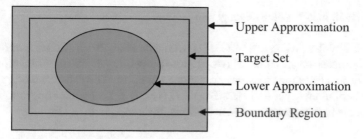

Upper Approximation

Target Set

Lower Approximation

Boundary Region

Fig. 2 Illustrate the boundary region of a target set

Obviously, from the definition of the positive region, equivalence class in $\mathbb{U}/\mathcal{P}$ that is not a subset of $\mathcal{X}$ in $\mathbb{U}/\mathcal{D}$, falls in the boundary region. If more falls in the boundary region, the dependency value will decrease. An equivalence class may go down because of some objects do not belong to $\mathcal{X}$. If very few objects of are responsible for placing it in the boundary region, then the class almost agrees to the target set $\mathcal{X}$, i.e., a class in $\mathbb{U}/\mathcal{D}$ So attribute dependency should not be the only criterion for *reduct* generation. To overcome this shortcoming, boundary region is explored by computing similarity factor of set (classes of $\mathbb{U}/\mathcal{P}$ whose objects lie in the boundary region, formed using Eq. (8)) to $\mathbb{U}/\mathcal{D}$, formulated using Eq. (9).

$$CB_{\mathcal{P}}(\mathcal{D}) = [x]_{\mathcal{P}} | \left([x]_{\mathcal{P}} \in \frac{\mathbb{U}}{\mathcal{P}}\right) \wedge (\mathcal{X} \in POS_{\mathcal{P}}(\mathcal{D})) \tag{8}$$

$$\delta_{\mathcal{P}}(\mathcal{D}) = \frac{1}{BND_{\mathcal{P}}(\mathcal{D})} \sum_{[x]_{\mathcal{P}} \in CB_{\mathcal{P}}(\mathcal{D})} \max_{[x]_{\mathcal{D}} \in \mathbb{U}/\mathcal{D}} (|[x]_{\mathcal{P}} \cap [x]_{\mathcal{D}}|) \tag{9}$$

In Eq. (9), summation of maximum number of common objects is calculated between an element $[x]_{\mathcal{P}} \in CB_{\mathcal{P}}(\mathcal{D})$ and all elements $[x]_{\mathcal{P}}$ in $\mathbb{U}/\mathcal{D}$ and then it is divided by the total number of objects in $BND_{\mathcal{P}}(\mathcal{D})$. So, if very few objects of are responsible for placing it into the boundary region, then the class almost agrees with the target class, i.e., a class $[x]_{\mathcal{D}}$ in $\mathbb{U}/\mathcal{D}$ and similarity factor will increase, where $\delta_{\mathcal{P}}(\mathcal{D})$ in the same situation dependency value $\gamma_{\mathcal{P}}(\mathcal{D})$ decreases.

Since, for a decision system, these two factors namely, dependency value $\gamma_{\mathcal{P}}(\mathcal{D})$ and similarity factor $\delta_{\mathcal{P}}(\mathcal{D})$ need to be maximized, so the fitness function $F(ch)$ for chromosome ch of associated *GA*-based optimization problem is considered as the weighted average of these two factors, computed using Eq. (10).

$$F(ch) = \mathcal{W}.\gamma_{\mathcal{P}}(\mathcal{D}) + (1 - \mathcal{W}).\delta_{\mathcal{P}}(\mathcal{D}) \tag{10}$$

where $\mathcal{W}$ is the weight factor, which is set experimentally. Obviously, higher the fitness value $F(ch)$, better the quality of the chromosome (or encoded string) ch.

2.2.2 Kullback–Leibler Divergence

Kullback–Leibler divergence (*KLD*) (Harmouche et al. 2016; Kullback and Leibler 1951) computes the proximity of two or more probability distributions in information theory. It quantifies the measures how close up a probability distribution (p) is to a model distribution (q) in statistics. The *KLD* (i.e., D_{KL}), which is nonsymmetric and nonnegative in two probability distributions, is defined in Eq. (11).

$$D_{KL}(p\|q) = \sum_i P_i \log_2\left(\frac{P_i}{q_i}\right) \tag{11}$$

$D_{KL}(p\|q)$ is computed using Eq. (11) which evaluates pair to pair whose mean distance is used as another fitness function which is to be minimized that governs maximum similarity between p and q.

2.3 Multipoint Crossover

It expresses examine towards the best existing offspring although is not able to generate new offspring. To generate new offspring, crossover mechanism is mandatory. In environment, individual has two parents and inherit several of their characteristics. The crossover does the same matter. The two new offspring is created by the crossover with probability (c_p) from an identified pair of parents. In this methodology, two-point crossover has been utilized generating the two random positions in the chromosome. The substrings of the parent chromosomes, lying between the two randomly identified positions, are exchanged and two new offsprings are generated. The motivation for this operation is that, the produced new offspring are most similar to one of their quality full parents than they are in one point crossover. Consequently, convergence arises in advance that is expected.

2.4 Jumping Gene Mutation

A gene is randomly selected to be mutated and its value is changed depending on the encoding type used although it lacks the diversity in the chromosome pool as the initial (first) bit of the string commonly does not modify in single-bit mutation. Alternatively, in multi-bit mutation, multi-genes are randomly selected for mutation and their values changed depending on the encoding type used. So, both of the mutations are dependent on the random bit number generation with respect to mutation probability which is inefficient in high dimensional space. To overcome these demerits, jumping gene mutation methodology (Chaconas et al. 1996) is used here for mutation. The jumping genes are a subset of original genes (with mutation probability m_p), which can dynamically jump within the respective chromosome. This type of genes has an immense prospective for maintaining diversity, a crucial characteristic for evolutionary search throughout the entire population.

Example 2 Let, in the population, the original chromosome be $(x_1, x_2, ..., x_n)$. Randomly, a jumping gene of length z $(z<<n)$ say, $(y_1, y_2, ..., y_z)$ is identified. Subsequently, randomly the primary position from anywhere of the chromosome to be changed by the jumping genes which are selected. Let the initial location be w, therefore the muted chromosome is $(x_1, x_2, ..., x_{w-1}, y_1, y_2, ..., y_z, x_{z+1}, x_{z+2}, ..., x_n)$ after mutation. This mutation mechanism offers much promising outcome compared to single-bit mutation operation.

2.5 Replacement Strategy

In this optimization strategy, the two objective functions are conflicting in nature and not possible to optimize simultaneously. Consequently, it is necessary to have a trade-off between them in decision-making during the replacement of chromosomes in population. The replacement strategy of *GSBOGA* is based on Pareto optimality concept (Zitzler and Thiele 1999; Knowles and Corne 2000; Shelokar et al. 2013). Figure 3 demonstrates the measurement of dominance regarding the Pareto optimality concept, which is defined below.

Definition 1 (*Strongly dominated solution*) A solution X_1 is strongly dominated by the other solution X_2, if the X_1 is strictly better compared to X_2 with respect to all fitness functions.

Definition 2 (*Nondominated solution*) The X_1 and X_2 are supposed to nondominated solutions to each other if some objectives of one solution are higher than that of the other.

Definition 3 (*Dominated solution*) A solution $X1$ is dominated by another solution $X2$, if the $X1$ is strictly worse compared to $X2$ regarding all objective function values.

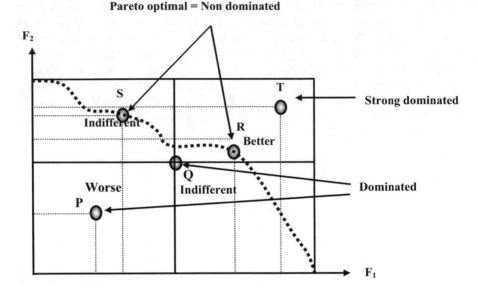

Fig. 3 The concept of Pareto optimality

Figure 3 demonstrates the nondominated and dominated solutions with respect to two fitness functions F_1 and F_2. Here, better means an outcome is not inferior in any fitness values. The solution symbolized by the point P is worse than the solution symbolized by the point Q and the solution at point R is better compared to that at Q. But, it cannot be stated that the R is better than the S or vice versa since the fitness value (single) of every point is better compared to the other solution. These are called nondominated or Pareto-optimal solution represented by the dotted line in the figure. The solution T is strongly dominated compared to all other solutions with respect to both objective values. So, after mutation, both fitness values are evaluated for offspring and the elitism property is maintained replacing parent with its offspring according to strong dominated or non-dominated property.

2.6 The GSBOGA Algorithm

The algorithm of the *GSBOGA* methodology is represented below

The algorithm of the *GSBOGA* methodology is represented below-

Procedure: *GSBOGA*

Input: Size of the Population: *M*

Maximum generation numbers: *G*

Probability of crossover: *cp*

Probability of mutation: *mp*

Output: Strongly dominated and non-dominated solutions

Begin

    Create initial population *P* of the size *M* utilizing nonlinear cellular automata;

    Evaluate fitness values of all chromosomes;

    Set, $c = 0$;

    Repeat

        For $i=1$ to *M* **do**

            First_ parent = *Pi*;

            Select another parent randomly from the remaining population;

            Apply multipoint crossover with probability *cp* to produce two offspring;

            Use jumping gene mutation to the offspring with mutation rate *mp*;

            Evaluate fitness values of the offspring;

            If (both offspring either strongly or non-dominate with the parents) **Then**

                Both the parents are replaced by the offspring;

            Else If (one offspring strongly dominates the parents) **Then**

                This offspring replaces the dominated parent;

            Else

                Both the offspring are discarded;

        End-for

        $c = c+1$;

    Until $(c<=G)$;

    Return (strongly dominated and non-dominated chromosomes);

End.

3 Experimental Results of *GSBOGA* Method

Extensive experiments are done to evaluate the *GSBOGA* method using experimental microarray data ("Kent Ridge Biomedical Dataset Repository"(n.d)) describes as follows.

3.1 Microarray Dataset Description

I. Leukemia Dataset

Publication: This is a well-understood gene database published in Golub et al. (1999).

Raw Data: The raw data is available in http://www-genome.wi.mit.edu/cgi-bin/ cancer/datasets.cgi.

Description: This data contains 7129 human genes over 38 samples (27 *ALL* and 11 *AML*) of bone marrow.

II. Lung Cancer Dataset

Publication: This is a well-understood gene database published in Gordon et al. (2002).

Raw Data: The raw data is available in http://www.chestsurg.org/microarray. htm.

Description: The data consists of 32 lung samples (16 are malignant pleural mesothelioma (*MPM*) and 16 are adenocarcinoma (*ADCA*)) with 12553 genes.

III. Prostate Cancer Dataset

Publication: The dataset published in Singh et al. (2002).

Raw Data: The raw dataset is available in http://www-genome.wi.mit.edu/mpr/ prostate.

Description: The data consists of 52 *tumor* samples and 50 *non-tumor* samples with 12600 prostate genes.

IV. Breast Cancer Dataset

Publication: The experimental dataset is published in Veer et al. (2002).

Raw Data: The raw data is available in http://www.rii.com/publications/2002/v antveer.htm.

Description: The data consists of 78 samples (34 samples (patient) labeled as "*relapse*" and the remaining 44 samples labeled as "*non-relapse*"). The total number of gene is 24481.

V. Diffuse Large B-cell Lymphoma (DLBCL) Data

Publication: This data (DLBCL) is available in Shipp et al. (2002).

Table 1 Initial parameters of *GSBOGA* methodology

Parameter	Value
Size of population (*M*)	100
Generation number (*G*)	300
Crossover probability (p_c)	0.90
Mutation probability (p_m)	0.05

Raw Data: The raw data available in http://www-genome.wi.mit.edu/cgi-bin/cancer/datasets.cgi.

Description: The classification about data addressed in the publication is *DLBCL* against Follicular Lymphoma morphology (*FL*). The data consists of 58 *DLBCL* and 19 *FL* samples over 6817 genes.

VI. Colon Cancer Dataset

Publication: This gene database is published in Alon et al. (1999).

Raw Data: The raw data available in http://microarray.princeton.edu/oncology/affydata/index.html.

Description: The data consists of 62 samples (40 tumor samples labeled as "*negative*" and rest 22 normal samples labeled as "*positive*"). The expression profile contains 2000 genes.

3.2   Parameter Setup and Preprocessing

The parameters used in *GSBOGA* are presented in Table 1. The parameters are identified after various test evaluation of the proposed methodology and data instances until achieve to the best arrangement in terms of the excellence of results and the computational efficiency.

Here, the ChiMerge discretization (Kerber 1992) algorithm is used to convert continuous attributes into discretized values.

3.3   Performance Measurement

The proposed *GSBOGA* method uses cellular automata for initial population generation instead of random initialization, multipoint crossover for new offspring generation instead of single-point crossover, and jumping gene mutation instead of multi-bit mutation to progress the efficiency of the created system. Table 2 shows the comparison between *GSBOGA* and, normal *MOGA* (i.e., *NMOGA*). The *NMOGA* uses random initialization, single-point crossover, two-point mutation, and two objective functions, such as *LBA* and *KLD*. It is observed that the *GSBOGA* method is more efficient than *NMOGA* with respect to the execution time.

Table 2 Comparison between *GSBOGA* and *NMOGA* regarding time (min)

Method	Dataset					
	Leukemia	Lung	Prostate	Breast	DLBCL	Colon
NMOGA	46.48	66.5	75.30	86.25	62.35	43.27
GSBOGA	39.53	43.32	64.55	77.50	53.30	35.45

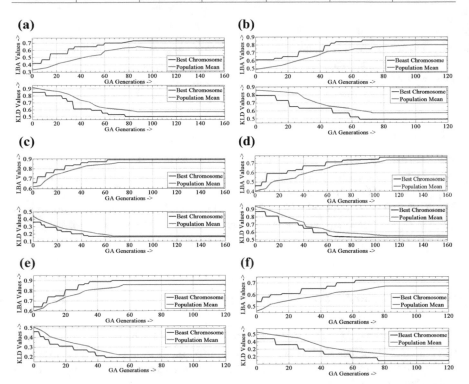

Fig. 4 Converging nature of best chromosome for **a** Leukemia **b** Lung **c** Prostate **d** Breast **e** DLBCL and **f** Colon cancer dataset

The *GSBOGA* is terminated after 300 generations for all data, in spite of the fact that for most of the cases, convergence is expected soon. Figure 4 represents behavior of the best chromosome and population mean of all chromosomes with respect to individual fitness values by the *GSBOGA* method. The usual examination is that the best chromosome discovers the top performing fitness values relatively quicker compared to the mean population.

The *GSBOGA* method runs several times and obtains the Pareto front approximations with respect to each of the fitness function for all datasets after final generation. For each fitness function, the statistical measures like minimum (*Min.*), maximum (*Max.*), mean (*Avg.*), and standard deviation (*Std.*) are computed among all chromo-

Table 3 Statistical measures of the population after final generation for a run

Dataset	LBA				KLD			
	Min.	Max.	Avg.	Std.	Min.	Max.	Avg.	Std.
Leukemia	0.5903	0.7317	0.6337	0.1714	0.5008	0.6470	0.5792	0.1972
Lung	0.7611	0.8601	0.8040	0.1618	0.4927	0.5605	0.5716	0.1296
Prostate	0.8681	0.8943	0.8841	0.2027	0.1570	0.2431	0.2226	0.0113
Breast	0.6355	0.7602	0.7373	0.2094	0.5382	0.5905	0.5514	0.1138
DLBCL	0.8375	0.9002	0.8662	0.1063	0.1922	0.2648	0.2207	0.0138
Colon	0.6229	0.7422	0.6770	0.2172	0.1400	0.2917	0.2262	0.2226

Table 4 Average statistical measures of the population for 50 runs

Dataset	LBA				KLD			
	Min.	Max.	Avg.	Std.	Min.	Max.	Avg.	Std.
Leukemia	0.5718	0.7217	0.6199	0.1857	0.5119	0.6651	0.5817	0.2007
Lung	0.7521	0.8581	0.7968	0.2600	0.5008	0.5596	0.5700	0.1218
Prostate	0.8172	0.8514	0.8266	0.3738	0.1853	0.3005	0.2374	0.2849
Breast	0.6004	0.7576	0.7204	0.2189	0.5399	0.6182	0.5629	0.2086
DLBCL	0.8144	0.8918	0.8417	0.1240	0.2134	0.2860	0.2397	0.1129
Colon	0.6162	0.7409	0.6721	0.3284	0.1511	0.3007	0.2212	0.2420

somes for a particular run, as listed in Table 3. Also, the average values of 50 runs for all datasets are presented in Table 4.

3.4 Comparative Study

The *GSBOGA* method produces some Pareto-optimal solutions after final generation from where all strongly dominated solutions and some best non-dominated outcomes regarding the fitness functions are considered for the comparative study. The average classification accuracies are computed with consideration of all base classifiers for all selected solutions and best feature subset is selected with respect to average classification accuracy, which is used as comparison purpose for *GSBOGA*.

The efficiency of *GSBOGA* is compared with some existing state-of-the-art feature selection methodologies like *PCA* (Jackson 1991), *SVD* (Petrou and Bosdogianni 2000), *CON* (Yang and Pedersen 1997), *CFS* (Hall 1999), *NSGA-II* (Deb et al. 2002), and *MOEA/D* (Zhang and Li 2007). The accuracies of some base classifiers and average accuracy are listed in Table 5. From the multiple gene subsets (i.e., chromosomes after final generation), generated by the *GSBOGA*, the result of the best subset is shown in Table 5.

Table 5 Performance comparison of *GSBOGA* and other gene selection methods

Dataset	Methods (#genes)	Classifiers (%)						Avg. acc.
		NB	J48	SVM	KNN	MLP	RBF	(%)
Leukemia	PCA (89)	78.95	76.32	76.32	73.68	76.32	81.58	77.20
	SVD (89)	81.58	78.95	76.32	81.58	76.32	81.58	79.39
	CON (159)	78.95	76.32	81.58	76.32	81.58	78.95	78.95
	CFS (147)	84.21	81.58	84.21	78.95	84.21	81.58	82.46
	NSGA-II (99)	86.84	**89.47**	92.11	**86.84**	86.84	84.21	87.72
	MOEA/D (100)	84.21	86.84	92.11	84.21	84.21	86.84	86.40
	GSBOGA (89)	**92.11**	**89.47**	**94.74**	**86.84**	**94.74**	**92.11**	**91.67**
Lung	PCA (95)	84.38	78.12	84.38	84.38	81.25	84.38	82.82
	SVD (95)	87.50	84.38	87.50	84.38	84.38	81.25	84.90
	CON (190)	87.50	90.62	87.50	87.50	90.62	87.50	88.54
	CFS (205)	90.62	87.50	84.37	87.50	90.62	84.37	87.50
	NSGA-II (110)	96.88	90.62	**100**	93.75	**100**	90.62	95.31
	MOEA/D (96)	96.88	**96.88**	**100**	93.75	87.50	90.62	94.27
	GSBOGA (95)	**100**	**96.88**	**100**	**96.87**	**100**	**96.87**	**98.44**
Prostate	PCA (74)	85.29	83.33	86.27	83.33	84.31	83.33	84.31
	SVD (74)	86.27	84.31	82.35	86.27	82.35	84.31	84.29
	CON (138)	84.31	82.35	88.24	83.33	86.27	84.31	85.64
	CFS (127)	86.27	84.31	90.20	88.24	91.18	88.24	88.07
	NSGA-II (107)	91.18	90.20	91.18	89.22	92.16	91.18	90.85
	MOEA/D (89)	88.24	83.33	89.22	86.27	91.18	84.31	87.09
	GSBOGA (74)	**94.12**	**93.14**	**95.10**	**93.14**	**96.08**	**94.12**	**94.28**

(continued)

Table 5 (continued)

Dataset	Methods (#genes)	Classifiers (%)						Avg. acc.
		NB	J48	SVM	KNN	MLP	RBF	(%)
Breast	PCA (85)	70.51	67.95	73.08	71.79	73.08	70.51	71.15
	SVD (85)	71.79	70.51	74.36	71.79	73.08	69.23	71.79
	CON (168)	73.08	71.79	75.64	70.51	74.36	71.79	72.86
	CFS (194)	73.08	70.51	73.08	76.92	75.64	73.08	73.72
	NSGA-II (117)	78.21	75.64	80.77	78.21	79.49	76.92	78.21
	MOEA/D (91)	82.05	83.33	85.90	**84.61**	**85.90**	**83.33**	84.19
	GSBOGA (85)	**83.33**	**84.61**	**87.18**	**84.61**	**85.90**	**83.33**	**84.61**
DLBCL	PCA (68)	75.32	72.73	80.52	74.02	77.92	72.73	75.54
	SVD (68)	80.52	76.62	81.82	79.22	80.52	77.92	79.44
	CON (145)	79.92	77.92	81.82	81.48	80.52	77.92	79.93
	CFS (180)	80.52	79.92	83.12	80.52	83.12	83.12	81.72
	NSGA-II (89)	84.41	83.12	84.41	81.82	84.41	83.12	83.55
	MOEA/D (83)	84.41	85.71	87.01	83.31	85.71	84.41	85.09
	GSBOGA (68)	**92.21**	**90.91**	**93.51**	**89.61**	**93.51**	**92.21**	**91.99**
Colon	PCA (57)	75.81	74.19	79.03	77.42	79.03	77.42	77.15
	SVD (57)	72.58	75.81	77.42	75.81	74.19	75.81	75.27
	CON (134)	80.64	79.03	82.26	79.03	80.64	82.26	80.64
	CFS (140)	77.42	79.03	80.64	75.80	82.26	75.80	78.49
	NSGA-II (93)	83.87	82.26	85.48	82.26	83.87	82.26	83.33
	MOEA/D (81)	80.64	79.03	82.26	80.64	82.26	80.64	80.91
	GSBOGA (57)	**85.48**	**83.87**	**88.71**	**85.48**	**87.10**	**85.48**	**86.02**

From Table 5, it has been observed that the *GSBOGA* is superior to most of the other methodologies with respect to both the number of features and the average classification accuracy. For example, in case of Leukemia dataset, *J*48 and *KNN* give the best results both for *NSGA-II* and *GSBOGA* methods; but for other classifiers, only *GSBOGA* gives the best results. Similarly, for remaining datasets, the other methods give better results for some classifiers but in most of the cases, the proposed *GSBOGA* gives the best results, which are marked by the bold font in Table 5. The *GSBOGA* represents very promising solution with lesser time complexity compared to *NSGA*-based methods since there is not require of global computation usual of other Pareto optimality concept based MOEA methods. It utilizes a stable state identification methodology, not required for fitness sharing parameters, which is employed in *NSGA* method or crowding distance employed in *NSGA-II* method.

4 Summary

A novel gene selection (GSBOGA) methodology regarding bi-objective GA is proposed in this chapter. The GSBOGA represents very promising solution with lesser time complexity compared to NSGA based methods, since there is not require any of global computation of other Pareto optimality based MOEA methods. The methodology also explores the boundary region of the rough set to allow some vagueness which ultimately helps to find the optimal solutions from the searching space. It utilizes a stable selection technique, not required for fitness sharing parameters, which is employed in NSGA method or crowding distance utilized in NSGA-II method. Jumping gene mutation in proposed method moreover conquers the lack of diversity of the chromosome pool that might occur in the sense of single-bit mutation.

References

U. Alon, N. Barkai, D.A. Notterman, K. Gish, S. Ybarra, D. Mack, A.J. Levine, Broad patterns of gene expression revealed by clustering analysis of tumor and normal colon tissues probed by oligonucleotide arrays. PNAS **96**, 6745–6750 (1999)

H.C. Causton, J. Quackenbush, A. Brazma, Microarray gene expression data analysis: a beginner's guide. Genet. Res. **82**, 151–153 (2003)

G. Chaconas, B.D. Lavoie, M.A. Watson, DNA transposition: jumping gene machine. Curr. Biol. **6**(7), 817–820 (1996)

K. Deb, A. Pratap, S. Agarwal, T.A. Meyarivan, A fast and elitist multi objective genetic algorithm: NSGA-II. IEEE Trans. Evol. Comp. **6**(2), 182–197 (2002)

T.R. Golub, D.K. Slonim, P. Tamayo, C. Huard, M. Gaasenbeek, J.P. Mesirov, H. Coller, M.L. Loh, J.R. Downing, M.A. Caligiuri, C.D. Bloomfield, E.S. Lander, Molecular classification of cancer: class discovery and class prediction by gene expression monitoring. Science **286**, 531–537 (1999)

D.E. Goldberg, J.H. Holland, Genetic algorithms and machine learning. Mach. Learn. **3**(2), 95–99 (1988)

D. Gong, G. Wang, X. Sun, Y. Han, A set-based genetic algorithm for solving the many-objective optimization problem. Soft Comput. **19**(6), 1477–1495 (2015)

G.J. Gordon, R.V. Jensen, L.L. Hsiao, S.R. Gullans, J.E. Blumenstock, S. Ramaswamy, W.G. Richards, D.J. Sugarbaker, R. Bueno, Translation of microarray data into clinically relevant cancer diagnostic tests using gene expression ratios in lung cancer and mesothelioma. Cancer Res. **62**, 4963–4967 (2002)

F. Gu, H.L. Liu, K.C. Tan, A hybrid evolutionary multi-objective optimization algorithm with adaptive multi-fitness assignment. Soft Comput. **19**(11), 3249–3259 (2015)

A.M. Hall, Correlation-based feature selection for machine learning, The University of Waikato, 1999

J. Harmouche, C. Delpha, D. Diallo, Y.L. Bihan, Statistical approach for non-destructive incipient crack detection and characterization using Kullback-Leibler divergence. IEEE Trans. Reliab. **65**(3), 1360–1368 (2016)

J.E. Jackson, *A User's Guide to Principal Components* (Wiley, New York, 1991), ISBN 0-471-62267-2

S.Y. Jing, A hybrid genetic algorithm for feature subset selection in rough set theory. Soft Comput. **18**(7), 1373–1382 (2014)

Kent Ridge Biomedical Dataset Repository, (n.d), http://datam.i2r.a-star.edu.sg/datasets/krbd/

R. Kerber, ChiMerge: discretization of numeric attributes. in *National Conference on Artificial Intelligence*, pp. 123–128 (1992)

J.D. Knowles, D.W. Corne, M-PAES: a memetic algorithm for multi-objective optimization. in *Proceedings of IEEE Congress on Evolutionary Computation*, pp. 325–332 (2000)

S. Kullback, R.A. Leibler, On information and sufficiency. Ann. Math. Stat. **22**(1), 79–86 (1951)

Y. Leung, Y. Hung, A multiple-filter-multiple-wrapper approach to gene selection and microarray data classification. IEEE/ACM Trans. Comput. Biol. Bioinform. **7**(1), 108–117 (2010)

H. Maaranen, K. Miettinen, M.M. Makela, A quasi-random initial population for genetic algorithms. Comput. Math. Appl. **47**(12), 1885–1895 (2004), Elsevier

J.V. Neumann, in *Theory of Self-reproducing Automata*, ed. by A.W. Burks (Univer. of Illinois Press, USA, 1996)

Z. Pawlak, Rough set theory and its applications to data analysis. Cybern. Syst. **29**, 661–688 (1998)

M. Petrou, P. Bosdogianni, An example of SVD. in *Image Processing: The Fundamentals* (Wiley, 2000), pp. 37–44

K. Price, R.M. Storn, J.A. Lampinen, in *Differential Evolution: A Practical Approach to Global Optimization*, Natural Computing Series (Springer, 2005), ISBN: 3540209506

L.S. Santana, A.M. Canuto, Filter-based optimization techniques for selection of feature subsets in ensemble systems. Expert Syst. Appl. **41**(4), 1622–1631 (2014)

G. Schaefer, Data mining of gene expression data by fuzzy and hybrid fuzzy methods. IEEE Trans. Inf. Technol. Biomed. **14**(1), 23–29 (2010)

P. Shelokar, A. Quirin, O. Cordón, MOSubdue: a Pareto dominance-based multi objective Subdue algorithm for frequent sub graph mining. Knowl. Inf. Syst. **34**(1), 75–108 (2013)

M.A. Shipp, K.N. Ross, P. Tamayo, A.P. Weng, J.L. Kutok, R.C.T. Aguiar, M. Gaasenbeek, M. Angelo, M. Reich, T.R. Golub, Diffuse large B-cell lymphoma outcome prediction by gene expression profiling and supervised machine learning. Natl. Med. **8**(1), 68–74 (2002)

D. Singh, P.G. Febbo, K. Ross, D.G. Jackson, J. Manola, C. Ladd, P. Tamayo, A.A. Renshaw, J.P. Richie, E.S. Lander, M. Loda, T.R. Golub, W.R. Sellers, Gene expression correlates of clinical prostate cancer behavior. Cancer Cell **1**, 203–209 (2002)

L.J. Veer, H. Dai, M.J. Vijver, Y.D. He, Y.D. He, A.A.M. Hart, Gene expression profiling predicts clinical outcome of breast cancer. Nature **415**, 530–536 (2002)

D.P. Waters, Von Neumann's theory of self-reproducing automata: a useful framework for biosemiotics? Biosemiotics **5**(1), 5–15 (2012)

Y. Yang, J.O. Pedersen, A comparative study on feature selection in text categorization. ICML **97**, 412–420 (1997)

Q. Zhang, H. Li, MOEA/D: a multi-objective evolutionary algorithm based on decomposition. IEEE Trans. Evol. Comput. **11**(6), 712–731 (2007)

E. Zitzler, L. Thiele, Multi-objective evolutionary algorithms: a comparative case study and the strength Pareto approach. IEEE Trans. Evol. Comput. **3**(4), 257–271 (1999)

Multi-Objective Optimization and Cluster-Wise Regression Analysis to Establish Input–Output Relationships of a Process

Amit Kumar Das, Debasish Das and Dilip Kumar Pratihar

1 Introduction

We, human beings, have a natural tendency to know input–output relationships of a process. Goldberg (2002) claimed that a genetic algorithm (GA) is competent for yielding innovative solutions in a single-objective optimization (SOO) problem domain. An SOO is generally used for finding out a single optimal solution out of several possibilities. However, a multi-objective optimization (MOO) problem involves at least two conflicting objective functions and a Pareto-optimal front of solutions can be obtained for the same. This present chapter deals with an application of MOO.

For the last few decades, MOO had been applied for solving various research and industrial problems (Ahmadi et al. 2015, 2016; Aghbashlo et al. 2016; Jarraya et al. 2015; Khoshbin et al. 2016; Marinaki et al. 2015; Ahmadi and Mehrpooya 2015; Sadatsakkak et al. 2015). Making large data intelligent transportation system (Wang et al. 2016), fabrication and optimization of 3D structures in bone tissue engineering area (Asadi-Eydivand et al. 2016), portfolio optimization with functional constraints (Lwin et al. 2014), optimization of building design (Brownlee and Wright 2015), disaster relief operations (Zheng et al. 2015), etc. are some of the worth-mentioning examples, where MOO has been utilized successfully in the recent times. By using a multi-objective evolutionary algorithm (MOEA), a Pareto-front of solutions is obtained. These solutions are not generated randomly (Askar and Tiwari 2009) and

A. K. Das · D. Das · D. K. Pratihar (✉)
Indian Institute of Technology Kharagpur, Kharagpur 721302, India
e-mail: dkpra@mech.iitkgp.ac.in

A. K. Das
e-mail: amit.besus@gmail.com

D. Das
e-mail: dd4311@gmail.com

© Springer Nature Singapore Pte Ltd. 2018
J. K. Mandal et al. (eds.), *Multi-Objective Optimization*,
https://doi.org/10.1007/978-981-13-1471-1_14

they must satisfy mathematically Karush–Kuhn–Tucker (KKT) conditions (Mietti-nen et al. 1999). It is argued (Deb and Srinivasan 2008) that there is a high chance of finding some commonalities among these high-performing solutions of the Pareto-front. Nevertheless, it is also debated that this commonality may exist for either the whole Pareto-front or different subsets of the same. However, if the said commonal-ities are embedded in those optimal or near-optimal solution sets, then it is expected to obtain the design principles for that process after some numerical analysis of the solutions.

The said fact can have an immense effect in designing a product or process. If a designer does have this type of information a priori, then it will help him to do his job more efficiently. For example, let us take the case of metal cutting operation with the inputs, such as cutting speed, feed, and depth of cut, and take two conflicting outputs as minimizing the machining time and minimizing the surface roughness of the machined product. By using the MOEA, if it is found that one or more parameters are varying in a particular fashion within the Pareto-front or the objective functions are changing with the inputs in a certain manner, then this information can provide an extra advantage to the process designer to set the same in a more efficient way. Similar things are applicable to other processes as well.

This fact of obtaining various design principles has another significance for the manufacturing industries. Reduction of cost without compromising the safety and quality of the products has always been the primary concern of any industry. The scope of cost minimization through proper inventory and manpower management would be possible, if the facts discussed above are available. Moreover, similar prior information may assist the process designer to establish and stabilize the process of interest with ease.

To use an MOEA, input–output relationships are required to define the objective functions. In many instances, these equations are not obtained from the literature and we have to derive these using the statistical tools on the experimental data. Now, generation of a wide range of experimental data requires not only proper facilities, but also a sequence of several tedious steps, and it demands time, cost, and effort. Moreover, noticeable variations in output data of any experiments are likely to be observed for a given set of input parameters. This results in fuzziness and inaccuracy in the experimental dataset, as shown by Gil and Gil (1992). The inaccuracy of the data can be minimized through multiple repetitions of the experiments, thereby providing the upper and lower limits of variation in the developed dataset, for a given set of input parameters. However, it again becomes difficult due to time and resource constraints in most of the cases. Gil (1987) also argued that experimental data may suffer from the loss of information about the state or parameter space owing to the fuzziness in it. Therefore, the generated Pareto-optimal solutions using an MOEA may also be affected by the fuzziness and inaccuracy of the experimental data, and we may end up with largely inaccurate design principles for the analyzed process. Due to this issue, it is desirable to tackle inherent fuzziness of the experimental data to establish input–output relationships of the process more accurately.

2   Literature Survey

Evolutionary multi-objective optimization has been used to establish input–output relationships by several researchers. The existence of the resemblance among the Pareto-optimal front of solutions was highlighted by Deb (2003). He also suggested that these commonalities could be revealed through regression and manual plotting. Obayashi and Sasaki (2003) used the self-organizing map to view the higher dimensional objective space and design variable space on a lower dimensional map. They also used a clustering technique to make the clusters of decision parameters, which showed the role of the variables in improving the design and trade-offs. Taboada and Coit (2006) suggested applying k-means clustering technique on the Pareto-optimal solutions for ease of further analysis. Deb and Srinivasan (2008) used Benson's method to obtain a modified Pareto-front from the initial one and after that, they used statistical regression analysis to get different design principles. This method had also been adopted and implemented by various researchers in their works (Deb et al. 2009; Deb and Jain 2003; Deb and Sindhya 2008). An analytical approach was suggested by Askar and Tiwari (2011) to get a Pareto-front for multi-objective optimization problem, and the obtained Pareto-optimal solutions were analyzed to derive several innovative commonality principles. Deb et al. (2014) proposed the method of automated innovization to decode several important relationships through the extensive use of an evolutionary algorithm. They used the word, innovization, which means the act of obtaining innovative solutions through optimization. In this case, they had not used regression tools to decipher the said principles. Later, a concept of higher level innovization was introduced by Bandaru et al. (2011) and Bandaru et al. (2015) in a generalized form. Also, a simulation-based innovization procedure was developed by Dudas et al. (2011). In this approach, they tried to evaluate the effect of variables on the performance of the process and they showed a method of getting in-depth knowledge about a process after these analyses.

Among all these stated approaches, no one adopted any method to model the inherent fuzziness of the Pareto-optimal solutions. As already discussed, if this work is not done, then there will be a high chance that we shall obtain imprecise design principle for the process of interest. Therefore, a method has been developed here to obtain different input–output relationships after modeling the fuzziness in the Pareto-optimal dataset. This method has been applied for an electron beam welding process on SS304 plates, and the obtained results clearly show the significance of this developed approach. The rest of the text has been arranged as follows:

The developed method has been described in Sect. 3, whereas Sect. 4 deals with the experimental data collection procedure. The results and discussion are provided in Sect. 5, and some concluding remarks are made in Sect. 6.

3 Developed Approach

In the proposed approach, an attempt has been made to establish the input–output relationships of a process through the extensive use of a multi-objective evolutionary algorithm (MOEA). The approach has been explained in the following steps:

Step 1: Develop initial Pareto-optimal front

In this step, an initial Pareto-front is obtained using an MOEA, where the input–output relationships are used to determine the fitness values of the objective functions. The obtained Pareto-front is subjected to some inherent fuzziness in it, which is going to be removed in further steps.

Several strategies for the MOEA are available in the literature, such as niched Pareto genetic algorithm (NPGA) (Horn et al. 1994), strength Pareto evolutionary algorithm (SPEA) (Zitzler and Thiele 1998), Pareto-archived evolution strategy (PAES) (Knowles and Corne 1999), non-dominated sorting genetic algorithm-II (NSGA-II) (Deb et al. 2002), multi-objective algorithm based on decomposition (MOEA/D) (Zhang and Li 2007), multiple populations for multiple objectives (MPMO) (Zhan et al. 2013), and others. In our study, NSGA-II has been used as the MOEA and its working principle is described in Fig. 1.

Step 2: Train a neuro-fuzzy system

Using the initial Pareto-front of solutions, a neuro-fuzzy system (NFS) is trained. In NFS, a fuzzy logic controller (FLC) is expressed as the form of a neural network. During the training, an evolutionary optimization technique is used to tune the NFS. The initial Pareto-optimal solutions are clustered using a clustering algorithm, and the number of rules of the NFS is kept as the same with that of a total number of clusters obtained. The data obtained through experiments are subjected to inaccuracy and fuzziness. This may be due to various reasons, such as experimental inaccuracy, error due to the unskilled operator, instrumental inaccuracy, and others. To take care of this inherent fuzziness, NFS is an efficient tool (Mitra and Pal 1996), which works based on the principles of fuzzy sets. In NFS, the advantages of both fuzzy logic controller and neural network are clubbed together to design and remove the uncertainty and imprecision of a set of data. The NFS has been used successfully to solve a variety of problems related to several fields of research (Takagi and Hayashi 1991; Takagi et al. 1992; Keller et al. 1992; Berenji and Khedkar 1992; Jang 1993; Ishibuchi et al. 1994). Here, an NFS with Mamdani approach (Mamdani and Assilian 1975) is used to model fuzziness in the initial Pareto-optimal solutions. The said NFS has mainly five layers, namely, input layer, fuzzification layer, And operation layer, fuzzy inference layer, and defuzzification layer. Gaussian type of membership functions, which has been used in input and output layers of the NFS, can be expressed using Eq. (1):

$$\mu_{Gaussian} = e^{-\left[\frac{(x-m)^2}{2\sigma^2}\right]}, \tag{1}$$

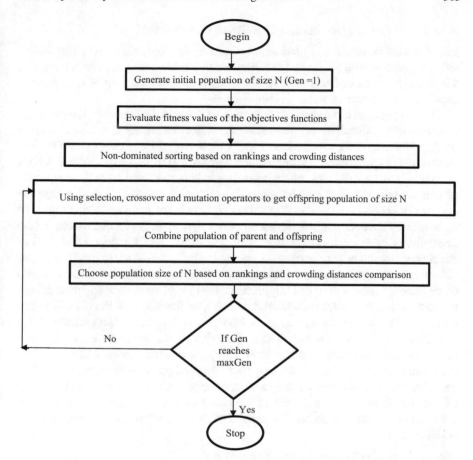

Fig. 1 Flowchart of NSGA-II algorithm

where σ and m are the standard deviation and mean of the Gaussian distribution, respectively. During the training, the NFS is evolved with the help of a genetic algorithm, where the Gaussian parameters (σ and m) are used as the design variables and the objective is to minimize the root-mean-square error (RMSE) value (here, error is the amount of deviation in prediction).

Step 3: Obtain a modified Pareto-front

In this step, the trained NFS is used in another MOEA to evaluate the fitness values of the objective functions. As NFS is a strong tool to take care the fuzziness of the used data, there is every possibility to get a better quality of Pareto-front compared to that of the initial one. In this way, the imprecision of the initial Pareto-optimal solutions is removed using the NFS and a modified and better Pareto-front in terms of both the objective functions' values is obtained.

Step 4: Clustering of the obtained modified Pareto-front

The modified Pareto-optimal dataset, obtained in Step 3, is clustered using a clustering algorithm. The purpose of carrying out this operation is to find that whether the design principles of the process are varying for the different clusters, or they are kept the same for the whole set of Pareto-front.

There are several techniques of clustering available in the literature. However, we have considered three popular clustering algorithms like fuzzy C-means clustering (FCM) (Bezdek 1973), entropy-based fuzzy clustering (EFC) (Yao et al. 2000), and density-based spatial clustering application with noise (DBSCAN) (Ester et al. 1996), in our study. In case of FCM, a data point may belong to several predefined numbers of clusters with different membership values. It is an iterative method, where the cluster centers and the membership values of the cluster members are going to be updated with the iterations. In every step of the algorithm, the focus is to reduce the dissimilarity measurement (which is evaluated in terms of Euclidean distance) of the cluster members. The members of the same cluster are likely to have the higher values of similarity, whereas the data points belonging to different clusters should be dissimilar in nature. The EFC algorithm works based on the entropy measures of the data points, which are calculated depending on the values of the distances (say Euclidean distances) among them. A data point is declared as a cluster center, if it is found with the minimum entropy value. The other points, which have the similarity measures with the cluster center greater than a user-defined value, will be put into the cluster. In case of DBSCAN, two user-specified parameters are used, such as neighborhood radius and minimum number of neighborhood points. The philosophy of this algorithm is that the density of the members lying within a cluster is higher than that of the outsider points. The details of these algorithms can be found out in (Pratihar 2007).

Step 5: Analysis of the modified Pareto-front

The obtained modified Pareto-optimal solutions are statistically analyzed cluster-wise to get several input–output relationships for the problem. Moreover, any relationship, which is common to all the clusters, has to be checked. The developed approach has been described through a flowchart, as shown in Fig. 2. Generally, the outputs of a natural process vary nonlinearly with the input parameters. Keeping this idea in mind, a nonlinear regression tool (using MINITAB 16.0 software (http://www.minitab.com)) has been used to determine various input–output relationships in power form.

4   Experimental Data Collection

To explain the proposed approach in more details, an engineering problem, namely, electron beam welding (EBW), has been selected and the developed approach has been implemented for the said process. The details of the experimental procedure, along with the setup information, have been provided in this section.

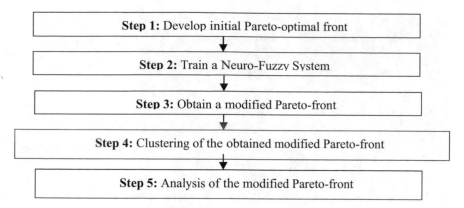

Fig. 2 Flowchart of the developed approach

4.1 Experimental Setup and Procedure

Bead-on-plate welding was carried out on EBW facility, developed by Bhabha Atomic Research Centre (BARC), Mumbai, at IIT Kharagpur (refer to Fig. 3). The machine has a maximum power rating of 12 kW. The beam was kept stationary, while the table containing the workpiece, fixture, and other arrangements traveled in the horizontal plane at a predefined welding speed. A vacuum was provided in the work chamber and gun chamber with the help of vacuum pumps before the initiation of the welding process. The stream of highly accelerated electrons was made incident on 20-mm-thick AISI 304 stainless steel workpiece in the vacuum environment. The chemical composition of the used material can be found out in (Das et al. 2016). The EBW experiments had been carried out following a multilevel full-factorial design. This study aims to investigate the effects of beam power and welding speed on the depth of penetration and bead width of the weld.

4.2 Data Collection

Two input parameters, namely, beam power (P in W) and welding speed (S in mm/min), were considered in this study. Considering four levels of the input parameters, the experiments had been carried out according to the multilevel full-factorial design with $2^4 = 16$ combinations of design variables. For each combination of input variables, welding was carried out three times in order to ensure repeatability.

These samples were sectioned, polished, etched, and observed under the microscope in order to obtain the desired measurements. The average values of the bead width (BW in mm) and depth of penetration (DP in mm) were calculated and used in this study. The details of the experimental data used for developing the model and testing the same are shown in Appendices 1 and 2, respectively.

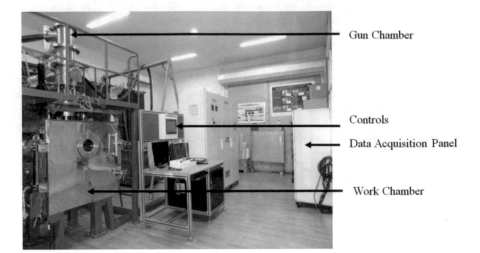

Fig. 3 Electron beam welding (EBW) setup, IIT Kharagpur, India

5 Results and Discussion

As like other natural processes, electron beam welding process has also nonlinear input–output relationships (Jha et al. 2014). These relationships were obtained from the experimental data using a nonlinear regression tool. After this, the developed approach was used to get the relationships among the responses and design variables, and other important information regarding the said process.

5.1 Obtaining Nonlinear Input–Output Relationships from the Experimental Data

Using the statistical software Minitab 16.0, a nonlinear regression analysis had been carried out to obtain the input–output relationships from the experimental data (refer to Appendix 1) collected within the upper and lower limits of the input variables, as provided in Table 1.

The following expression was obtained for depth of penetration:

Table 1 Input variables and their ranges

Sl. no	Input variables	Symbols	Minimum value	Maximum value
1	Beam power (W)	P	3200	5600
2	Welding speed (mm/min)	S	900	1800

$$DP = -2.31965 + 0.00443709 \times P - 0.00486308 \times S - 2.08008E - 07 \times P^2$$
$$+ 2.67361E - 06 \times S^2 - 1.16208E - 06 \times P \times S. \tag{2}$$

The model was found to be capable of predicting accurate results because of a high value of regression coefficient of 0.97. A confidence level of 95% was considered. The other response, BW was also expressed in terms of the input parameters as follows:

$$BW = 18.0029 - 0.00554238 \times P + 0.000857167 \times S + 6.32813E - 07 \times P^2$$
$$- 5.69444E - 07 \times S^2 - 2.61667E - 07 \times P \times S. \tag{3}$$

The regression coefficient for this case was found to be 0.8.

5.2 Formulation of the Optimization Problem

The objective was to maximize the depth of penetration (DP), while keeping the bead width (BW) at a lower value. These said goals are contradictory to each other, as with the increase of DP, and output parameter BW increases. Therefore, this is an ideal problem for MOO and it can be expressed as follows:

$$\begin{aligned} &\text{Minimize} && 1/DP \\ &\text{Minimize} && BW \\ &\text{subject to} \\ & && 3200 \le P \le 5600 \\ & && 900 \le S \le 1800 \end{aligned} \tag{4}$$

5.3 Obtaining Initial Pareto-Front

Using NSGA-II, an initial Pareto-front was obtained (refer to Fig. 4). In this case, Eqs. (2) and (3) were used to evaluate the numerical values for the outputs DP and BW, respectively. The user-defined parameters for the NSGA-II, such as probability of crossover (p_c), probability of mutation (p_m), population size (N), and maximum number of generations (G_{max}), were selected through a detailed parametric study. This study had been carried out by varying parameters one at a time and keeping the others fixed. The sequence for varying the parameters was taken as follows: p_c, p_m, N and G_{max}, as suggested in Pratihar (2007). The best parameters were chosen based on the maximum spread of the Pareto-front. The details of this parametric study have been provided in Table 2, where the selected parameters are written in bold.

Table 2 Results of the parametric study to select NSGA-II parameters

Exp no.	Fixed GA parameters' values	Varying GA parameter value	Spread of Pareto-optimal data set (X- and Y-coordinates represent 1/MRR and SR values, respectively)	Is the parameter selected?
1	$p_m = 0.06$, $N = 300$ and $G_{max} = 1000$	$p_c = 0.6$	(0.125994, 5.80199) to (0.247707, 3.41509)	No
2		$p_c = 0.7$	(0.125994, 5.80199) to (0.247708, 3.41509)	No
3		$\boldsymbol{p_c = 0.8}$	(0.125994, 5.80199) to (0.247716, 3.41509)	Yes
4		$p_c = 0.9$	(0.125994, 5.80199) to (0.247704, 3.41509)	No
5		$p_c = 1.0$	(0.125994, 5.80199) to (0.247707, 3.41509)	No
6	$p_c = 0.8$, $N = 300$ and $G_{max} = 1000$	$p_m = 0.02$	(0.125994, 5.80199) to (0.247703, 3.41509)	No
7		$p_m = 0.04$	(0.125994, 5.80199) to (0.247715, 3.41509)	No
8		$\boldsymbol{p_m = 0.06}$	(0.125994, 5.80199) to (0.247716, 3.41509)	Yes
9		$p_m = 0.08$	(0.125994, 5.80199) to (0.247711, 3.41509)	No
10		$p_m = 0.1$	(0.125994, 5.80199) to (0.247708, 3.41509)	No
11	$p_c = 0.8$, $p_m = 0.06$ and $G_{max} = 1000$	$N = 100$	(0.125994, 5.80199) to (0.247701, 3.41509)	No
12		$N = 200$	(0.125994, 5.80199) to (0.2477, 3.41509)	No
13		$\boldsymbol{N = 300}$	(0.125994, 5.80199) to (0.247716, 3.41509)	Yes
14		$N = 400$	(0.125994, 5.80199) to (0.247709, 3.41509)	No
15		$N = 500$	(0.125994, 5.80199) to (0.247708, 3.41509)	No
16	$p_c = 0.8$, $p_m = 0.06$ and $N = 300$	$\boldsymbol{G_{max} = 1000}$	(0.125994, 5.80199) to (0.247716, 3.41509)	Yes
17		$G_{max} = 2000$	(0.125994, 5.80199) to (0.247708, 3.41509)	No
18		$G_{max} = 3000$	(0.125994, 5.80199) to (0.247705, 3.41509)	No
19		$G_{max} = 4000$	(0.125994, 5.80199) to (0.247709, 3.41509)	No
20		$G_{max} = 5000$	(0.125994, 5.80199) to (0.247709, 3.41509)	No

Fig. 4 Obtained initial
Pareto-front of solutions

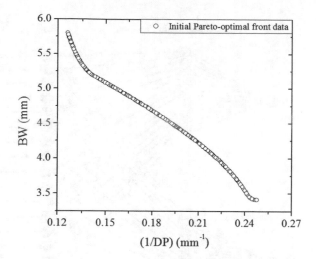

In NSGA-II, different genetic operators, such as tournament selection, arithmetic crossover, and Gaussian mutation, were used. Now, the initial Pareto-front of optimal solutions was obtained utilizing the selected parameters, as shown in Fig. 4.

5.4 Training of an NFS

Using the initial Pareto-front dataset, a neuro-fuzzy system (NFS) had been trained. The training dataset had been clustered using a clustering technique, namely, fuzzy C-means clustering and in this case, the level of cluster fuzziness was considered as 2.0. To obtain the best results, the said data set was clustered into 16 different clusters. Therefore, the total number of rules for the NFS became equal to the number of clusters made. The structure of the developed NFS has been shown in Fig. 5.

In the used NFS, a supervised learning with a batch mode of training method had been applied. In the input and output layers, the Gaussian type of membership functions had been used. So, the total number of unknown parameters of this model was found to be equal to $(16 \times 4 \times 2 =) 128$ (as there are two inputs and two outputs each in the model, and each Gaussian function has two unknown parameters, that is, σ and m). For the training purpose, 300 input data points of the initial Pareto-front had been used and a root-mean-square error (RMSE) was calculated each time. Here, the error is nothing but the deviation in prediction. This NFS was evolved using a genetic algorithm, where the objective was set to minimize the RMSE value and the design variables were those 128 numbers of unknown parameters of the model. Different genetic operators, such as roulette wheel selection, linear crossover, and random mutation, had been utilized in the GA, and to obtain the best results, the selected GA parameters were as follows: crossover probability ($p_c = 0.9$), mutation probability ($p_m = 0.1$), and population size ($N = 60$).

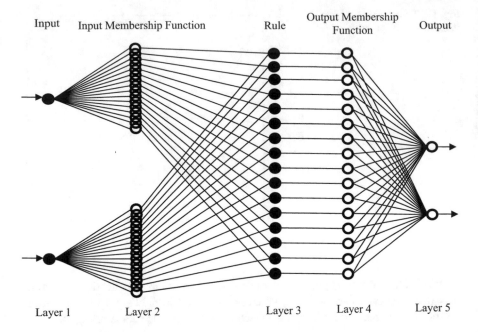

Input   Input Membership Function   Rule   Output Membership Function   Output

Layer 1   Layer 2   Layer 3   Layer 4   Layer 5

Fig. 5 Structure of the neuro-fuzzy system

5.5 Obtaining Modified Pareto-Front

In this step, the trained NFS had been used in an NSGA-II for determining the fitness values of the objectives. It is important to note that Eqs. (2) and (3) were used previously in the NSGA-II for obtaining the initial Pareto-front of solutions. The parameters for this NSGA-II were kept the same as in that of the previous case described in Sect. 5.3. We could get a modified Pareto-front, as shown in Fig. 6, and the quality of this Pareto-front had been improved in terms of the objective function values compared to that of the initial one.

5.6 Clustering of the Modified Pareto-Front Data Set

The obtained modified Pareto-optimal solutions were clustered using three different algorithms, namely, fuzzy C-means clustering (FCM), entropy-based fuzzy clustering (EFC), and density-based spatial clustering application with noise (DBSCAN). By using FCM algorithm with the level of cluster fuzziness kept equal to 1.25, two clusters were obtained, as shown in Fig. 7.

Fig. 6 Initial and modified
Pareto-front of solutions

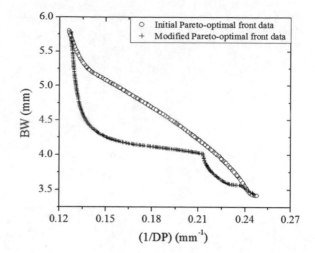

Fig. 7 Clustering of the
modified Pareto-front using
FCM

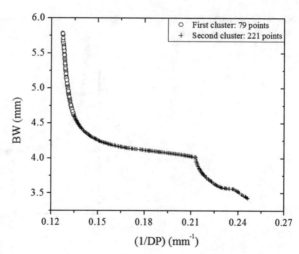

In case of EFC, two distinct clusters (refer to Fig. 8) were obtained with the user-defined parameters like the constant of similarity ($\alpha = 0.12$) and threshold value of similarity ($\beta = 0.9$).

In another case, where the clustering was done using DBSCAN algorithm, three distinct clusters were obtained, as shown in Fig. 9. In this algorithm, a point was considered to form a cluster, when a minimum of three other points were found to be present in a neighborhood radius of 0.032.

For the different clusters obtained using the said three clustering algorithms, the respective ranges of variation for the two outputs, such as DP and BW, are given in Table 3.

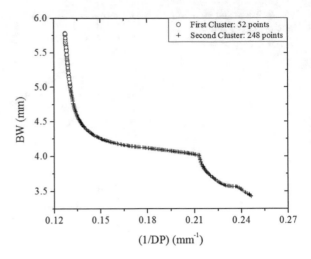

Fig. 8 Clustering of the modified Pareto-front using EFC

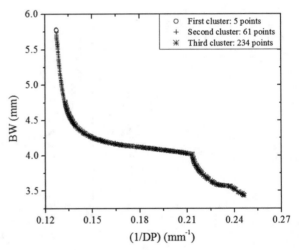

Fig. 9 Clustering of the modified Pareto-front using DBSCAN

The clustered solutions were analyzed using nonlinear regression analysis. For this purpose, a Gauss–Newton approach was used with 95% confidence level for all intervals and the convergence tolerance was assumed to be equal to 0.00001. In Table 4, cluster-wise obtained different input–output relationships are provided.

The two extreme points on the modified Pareto-front correspond to the maximum and minimum values of the two outputs of the EBW process. One of these points shows the highest values of DP and BW with the input parameters as follows: $P = 5600\,W$, $S = 900\,mm/min$. On the other hand, the other point provides the information regarding the lowest values of the outputs with an input variables setting as $P = 4750.98\,W$, $S = 1800\,mm/min$. It is observed (refer to Fig. 10) that for increasing the depth of penetration (DP), beam power (P) has to be increased and

Table 3 Ranges of DP and BW in different clusters

Cluster algorithm	Cluster number	Output parameters			
		DP (mm)		BW (mm)	
		Minimum	Maximum	Minimum	Maximum
FCM	1	7.4441	7.8581	4.6191	5.7872
	2	4.0632	7.4252	3.4281	4.6039
EFC	1	7.6593	7.8581	5.0077	5.7872
	2	4.0632	7.6535	3.4281	4.9897
DBSCAN	1	7.8502	7.8581	5.7519	5.7872
	2	7.5697	7.8472	4.7989	5.7124
	3	4.0632	7.5481	3.4281	4.7629

Table 4 Relationships among decision variables and responses for EBW problem

Clustering algorithm	Cluster number	Relationships among design variables and objectives
FCM	1	$DP = P^{0.443585} \times S^{-0.257345}$
	1	$BW = P^{1.13416} \times S^{-1.17341}$
	2	$DP = P^{0.790491} \times S^{-0.713536}$
	2	$BW = P^{0.40339} \times S^{-0.290404}$
EFC	1	$DP = P^{0.351876} \times S^{-0.14233}$
	1	$BW = P^{0.848204} \times S^{-0.814444}$
	2	$DP = P^{0.880063} \times S^{-0.818614}$
	2	$BW = P^{0.450339} \times S^{-0.345143}$
DBSCAN	1	$DP = P^{0.228065} \times S^{0.013738}$
	1	$BW = P^{0.228118} \times S^{-0.0315064}$
	2	$DP = P^{0.454036} \times S^{-0.26975}$
	2	$BW = P^{1.23061} \times S^{-1.29156}$
	3	$DP = P^{0.835263} \times S^{-0.765954}$
	3	$BW = P^{0.422925} \times S^{-0.313142}$

welding speed (S) should be at its lower value. In other situation, where a user requires a lower value of bead width (BW), the input parameter, P, has to be decreased and S is needed to be increased. Moreover, both DP and BW are found to be proportional to the heat input, which is a unified effect of beam power and welding speed on the weld geometries. This trend is in accordance with the literature (Das et al. 2017; Kar et al. 2015). Therefore, a user may be recommended to choose input parameters setting to avail the high depth of penetration (7.01 mm) and low bead width (4.36 mm) as follows: $P = 5548.8\,\text{W}$, $S = 1096.46\,\text{mm/min}$.

The obtained input–output relationships were used on some test data (refer to Appendix 2), and an average absolute percentage error (AAPE) was calculated for each of the cases of clustering techniques. These were compared to the results

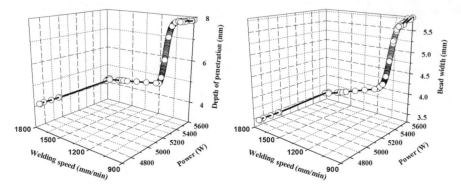

Fig. 10 Variations of DP and BW with the inputs, P and S, in the modified Pareto-front

Table 5 Comparison of results using the relationships obtained from developed approach and experimental data

Output	AAPE in case of FCM (%)	AAPE in case of EFC (%)	AAPE in case of DBSCAN (%)	AAPE in case of Eqs. from experimental data (%)
DP	2.9822	3.2329	2.9903	3.7288
BW	5.4764	5.4009	5.6698	6.3340

obtained using Eqs. (2) and (3). The comparison (refer to Table 5) clearly indicates the fact that the relationships derived by the developed approach could predict more accurately compared to the regression equations did. In addition, FCM algorithm could perform a slightly better compared to the other two algorithms of clustering.

Another interesting fact to observe here is that the range for the input variable power (P) had been squeezed from (3200, 5600 W) to (4750.98, 5600 W) in the modified Pareto-optimal dataset. This fact denotes that the effective range of the input parameter P for this process has been shortened and it is advisable to operate only in this squeezed range to get the best results. This information will surely help the designers to design and establish the process efficiently.

6 Conclusion

An approach was developed to obtain different input–output relationships of the EBW process by the extensive use of a multi-objective optimization technique and a neuro-fuzzy system. It is quite different from the approaches available, because it adopts a method to model the inherent fuzziness in the experimental data and at the end, it could generate more accurate input–output relationships for a process. The approach was applied for an EBW process, and the results obtained were superior to that of the other methods in terms of precision and accuracy. This happens due to the fact that in the developed approach, an NFS, which is capable of handling uncertainty and inaccuracy of the data, is used and the imprecision in the data is removed to come up with the more accurate relationships.

The solutions of the modified Pareto-front were clustered and analyzed to obtain input–output relationships cluster-wise. A comparison has also been made among the relationships obtained in different cases of clustering algorithms based on the results of the test cases. Another interesting fact has been found out that for an input parameter, the effective range to get the best results has been squeezed. This prior information may increase the opportunities to reduce the operating cost and make the process more stable. Moreover, some physical aspects of the process are derived after analyzing the modified Pareto-optimal dataset, and the conclusions are seen to be inline with those made by other researchers for the process. Therefore, the developed approach can be applied to any process for obtaining input–output relationships and other important facts of the same.

Appendices

Appendix 1 collected experimental data

Sl. no.	Power (W)	Speed (mm/min)	Depth of penetration (mm)	Bead width (mm)
1	3200	1800	2.73	4.82
2	3200	1500	3.27	5.36
3	4000	1800	3.43	4.46
4	3200	1200	4	3.4
5	4000	1500	4.13	4.4
6	4800	1800	3.41	5.54
7	5600	1800	4.55	3.4
8	4800	1500	4.5	3.5
9	4000	1200	4.6	4.7
10	3200	900	3.9	6.92
11	5600	1500	4.8	5.1
12	4800	1200	5.29	5.31
13	4000	900	5.69	4.99
14	5600	1200	5.8	5.5
15	4800	900	7.15	5
16	5600	900	8.2	5.6

Appendix 2 Experimental data collected for testing the performance of the developed approach

Sl. no.	Power (W)	Speed (mm/min)	Depth of penetration (mm)	Bead width (mm)
1	4800	1650	4.37	3.38
2	4800	1325	5.03	3.87
3	5200	1650	4.39	3.86
4	5200	1325	5.27	4.51
5	5600	1000	7.91	5.28

References

M. Aghbashlo, S. Hosseinpour, M. Tabatabaei, H. Younesi, G. Najafpour, On the exergetic optimization of continuous photobiological hydrogen production using hybrid ANFIS–NSGA-II (adaptive neuro-fuzzy inference system–non-dominated sorting genetic algorithm-II). Energy **96**, 507–520 (2016)

M.H. Ahmadi, M. Mehrpooya, Thermo-economic modeling and optimization of an irreversible solar-driven heat engine. Energy Convers. Manag. **103**, 616–622 (2015)

M.H. Ahmadi, M.A. Ahmadi, S.A. Sadatsakkak, Thermodynamic analysis and performance optimization of irreversible Carnot refrigerator by using multi-objective evolutionary algorithms (MOEAs). Renew. Sustain. Energy Rev. **51**, 1055–1070 (2015)

M.H. Ahmadi, M.A. Ahmadi, A. Mellit, F. Pourfayaz, M. Feidt, Thermodynamic analysis and multi objective optimization of performance of solar dish Stirling engine by the centrality of entransy and entropy generation. Int. J. Electr. Power Energy Syst. **78**, 88–95 (2016)

M. Asadi-Eydivand, M. Solati-Hashjin, A. Fathi, M. Padashi, N.A.A. Osman, Optimal design of a 3D-printed scaffold using intelligent evolutionary algorithms. Appl. Soft Comput. **39**, 36–47 (2016)

S.S. Askar, A. Tiwari, Multi-objective optimisation problems: a symbolic algorithm for performance measurement of evolutionary computing techniques. in *Proceedings of EMO 2009* (Springer, 2009), pp. 169–182

S. Askar, A. Tiwari, Finding innovative design principles for multiobjective optimization problems. IEEE Trans. Syst. Man Cybern. Part C Appl. Rev **41**(4), 554–559 (2011)

S. Bandaru, C.C. Tutum, K. Deb, J.H. Hattel, Higher-level innovization: a case study from friction stir welding process optimization. Evol. Comput. (CEC) **2011**, 2782–2789 (2011)

S. Bandaru, T. Aslam, A.H. Ng, K. Deb, Generalized higher-level automated innovization with application to inventory management. Eur. J. Oper. Res. **243**(2), 480–496 (2015)

J.C. Bezdek, Fuzzy mathematics in pattern classification. Ph.D. thesis, Applied Math. Center, Cornell University, 1973

H.R. Berenji, P. Khedkar, Learning and tuning fuzzy logic controllers through reinforcements. IEEE Trans. Neural Netw. **3**(5), 724–740 (1992)

A.E. Brownlee, J.A. Wright, Constrained, mixed-integer and multi-objective optimisation of building designs by NSGA-II with fitness approximation. Appl. Soft Comput. **33**, 114–126 (2015)

D. Das, D.K. Pratihar, G.G. Roy, Electron beam melting of steel plates: temperature measurement using thermocouples and prediction through finite element analysis. *CAD/CAM, Robotics and Factories of the Future* (Springer, 2016), pp. 579–588

D. Das, D.K. Pratihar, G.G. Roy, A.R. Pal, Phenomenological model-based study on electron beam welding process, an input-output modeling using neural networks trained by back-propagation algorithm, genetic algorithm, particle swarm optimization algorithm and bat algorithm. Appl. Intell. (2017). https://doi.org/10.1007/s10489-017-1101-2

K. Deb, Unveiling innovative design principles by means of multiple conflicting objectives. Eng. Optim. **35**(5), 445–470 (2003)

K. Deb, S. Jain, Multi-speed gearbox design using multi-objective evolutionary algorithms. Trans. Am. Soc. Mech. Eng. J. Mech. Des. **125**(3), 609–619 (2003)

K. Deb, A. Pratap, S. Agarwal, T. Meyarivan, A fast and elitist multiobjective genetic algorithm: NSGA-II. IEEE Trans. Evol. Comput. **6**(2), 182–197 (2002)

K. Deb, A. Srinivasan, Innovization: Discovery of innovative design principles through multiobjective evolutionary optimization. *Multiobjective Problem Solving from Nature*, pp. 243–262 (2008)

K. Deb, S. Gupta, D. Daum, J. Branke, A.K. Mall, D. Padmanabhan, Reliability-based optimization using evolutionary algorithms. IEEE Trans. Evol. Comput. **13**(5), 1054–1074 (2009)

K. Deb, K. Sindhya, Deciphering innovative principles for optimal electric brushless DC permanent magnet motor design. in *IEEE World Congress on Computational Intelligence Evolutionary Computation 2008*, pp. 2283–2290 (2008)

K. Deb, S. Bandaru, D. Greiner, A. Gaspar-Cunha, C.C. Tutum, An integrated approach to automated innovization for discovering useful design principles: case studies from engineering. Appl. Soft Comput. **15**, 42–56 (2014)

C. Dudas, M. Frantzén, A.H. Ng, A synergy of multi-objective optimization and data mining for the analysis of a flexible flow shop. Robot. Comput. Integr. Manuf. **27**(4), 687–695 (2011)

M. Ester, H.-P. Kriegel, J. Sander, X. Xu, A density-based algorithm for discovering clusters in large spatial databases with noise. in *Proceedings of KDD 1996*, vol. 34, pp. 226–231 (1996)

M.A. Gil, Fuzziness and loss of information in statistical problems. IEEE Trans. Syst. Man Cybern. **17**(6), 1016–1025 (1987)

M.A. Gil, P. Gil, Fuzziness in the experimental outcomes: comparing experiments and removing the loss of information. J. Stat. Plan. Inference **31**(1), 93–111 (1992)

D.E. Goldberg, *The Design of Innovation: Lessons from and for Competent Genetic Algorithms*, vol. 7 (Springer Science & Business Media, 2002)

http://www.minitab.com

J. Horn, N. Nafpliotis, D.E. Goldberg, A niched Pareto genetic algorithm for multiobjective optimization. in *IEEE World Congress on Computational Intelligence Evolutionary Computation 1994*, pp. 82–87 (1994)

H. Ishibuchi, H. Tanaka, H. Okada, Interpolation of fuzzy if-then rules by neural networks. Int. J. Approx. Reason. **10**(1), 3–27 (1994)

J.-S. Jang, ANFIS: adaptive-network-based fuzzy inference system. IEEE Trans. Syst. Man Cybern. **23**(3), 665–685 (1993)

Y. Jarraya, S. Bouaziz, A.M. Alimi, A. Abraham, Evolutionary multi-objective optimization for evolving hierarchical fuzzy system. Evol. Comput. (CEC) **2015**, 3163–3170 (2015)

M. Jha, D.K. Pratihar, A. Bapat, V. Dey, M. Ali, A. Bagchi, Modeling of input-output relationships for electron beam butt welding of dissimilar materials using neural networks. Int. J. Comput. Intell. Appl. **13**(03), 1450016 (2014)

J. Kar, S. Mahanty, S.K. Roy, G. Roy, Estimation of average spot diameter and bead penetration using process model during electron beam welding of AISI 304 stainless steel. Trans. Indian Inst. Met. **68**(5), 935–941 (2015)

J.M. Keller, R.R. Yager, H. Tahani, Neural network implementation of fuzzy logic. Fuzzy Sets Syst. **45**(1), 1–12 (1992)

F. Khoshbin, H. Bonakdari, S.H. Ashraf Talesh, I. Ebtehaj, A.H. Zaji, H. Azimi, Adaptive neurofuzzy inference system multi-objective optimization using the genetic algorithm/singular value decomposition method for modelling the discharge coefficient in rectangular sharp-crested side weirs. Eng. Optim. **48**(6), 933–948 (2016)

J. Knowles, D. Corne, The Pareto archived evolution strategy: a new baseline algorithm for Pareto multiobjective optimisation. CEC **99**, 98–105 (1999)

K. Lwin, R. Qu, G. Kendall, A learning-guided multi-objective evolutionary algorithm for constrained portfolio optimization. Appl. Soft Comput. **24**, 757–772 (2014)

E.H. Mamdani, S. Assilian, An experiment in linguistic synthesis with a fuzzy logic controller. Int. J. Man Mach. Stud. **7**(1), 1–13 (1975)

M. Marinaki, Y. Marinakis, G.E. Stavroulakis, Fuzzy control optimized by a multi-objective differential evolution algorithm for vibration suppression of smart structures. Comput. Struct. **147**, 126–137 (2015)

K. Miettinen, *Nonlinear Multiobjective Optimization*, vol. 12 (Springer Science & Business Media, 1999)

S. Mitra, S.K. Pal, Neuro-fuzzy expert systems: relevance, features and methodologies. IETE J. Res. **42**(4–5), 335–347 (1996)

S. Obayashi, D. Sasaki, Visualization and data mining of Pareto solutions using self-organizing map. in *Proceedings of EMO 2003* (Springer, 2003), pp. 796–809

D.K. Pratihar, *Soft Computing* (Alpha Science International, Ltd, 2007)

S.A. Sadatsakkak, M.H. Ahmadi, M.A. Ahmadi, Optimization performance and thermodynamic analysis of an irreversible nano scale Brayton cycle operating with Maxwell-Boltzmann gas. Energy Convers. Manag. **101**, 592–605 (2015)

H.A. Taboada, D.W. Coit, Data mining techniques to facilitate the analysis of the Pareto-optimal set for multiple objective problems. in *Proceedings 2006, Institute of Industrial and Systems Engineers (IISE) IIE Annual Conference*, pp. 1–6 (2006)

H. Takagi, I. Hayashi, NN-driven fuzzy reasoning. Int. J. Approx. Reason. **5**(3), 191–212 (1991)

H. Takagi, N. Suzuki, T. Koda, Y. Kojima, Neural networks designed on approximate reasoning architecture and their applications. IEEE Trans. Neural Netw. **3**(5), 752–760 (1992)

C. Wang, X. Li, X. Zhou, A. Wang, N. Nedjah, Soft computing in big data intelligent transportation systems. Appl. Soft Comput. **38**, 1099–1108 (2016)

J. Yao, M. Dash, S. Tan, H. Liu, Entropy-based fuzzy clustering and fuzzy modeling. Fuzzy Sets Syst. **113**(3), 381–388 (2000)

Z.-H. Zhan, J. Li, J. Cao, J. Zhang, H.S.-H. Chung, Y.-H. Shi, Multiple populations for multiple objectives: a coevolutionary technique for solving multiobjective optimization problems. IEEE Trans. Cybern. **43**(2), 445–463 (2013)

Q. Zhang, H. Li, MOEA/D: a multiobjective evolutionary algorithm based on decomposition. IEEE Trans. Evol. Comput. **11**(6), 712–731 (2007)

Y.-J. Zheng, S.-Y. Chen, H.-F. Ling, Evolutionary optimization for disaster relief operations: a survey. Appl. Soft Comput. **27**, 553–566 (2015)

E. Zitzler, L. Thiele, An evolutionary algorithm for multiobjective optimization: the strength Pareto approach (1998)

Printed in the United States
By Bookmasters